Mechanisms and Machine Science

Volume 127

This book series establishes a well-defined forum for monographs, edited Books, and proceedings on mechanical engineering with particular emphasis on MMS (Mechanism and Machine Science). The final goal is the publication of research that shows the development of mechanical engineering and particularly MMS in all technical aspects, even in very recent assessments. Published works share an approach by which technical details and formulation are discussed, and discuss modern formalisms with the aim to circulate research and technical achievements for use in professional, research, academic, and teaching activities.

This technical approach is an essential characteristic of the series. By discussing technical details and formulations in terms of modern formalisms, the possibility is created not only to show technical developments but also to explain achievements for technical teaching and research activity today and for the future.

The book series is intended to collect technical views on developments of the broad field of MMS in a unique frame that can be seen in its totality as an Encyclopaedia of MMS but with the additional purpose of archiving and teaching MMS achievements. Therefore, the book series will be of use not only for researchers and teachers in Mechanical Engineering but also for professionals and students for their formation and future work.

The series is promoted under the auspices of International Federation for the Promotion of Mechanism and Machine Science (IFToMM).

Prospective authors and editors can contact Mr. Pierpaolo Riva (publishing editor, Springer) at: pierpaolo.riva@springer.com

Indexed by SCOPUS and Google Scholar.

Ioan Doroftei · Mircea Nitulescu · Doina Pisla · Erwin-Christian Lovasz
Editors

Proceedings of SYROM 2022 & ROBOTICS 2022

13th IFToMM International Symposium on Science of Mechanisms and Machines & XXV International Conference on Robotics

Springer

Editors
Ioan Doroftei
Department of Mechanical Engineering, Mechatronics and Robotics
"Gheorghe Asachi" Technical University of Iaşi
Iaşi, Romania

Mircea Nitulescu
Department of Computers and Electronics, Mechatronics and Robotics
Faculty of Automation
University of Craiova
Craiova, Romania

Doina Pisla
CESTER
Technical University of Cluj-Napoca
Cluj-Napoca, Romania

Erwin-Christian Lovasz
Department of Mechatronics
Polytechnic University of Timisoara
Timisoara, Romania

ISSN 2211-0984 ISSN 2211-0992 (electronic)
Mechanisms and Machine Science
ISBN 978-3-031-25654-7 ISBN 978-3-031-25655-4 (eBook)
https://doi.org/10.1007/978-3-031-25655-4

This Springer imprint is published by the registered company Springer Nature Switzerland AG
The registered company address is: Gewerbestrasse 11, 6330 Cham, Switzerland

Organization

Conference Chairman

Ioan Doroftei, "Gheorghe Asachi" Technical University of Iaşi, Mechanical Engineering, Mechatronics and Robotics Department, Iaşi, Romania

Conference Co-Chairmen

Mircea Nitulescu, University of Craiova, Faculty of Automation, Computers and Electronics, Mechatronics and Robotics, Craiova, Romania
Doina Pisla, CESTER, Technical University of Cluj-Napoca, Romania
Erwin-Christian Lovasz, University Politehnica Timisoara, Mechatronics, Timisoara, Romania

Program Committee Members

Stelian Alaci (Romania)
Mircea-Viorel Bara (Romania)
Yvan Baudoin (Belgium)
Karsten Berns (Germany)
Nicu-George Bizdoaca (Romania)
Theodor Borangiu (Romania)
Stelian Brad (Romania)
Adrian Burlacu (Romania)
Mehmet Isman Can Dede (Turkey)
Giuseppe Carbone (Italy)
Marco Ceccarelli (Italy)

Reviewers

Adrian Burlacu, "Gheorghe Asachi" Technical University of Iaşi, Romania
Giuseppe Carbone, University of Calabria, DIMEG, Rende, CS, Italy
Marco Ceccarelli, University of Rome Tor Vergata, Department of Industrial Engineering, Roma, Italy
Dorian Cojocaru, University of Craiova, Romania
Daniel Condurache, Technical University of Iaşi, Theoretical Mechanics, Iaşi, România
Burkhard Corves, RWTH Aachen University, Germany
Ioan Doroftei, "Gheorghe Asachi" Technical University of Iaşi, Mechanical Engineering, Mechatronics and Robotics Department, Iaşi, Romania
Jean-Christophe Fauroux, SIGMA Clermont; MecaBotiX Touluse, France
Alessandro Gasparetto, Università degli Studi di Udine, Italy
Corina Gruescu, Politehnica University of Timisoara, Romania
Mathias Hüsing, RWTH Aachen University, Germany
Erwin-Christian Lovasz, University Politehnica Timisoara, Mechatronics, Timisoara, Romania
Dan Mandru, Technical University of Cluj-Napoca, Mechatronics and Machine Dynamics, Cluj-Napoca, Romania
Mircea Neagoe, Transilvania University of Brasov
Mircea Nitulescu, University of Craiova, Faculty of Automation, Computers and Electronics, Mechatronics and Robotics Department, Craiova, Romania
Victor Petuya, University of the Basque Country (UPV/EHU), Spain
Doina Cester Pisla, Technical University of Cluj-Napoca, Romania
Ionel Staretu, Transilvania University of Brasov, Product Design, Mechatronics and Environment, Brasov, Romania
Daniela Tarnita, University of Craiova, Faculty of Mechanics, Applied Mechanics, Craiova, Romania
Mihai-Olimpiu Tatar, Technical University of Cluj-Napoca, Mechatronics and Machine Dynamics, Cluj-Napoca, România
Hidetsugu Terada, University of Yamanashi, Graduate School of Engineering, Kofu, Japan
Calin Vaida, CESTER, Technical University of Cluj-Napoca, Romania, Research Center for Industrial Robots Simulation and Testing (CESTER), Cluj-Napoca, Romania

Preface

The Joint International Conference of the 13th IFToMM International Symposium on Science of Mechanisms and Machines (SYROM 2022) and the XXV International Conference on Robotics (ROBOTICS 2022), was organized by Romanian Association for the Science of Mechanisms and Machines (ARoTMM), Robotics Society of Romania (SRR), and Mechanical Engineering, Mechatronics and Robotics Department at the Mechanical Engineering Faculty, "Gheorghe Asachi" Technical University of Iaşi, Romania, with the support of the International Federation for the Promotion of Mechanism and Machine Science (IFToMM) and the Technical Sciences Academy of Romania (ASTR).

The aim of the joint conference was to bring together researchers, scientists, industry experts and Ph.D. students involved in the general area of mechanisms, mechanical transmissions, robotics and mechatronics, to disseminate their latest research results and exchange views on the future research directions of these fields.

The topics of the symposium are related to mechanisms and robotics within aspects of theory, design, practice, and applications in engineering, including but not limited to: Theoretical and computational kinematics; Mechanism design; Experimental mechanics; Dynamics of machinery; Dynamics of multi-body systems; Mechanisms for biomechanics; Mechanical transmissions; Linkages and mechanical controls; Micro-mechanisms; History of mechanism science; Industrial and non-industrial applications; Serial and parallel robots; Mobile robots; Collaborative robots; Micro and Nano robots; Sensors and actuators; Advanced control; Artificial intelligence; Medical robots; Teleoperation, haptics, virtual reality.

We would like to express grateful thanks to IFToMM International Federation for the Promotion of Mechanism and Machine Science, to the Romanian IFToMM National Committee ARoTMM, to the National Committee of the Robotics Society of Romania, and to the members of the International Scientific Committee of SYROM and ROBOTICS 2022.

We appreciate the effort of the reviewers gathered in the International Scientific Committee of SYROM and ROBOTICS 2022. They spent time for a serious work of evaluation and improvement guidance, meant to assure a high quality of all papers.

We thank the authors who contributed valuable papers on different subjects, covering a wide scientific area of Mechanisms and Machine Science, and Robotics.

We thank the "Gheorghe Asachi" Technical University of Iaşi, the Mechanical Engineering Faculty, and the Mechanical Engineering, Mechatronics and Robotics Department for hosting the scientific event and supporting all associated activities. The Joint International Conference of the 13th IFToMM International Symposium on Science of Mechanisms and Machines (SYROM 2022) and the XXV International Conference on Robotics (ROBOTICS 2022) became a real event due to a hard-working local organizing team: Emil Budescu, Florentin Buium, Eugen Merticaru, Cristina-Magda Cazacu, and Robert Kristof.

Last but not least, we are grateful to the staff at Springer Publishers for their excellent technical and editorial support.

Iaşi, Romania
November 2022

Ioan Doroftei
Mircea Nitulescu
Doina Pisla
Erwin-Christian Lovasz

Contents

Part I
Mechanism Design and Analysis

Chapter 1
Experimental Energy Gain Assessment of a Photovoltaic System Equipped with a Biaxial Solar Tracking Mechanism

M. Moldovan, I. Visa, and B. Burduhos

Abstract Starting with January 2021, all new buildings and those undergoing major renovation must comply with the nearly Zero Energy Building standard imposed by the European Directive 31/2010. Besides specific indicators related to the energy performance and to the renewable energy share, the renewable energy systems must be implemented on or near the buildings. For solar energy conversion systems, this pre-requisite can be an issue for buildings with rather small available (opaque) areas, implemented in communities with high buildings density. To tackle this issue, solar tracking systems can be implemented to increase the amount of received solar energy and thus the solar energy convertor output (thermal or electrical energy). The experimental results obtained during a year of monitoring are comparatively presented in the paper for tracked and fixed photovoltaic modules implemented in the Renewable Energy Systems and Recycling R&D Center at Transilvania University of Brasov, Romania.

Keywords Solar energy · Photovoltaic system · Solar tracking mechanism · Energy gain

1.1 Introduction

The European Directive 31/2010 [1], imposes the standard nearly Zero Energy Building (nZEB) from January 1st, 2021 to all new or undergoing major renovation buildings. The buildings' energy consumption has to be minimized and covered

M. Moldovan (✉) · I. Visa
Renewable Energy Systems and Recycling Research Center, Transilvania University of Brasov, Brasov, Romania, Romanian Academy of Technical Sciences, Bucharest, Romania
e-mail: macedon.moldovan@unitbv.ro

I. Visa
e-mail: visaion@unitbv.ro

B. Burduhos
Renewable Energy Systems and Recycling Research Center, Transilvania University of Brasov, Brasov, Romania
e-mail: bogdan.burduhos@unitbv.ro

I. Doroftei et al. (eds.), *Proceedings of SYROM 2022 & ROBOTICS 2022*, Mechanisms and Machine Science 127, https://doi.org/10.1007/978-3-031-25655-4_1

in a large extent using renewable energy sources [2]. Moreover, the renewable energy systems must be implemented on or nearby the buildings to reduce the thermal and electrical energy loses [3].

The Agenda 2030 states that to develop a sustainable community, renewable energy systems must be implemented supporting affordable and clean energy for the buildings [4].

Among renewable energies, solar energy is an abundant source that can be harnessed in the built environment [5]. Its main disadvantage is represented by its low energy density requiring large implementation surfaces, especially in the case of photovoltaic (PV) systems [6]. To mitigate this issue, solar tracking systems represent a solution to increase the amount of the received solar energy and therefore the energy output of the solar energy conversion system. Several types of solar tracking systems are investigated [7], reporting an yearly output increases up to 35% for tracked solar thermal collectors [8], and up to 30% for tracked PV modules [9]. Literature reports usually present numerical simulations and experimental results over short periods of time [10].

This paper aims to experimentally assess the output increase of a PV system installed on a biaxial solar tracking system compared to an optimally tilted fixed PV system, both installed in a continental temperate climate in the Renewable Energy Systems and Recycling R&D Center of the Transilvania University of Brasov, Romania. Experimental results collected over a one year period (July 1, 2018 … June 30, 2019) are analysed and presented in the paper.

1.2 Methodology

To evaluate the advantages of a tracked PV platform versus a fixed one installed at a yearly optimal tilt angle (χ), the following relevant indicators are defined.

The available global solar energy is calculated using:

$$E_G = \sum_{t_1}^{t_2} \frac{G}{60} \left[\text{Wh/m}^2\right] \tag{1.1}$$

where:

t_1 and t_2 are the sunrise and sunset moments in the day;

G is the available global solar irradiance calculated each minute using:

$$G = B + D \left[\text{W/m}^2\right] \tag{1.2}$$

where:

B is the available direct solar irradiance measured each minute;

D is the available diffuse solar irradiance calculated based on the diffuse solar irradiance measured each minute in the horizontal plane (D_h) and on the solar altitude angle (α) using:

$$D = D_h \cdot \frac{1 + \sin\alpha}{2}\left[\mathrm{W/m^2}\right] \tag{1.3}$$

The solar altitude angle is calculated according to [4] using:

$$\alpha = arcsin\,(sin\delta \cdot sin\varphi + cos\delta \cdot cos\varphi \cdot cos\omega)\,[\circ] \tag{1.4}$$

where:

φ is the latitude of the location [°]

ω is the hour angle [°]

δ is the declination angle [°]

The received global solar energy is calculated using:

$$E_{Gn} = \sum_{t_1}^{t_2} \frac{G_n}{60}\left[\mathrm{W/m^2}\right] \tag{1.5}$$

where G_n is the received global solar irradiance calculated each minute using

$$G_n = B_n + D_n\left[\mathrm{W/m^2}\right] \tag{1.6}$$

where:

B_n is the received direct solar irradiance calculated based on the incidence angle (ν), defined as the angle between the sunray and the normal to the PV module [11], using:

$$B_n = B\cos\nu\left[\mathrm{W/m^2}\right] \tag{1.7}$$

D_n is the received diffuse solar irradiance calculated based on the diffuse solar irradiance measured each minute in the horizontal plane (D_h) and on the PV tilt angle (χ) using:

$$D_n = D_h \cdot \frac{1 + \cos\chi}{2}\left[\mathrm{W/m^2}\right] \tag{1.8}$$

The electrical energy gain of the tracked *vs.* the fixed PV modules is calculated using:

$$\Delta_E = E_t - E_f\,[\mathrm{Wh}] \tag{1.9}$$

where:

E_t is the electrical energy output of the tracked PV module, recorded daily, monthly or yearly by the monitoring system of the PV platform.

E_f is the electrical energy output of the fixed PV module, recorded daily, monthly or yearly by the monitoring system of the PV platform.

(a) (b)

Fig. 1.1 Pseudo-equatorial solar tracking system for three PV modules (**a**) and fixed PV platform (**b**) installed on the rooftop of the RESREC building

1.3 Experimental Setup

A biaxial solar tracking system for three PV modules (Fig. 1.1a) and a fixed PV platform, with three PV modules, (Fig. 1.1b) are installed on the rooftop of the Renewable Energy Systems and Recycling R&D Center of the Transilvania University of Brasov, Romania. The solar tracking system consists of a mobile frame that is rotated about a horizontal East–West axis by a linear actuator and three photovoltaic modules connected with rotational joints to the mobile frame. The diurnal angle of the photovoltaic modules relative to the mobile frame is adjusted with a linear actuator through a parallelogram mechanism.

Identical 250 W polycrystalline silicon PV modules are installed on the solar tracking system and on the fixed platform. As presented in Fig. 1.2a, each PV module (1) is serially connected to an inverter (2) through individual power optimizers (3) that monitor the voltage, current, power and energy for each PV module and transmit these parameters to the inverter. The inverter manages the data and transmits it to a web server where it is stored and can be downloaded for processing. The direct solar irradiance (B) and horizontal diffuse solar irradiance (D_h) are measured each minute with a CHP1 pyrheliometer and a CMP22 pyranometer installed on a Kipp & Zonen Solys 2 Sun Tracker (Fig. 1.2b).

1.4 Results and Discussion

The results recorded in four representative clear sky days during the monitoring period (July 1 2018 … June 30 2019) are further presented in terms of available and received solar energy and electrical energy output of the PV modules.

The sunniest days closest to the equinoxes and solstices were selected and the available global solar irradiance (G) and the global solar irradiance received in the plane of the fixed PV module (Gn) are comparatively presented in Fig. 1.3.

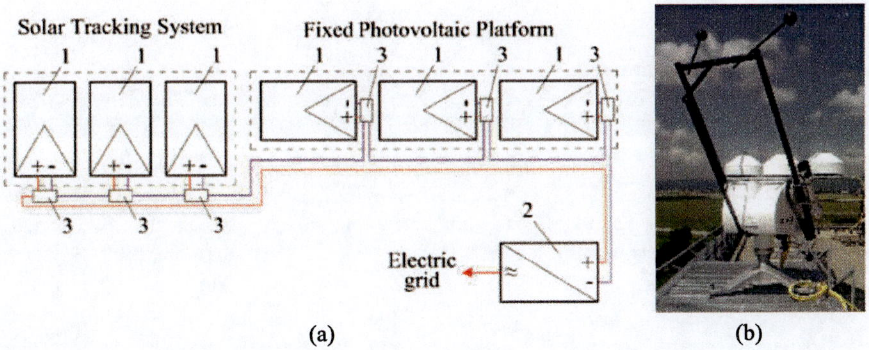

Fig. 1.2 Monitoring system of the PV modules (**a**) and Solys 2 Sun Tracker (**b**)

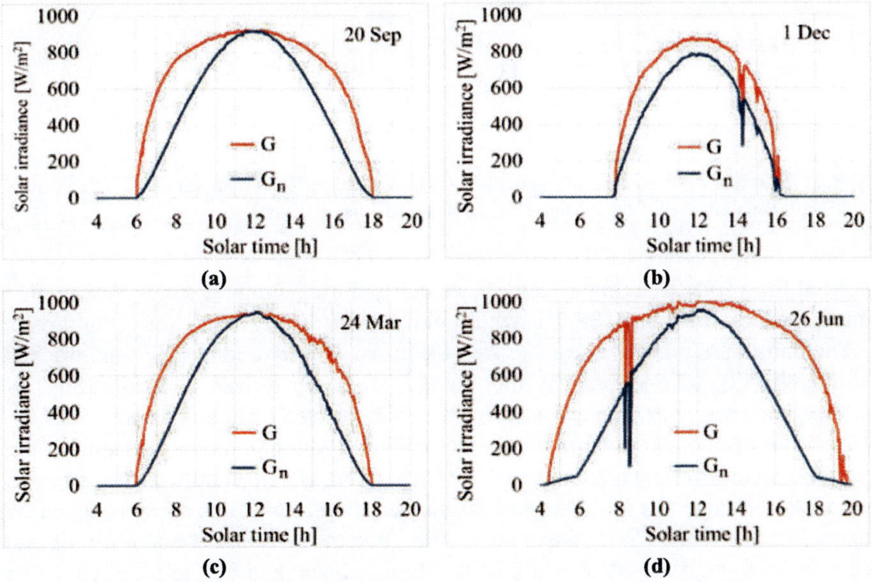

Fig. 1.3 Available global solar irradiance (G), global solar irradiance received in the plane of the fixed PV (G_n) in: **a** September 20, 2018, **b** December 1, 2018, **c** March 24, 2019 and **d** June 26, 2019

The daily available global solar energy ranges between 5.72 kWh/m^2 in December and 11.89 kWh/m^2 in June with almost similar values close to equinoxes (8.77 kWh/m^2 and 8.88 kWh/m^2). The global solar energy received in the plane of the fixed PV module ranges between 4.47 kWh/m^2 in December and 7.29 kWh/m^2 in June, with 6.5 kWh/m^2 and 6.66 kWh/m^2 in March and September. These values represent 78.1%, 61.3%, 74.1% and 75% of the available solar energy, thus justifying

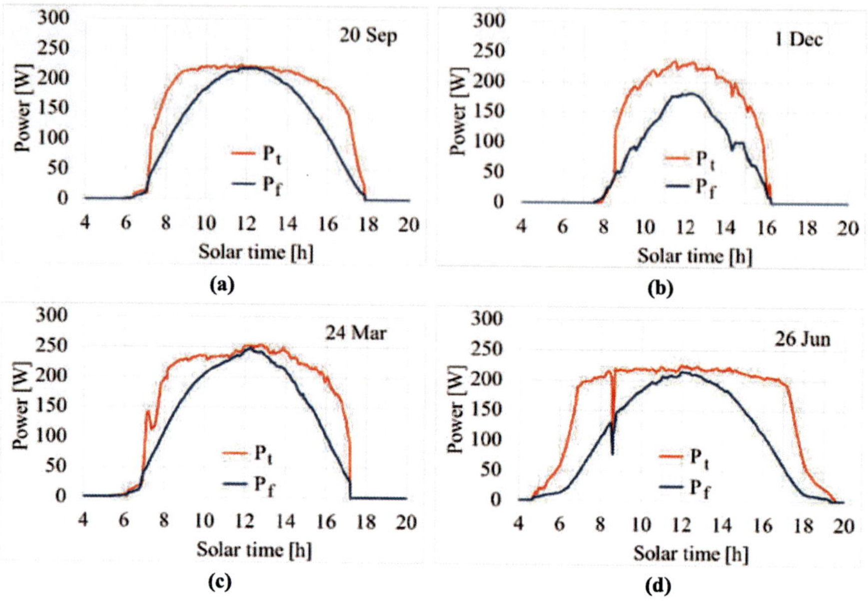

Fig. 1.4 Electrical power output of the tracked (P_t) versus fixed (P_f) PV modules in: **a** September 20, 2018, **b** December 1, 2018, **c** March 24, 2019 and **d** June 26, 2019

the need for a solar tracking system, particularly during summer sunny days with a potential maximum gain of 38.7%, when continuous tracking of the Sun is employed.

The daily variations of the measured electrical power produced by one tracked PV module (Pt) and one fixed PV module (Pf) are comparatively plotted in Fig. 1.4 with red and blue lines, respectively, for the four selected clear sky days.

In all cases, the electrical power of the tracked PV module is higher during the morning and the evening hours when the solar tracking system significantly decreases the incidence angle. At noon, except December, the electrical powers produced by the tracked and fixed PV modules are similar, due to their almost equal tilt angles. In December, the fixed PV module is visibly disadvantaged by the tilt angle (40°) generating a high incidence angle (27°).

The largest amount of electrical energy was obtained by the tracked PV module in June 26 (2.48 kWh), 49.4% higher than the electrical energy produced by the fixed PV module (1.66 kWh). On September 20 and on March 24 the output of the tracked PV module was 2.05 kWh and 2.23 kWh, 29.7% and 28.2% higher than the electrical energy produced by the fixed PV module (1.58 kWh and 1.74 kWh). On December 1, the electrical energy produced by the tracked PV module was of 1.47 kWh/day, 59.8% higher than 0.92 kWh produced by the fixed PV module; due to the reduced number of daylight hours in this day, the outputs of the analysed PV modules are significantly lower (59.3% for the tracked PV and 55.4% for the fixed PV) than during the summer day (June 26).

Extending the analysis over the entire monitoring period (July 1, 2018 … June 30, 2019) the electrical energy monthly produced by a tracked PV module (Et) vs. a fixed PV module (Ef) is presented in Fig. 1.4.

The electrical energy produced by the tracked PV module was of 414.68 kWh/year, 27.5% higher than 325.35 kWh/year produced by the fixed PV module. The monthly energy gain (ΔE) ranges between 3.73 kWh in December and 12.71 kWh in June resulting a yearly energy gain of 89.33 kWh for the tracked PV module, of 250 W.

Following these findings, a photovoltaic platform of 100 PV modules (25 kW) and a photovoltaic park of 4000 PV modules (1 MW) will record a yearly energy gain due to the use of a solar tracking system up to 8.9 MWh and 357.3 MWh respectively. Considering the actual price of electrical energy in Romania (0.2 Euro/kWh) the yearly energy gains correspond to a financial gain of 1780 Euro/year and 71,460 Euro/year respectively, reducing thus the payback time of the initial investment.

1.5 Conclusions

The electrical energy output of a tracked PV module compared to a fixed PV module was experimentally assessed in a temperate continental climate (Brasov, Romania) between July 1 2018 and June 30 2019.

During the four selected clear sky days (close to the solstices and equinoxes), the electrical energy gain ranges between 1.47 kWh (winter) and 2.48 kWh (summer).

The monthly energy gain ranges between 3.73 kWh in December and 12.71 kWh in June. The electrical energy produced yearly by the tracked PV module was of 414.68 kWh, 27.5% higher than 325.35 kWh produced by the fixed PV module, resulting a yearly energy gain of 89.33 kWh. Extrapolating this value from one PV module to a larger PV platform of 100 PV modules or to a PV park of 4000 PV modules, the yearly energy gain increases to 8.9 MWh and 357.3 MWh respectively, corresponding to increased incomes of about 1780 Euro/year and 71,460 Euro/year respectively.

References

1. European Parliament, The Directive 2010/31 of the European Parliament and of the Council of 19 May 2010 on the energy performance of buildings, Official Journal of the European Union 53 (2010)
2. Visa, I., Moldovan, M., Comsit, M., Duta, A.: Improving the renewable energy mix in a building towards the nearly zero energy status. Energy Build. **68**, 72–78 (2014)
3. Moldovan, M., Burduhos, B.-G., Visa, I.: Yearly Electrical Energy Assessment of a Photovoltaic Platform/Geothermal Heat Pump Prosumer–Energies **14**(13), 3776 (2021)
4. United Nations General Assembly, Transforming Our World: the 2030 Agenda for Sustainable Development, Resolution A/RES/70/1, New York (2015)
5. Kalogirou, S.A.: Solar Energy Engineering—Processes and Systems, 2nd edn. Academic Press, London, UK (2009)

6. Visa, I., Duta, A., Moldovan, M., Neagoe, M., Burduhos, B.: Solar Energy Conversion Systems in the Built Environment—Green Energy and Technology. Springer, Switzerland (2020)
7. Visa, I., Jaliu, C., Duta, A., Neagoe, M., Comsit, M., Moldovan, M., Ciobanu, D., Burduhos, B., Saulescu, R.: The Role of Mechanisms in Sustainable Energy Systems. Transilvania University Publishing House, Brasov, Romania (2015)
8. Fahl, P., Ganapathisubbu, S.: Tracking benefits for solar collectors installed in Bangalore. J. Renew. Sustain. Energy **3**, 1–12 (2011)
9. Moldovan, M.D., Visa, I., Burduhos, B.G.: Energetic autonomy for a solar house. Environ. Eng. Manag. J. **10**(9), 1283–1290 (2011)
10. Burduhos, B.G., Toma, C., Neagoe, M., Moldovan, M.D.: Pseudo-equatorial tracking optimization for small photovoltaic platforms from Toronto/Canada. Environ. Eng. Manag. J. **10**(8), 1059–1068 (2011)
11. Duffie, J., Beckmann, W.: Solar Engineering of Thermal Processes, Fourth Edition, Wiley Interscience, New York (2013)

Chapter 2
A Study of Feasibility of a Mechanism for Rib-Vertebra-Sternum Prosthesis

O. Ramirez, M. Ceccarelli, and C. R. Torres-San-Miguel

Abstract A mechanism with flexural joints is proposed as a prosthesis of a complex vertebra-rib-sternum. The flexible anatomic behavior is replicated by a mechanism structure concentrating the flexibility in suitable flexural joint designs. The feasibility of the proposed design is characterized by numeral performance values from quasi-static simulations.

Keywords Biomechanics · Biomedical devices · Design · Flexural joint · Simulation

2.1 Introduction

During breathing the inspiration process takes place when oxygen is captured in the lungs, the thorax expands in the transverse and anteroposterior diameter because of the contraction of the inspiratory muscles involved [1, 2]. The expiration process takes place when the inspiratory muscles relax completely because of the continuous expulsion of air caused by the balance of forces between the lungs and the chest wall. This action leads to decreased volume in the thorax [3].

Several investigations have been carried out to establish a characterization of human ribs subjected to different load scenarios. In the work carried out by Aguilar et al. [4], they describe the design of an artificial rib that is used to determine its behavior when it is subjected to a biological compression process, resulting in

O. Ramirez (✉) · C. R. Torres-San-Miguel
Escuela Superior de Ingeniería Mecánica y Eléctrica Sección de Estudios de Posgrado e Investigación, Instituto Politécnico Nacional, Unidad Profesional "Adolfo López Mateos" Zacatenco, Mexico City, Mexico
e-mail: oramirezj1400@alumno.ipn.mx

C. R. Torres-San-Miguel
e-mail: ctorress@ipn.mx

O. Ramirez · M. Ceccarelli
LARM2: Laboratory of Robotics and Mechatronics Department of Industrial Engineering, University of Rome Tor Vergata, 00133 Roma, Italy
e-mail: marco.ceccarelli@uniroma2.it

I. Doroftei et al. (eds.), *Proceedings of SYROM 2022 & ROBOTICS 2022*, Mechanisms and Machine Science 127, https://doi.org/10.1007/978-3-031-25655-4_2

behavior like a natural rib. In the work carried out by Ramirez et al. [5], the experimental characterization of a segmented rib is carried out when it is joined with an implant in order to know the behavior of this one when it is subjected to a normal breathing cycle of a human being. Rib designs are also used as stiffening elements that are used to characterize the dynamic response of the thorax in an automobile impact event, as in the case of a crash test dummy [6].

The present work focuses on the design of a rib-vertebra-sternum model with flexural joints. To reduce the complexity of the analysis of the flexural joints, the Pseudo-Rigid Body Model (PRBM) is used as formulated in [7]. A numerical analysis is performed to characterize the performance of the proposed mechanism during normal breathing. The development of this work can be used for its implementation with flexible prostheses of sternum [8] to better reproduce the ventilatory mechanics of the thorax.

2.2 The Vertebra-Rib-Sternum System

The normal breathing process comprises two stages, inhalation where air enters the lungs and then exhalation where gases are expelled from the lungs. For a healthy young adult, normal breathing occurs at a rate of about 14 breaths per minute, in such a way that the time necessary to carry out an inhalation and an expiration is about 2 s [9].

In the same way, during breathing, different movements are produced in the chest thorax. During inspiration, there is an increase in the anteroposterior diameter of the thorax, reaching a maximum sagittal displacement of about 30 mm [10].

The design proposal seeks to reproduce the two states of movement of the thorax during respiration, reproducing the anteroposterior expansion and contraction of the thorax.

A cross-sectional image of the chest for a healthy young adult and its coronal and sagittal dimensions are selected to determine the dimensions of the mechanism. Those dimensions are 358 mm in the coronal plane and 165 mm in the sagittal plane [11] which represents the workspace of the mechanism, as shown in Fig. 2.1.

The desired movement can be realized using a 6-bar mechanism, so that, a longitudinal displacement of the intermediate link can be generated as long as the same rotation is applied to the two links that are attached to the fixed one.

2.3 The Proposed Prosthesis Mechanism

The proposed design consists of 6 links with flexural joints, as in Fig. 2.2. Link 0 represents the vertebra, and it is the fixed link. Links 1 and 2 represent the left rib, in the same way links 4 and 5 represent the right rib and finally link 3 represents the sternum.

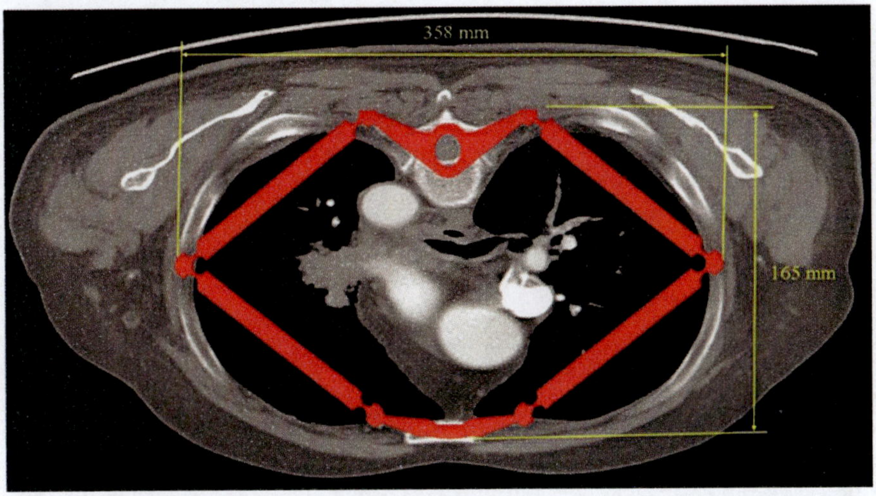

Fig. 2.1 A scheme for mechanism workspace within the thoracic area in a tomographic view

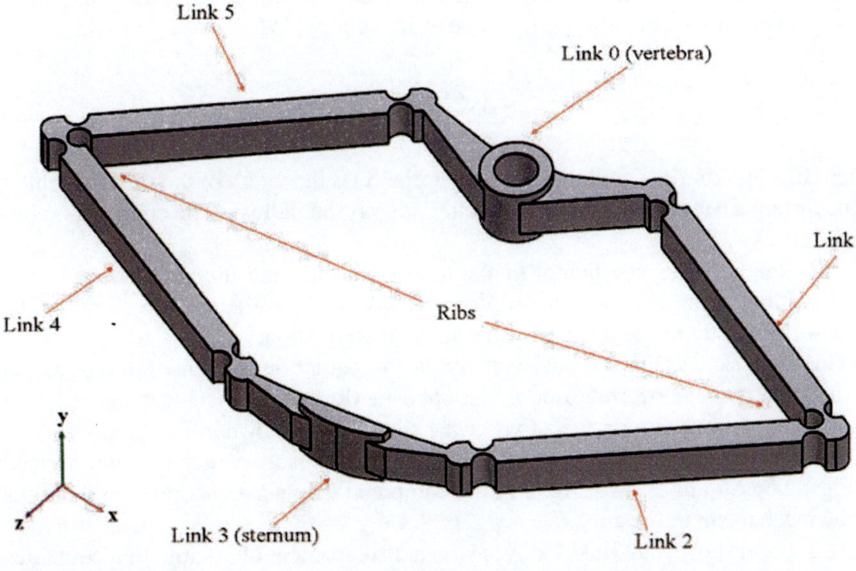

Fig. 2.2 The proposed 6 bar compliant mechanism as rib-vertebra-sternum prosthesis

Links are joined by circular notches that allow the movement of the mechanism. The flexural joints between link 0 and links 1 and 5 represent the flexibility of the costostenal joints. The flexural joints between links 2, 3 and 4 represent the flexibility of the costovertebral joints. Finally, the flexural joints between links 1, 2 and 4, 5 represent the flexibility of the rib.

When the same angular displacements are given, in opposite directions, on the costovertebral flexural joints, all the joints are simultaneously flexed in such a way that the longitudinal displacement of link 3 on the z axis is produced.

The mechanism was designed in SolidWorks software, as shown in Fig. 2.2. Determined lengths for the links to fit the thoracic volume as in Fig. 2.1 are: 108 mm for link 0, 149.8 mm for links 1 and 5, 153.1 mm for links 2 and 4, 100 mm for link 3.

2.4 Performance Analysis

An analysis of the compliant mechanism is performed using the PRBM. Using this technique, it is possible to model the compliant mechanism as a rigid body by replacing each flexural joint for a torsion spring and a rotatory joint [7]. On the other hand, a material is suitable for its application in a flexible mechanism when it has large ratios of yield strength to Young's modulus [12]. To design the flexible elements, Polylactic acid (PLA) was selected as material for a first construction design, since the ratio between its Young's modulus and its yield stress is equal to 20, because its Young's modulus is of 2346.5 MPa and a Yield stress is of 49.5 MPa [13].

The stiffness of a flexural joint can be evaluated by [14]

$$k = \frac{2Ew}{9\pi}\sqrt{\frac{t^5}{r}} \tag{2.1}$$

where E is the Young's modulus of the material, w is the width of the flexural joint, t is the distance between the circular notches and r is the radius of the circular notches (Fig. 2.3).

Once the stiffness coefficient of the torsion springs and the dimensions of the flexural joint have been determined, the ADAMS software is used to simulate the pseudo-rigid body model for performance characterization.

This analysis was carried out by applying an angular displacement of 6.5° as the input in the two costovertebral joints in opposite directions to achieve the required displacement of 30 mm on the z axis of the sternum, as shown in Fig. 2.4.

The analysis was carried out obtaining significant results which are summarized in Figs. 2.5, 2.6 and 2.7. In Fig. 2.5, the computed displacements refer to the links of the mechanism in the z axis.

This graph shows that link 3 moves with a displacement of 30 mm, that coincides with the maximum sagittal displacement of the assumed thorax during respiration. On the other hand, it can be observed that links 1, 5 have the same displacement of 7.2 mm. Similarly links 2, 4 have the same displacement for each other.

This is consistent since the pairs of links 1, 5 and 2, 4 have the same dimensions and perform the same movement as it is expected since they should move symmetrically as per the proposed design in Figs. 2.1 and 2.2.

Figure 2.6 shows the angular velocity that is generated in the links during the movement.

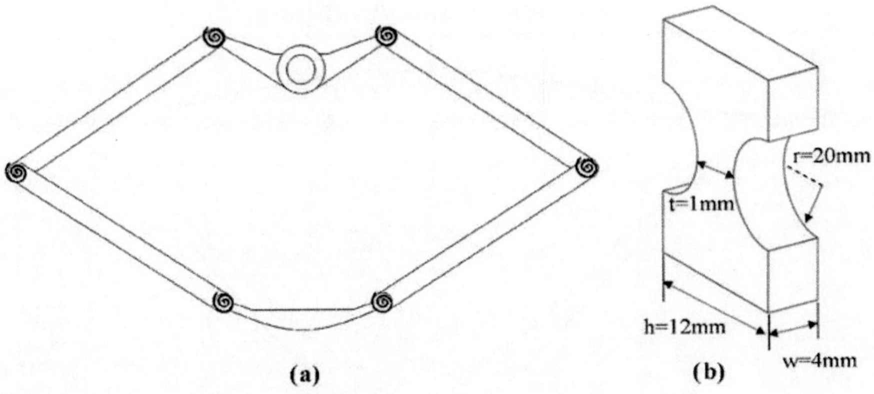

Fig. 2.3 A three-dimensional model of the compliant mechanism in Fig. 2.2: **a** PRBM of the compliant mechanism; **b** A scheme of the flexural joint with parameter sizes

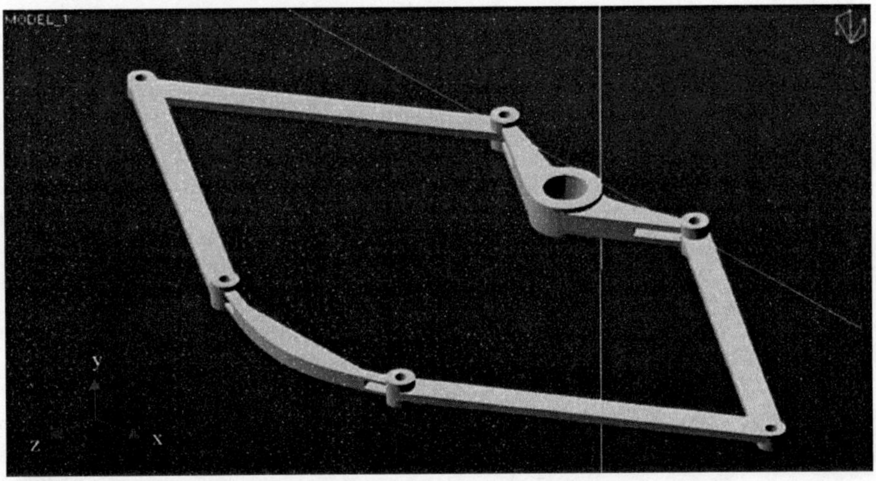

Fig. 2.4 PRBM of the mechanism model in Fig. 2.2, simulated in Adams software

It can be observed that link 3 does not generate any angular velocity, since it only moves about the y-axis and does not rotate at any time. This demonstrates that the movement of link 3 is the desired one as to represent the sternum, because the sternum only moves in the sagittal plane during respiration. It can be also observed that the 4 links have the same angular velocity, since link 3 does not rotate. Links 1, 4 and links 2, 5 generate the same motion because links 1, 4 rotate in the same direction, while links 2, 5 rotate in the opposite direction.

Figure 2.7 shows the angular accelerations that are generated in each of the links. Similarly, in link 3 no angular acceleration is generated. This is expected since it has no angular velocity, and it is not possible to generate any angular acceleration.

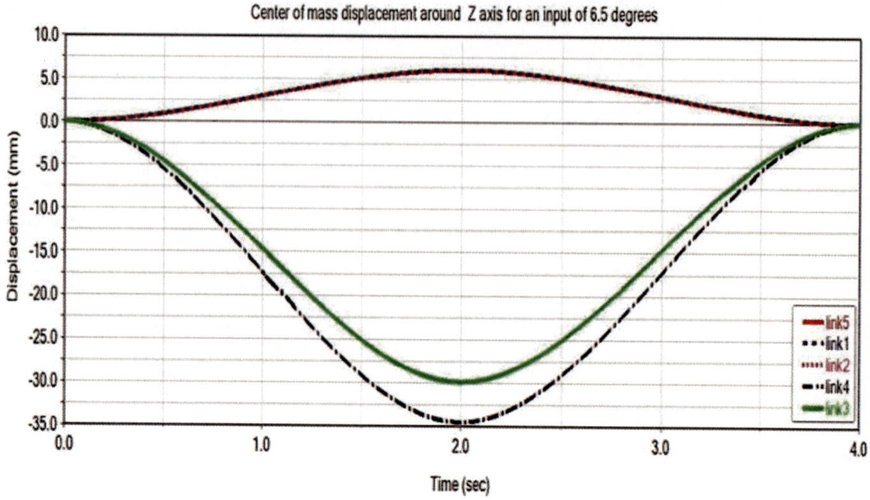

Fig. 2.5 Computed displacements of each link around Z axis during the normal breathing cycle as simulated in Fig. 2.4

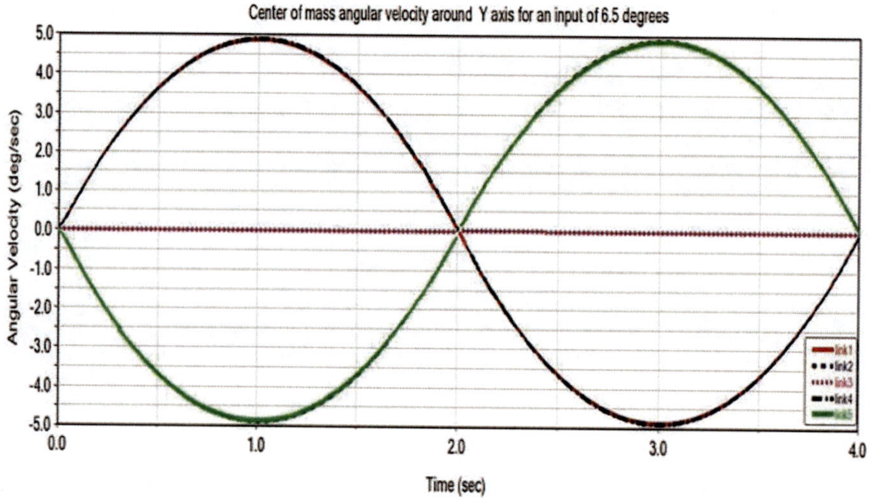

Fig. 2.6 Computed angular velocity of each link around Y axis during the normal breathing cycle as simulated in Fig. 2.4

In addition, from the simulation results, links 1, 4 and 2, 5 have the same trajectory but with opposite directions due to the direction they rotate. In Fig. 2.7 it is also possible to observe that the angular accelerations have a linear behavior with respect to time.

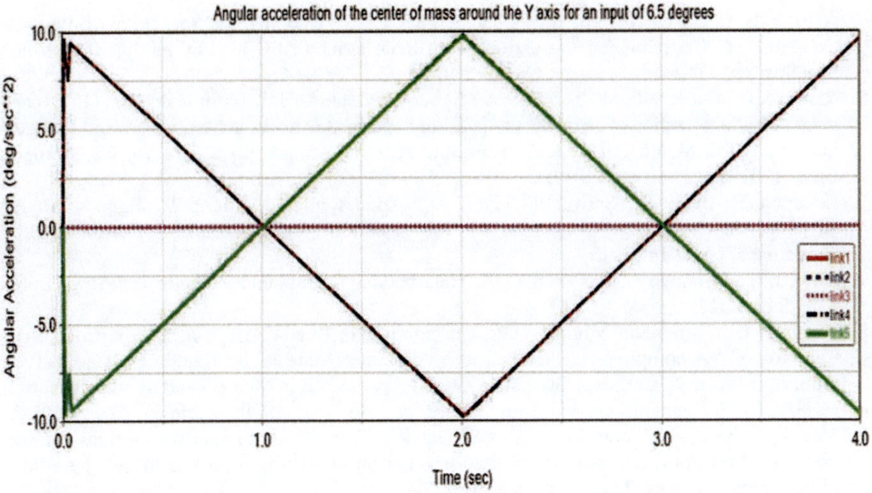

Fig. 2.7 Computed angular acceleration of each link around Y axis during the normal breathing cycle as simulated in Fig. 2.4

2.5 Conclusions

Results motion simulations are discussed to check the feasibility of a proposal for a compliant mechanism whose design includes the interaction between the pair of ribs, the vertebra, and the sternum that are represented as links. With the results from the simulations, it can be concluded that the mechanism is able to run the desired movements since the rotation of link 3 (sternum) is not allowed, producing only its displacement on the sagittal plane. This movement satisfactorily represents the increase and decrease of the sagittal diameter in the rib cage during normal breathing. Future work can be focused on conducting investigations now considering cough conditions. It is also sought to carry out experimental tests to corroborate the analytical results. This paper presents a first solution of designing and modelling a full rib system connecting the spine and sternum. The final goal will be to have the full ribcage modelled to investigate with proper simulation the behaviour of an artificial ribcage.

References

1. West, J.: Pulmonary Physiology. 7th edn. Panamericana, Buenos Aires (2005)
2. Ratnovsky, A., Elad, D., Halpern, P.: Mechanics of respiratory muscles. Respiratory Physology & Neurobiol. **163**, 82–89 (2008)
3. Ya-Ting, Y., Yasuyuki, Y., Tibor, H., Shuji, S.: Interaction between thorax, lumbar, and pelvis movements in the transverse plane during gait at three velocities. J. Appl. Biomech. **29**, 261–269 (2013)

4. Aguilar, L.A., Ceccarelli, M., Torres San Miguel, Ch.R., Urriolagoitia Sosa, G., Urriolagoitia Calderón, G.: Experimental evaluation of artificial human ribs. In: Multibody Mechatronic Systems, pp. 434–443. Springer, Cham (2017)
5. Ramirez, O., Ceccarelli, M., Torres-San-Miguel, C.R., and Urriolagoitia-Calderon, G.: Experimental characterization of an osteosynthesis implant. In: Advances in Mechanism and Machine Science. IFToMM WC 2019. Mechanisms and Machine Science, pp. 53–62, vol. 73, Springer, Cham (2019)
6. Lemmen, P., Been, B., Carroll, J., Hynd, D., Davidsson, J., Martinez, L., Eggers, A.: An advanced thorax–shoulder design for the THOR dummy. In: Proceedings of the 23rd ESV Conference, pp. 1–14 (2013)
7. Howell, L.L., Magleby, S.P., Olsen, B.M.: Handbook of Compliant Mechanisms. Wiley, United Kingdom (2013)
8. Ramirez, O., Torres-San-Miguel, C.R., Ceccarelli, M., Rueda Arreguín, J.L., Urriolagoitia-Calderón, G.: A compliant mechanism as a sternum prosthesis. In: Rauter, G., Cattin, P.C., Zam, A., Riener, R., Carbone, G., Pisla, D. (eds) New Trends in Medical and Service Robotics. MESROB. Mechanisms and Machine Science, pp. 143–151, vol. 93. Springer, Cham (2021)
9. Merrill, L., Greg, B., Scott, W., J.W., McWhorter, Claire, P.: A comparison of tidal volume, breathing frequency, and minute ventilation between two sitting postures in healthy adults. Physiotherapy Theory Practice **19**, 109–119 (2003)
10. Padasdao, B., Shahhaidar, E., Boric-Lubecke, O.: Measuring chest circumference change during respiration with an electromagnetic biosensor. In: 35th Annual International Conference of the IEEE Engineering in Medicine and Biology Society (EMBC), pp. 1936–1939 (2013)
11. Nathan, V., Prithishkumar, I.J., Livingstone, R.S.: Estimation of anatomical dimensions of the thorax from computed tomography images of the adult and pediatric Indian population for developing optimal radiological protocols. J. Med. Phys. **46**, 295–299 (2021)
12. Mattson, C.A., Howell, L.L., Magleby, S.P.: Development of commercially viable compliant mechanisms using the pseudo-rigid-body model: case studies of parallel mechanisms. J. Intell. Mater. Syst. Struct. **15**, 195–202 (2004)
13. PLA technical data sheet, https://docs.rs-online.com/7ade/0900766b81698006.pdf. Accessed 21 April 2022
14. Hun Kang, B., Wen, J.Y., Dagalakis, N., Gorman, J.J.: Analysis and design of parallel mechanisms with flexure joints. IEEE Trans. Rob. **21**, 1179–1185 (2005)

Chapter 3
A Review on Mechanisms Used for the Reconfigurable Wheelchairs

I.-E. Popescu, F. Florescu, C. Sticlaru, D. T. Mărgineanu, and E.-C. Lovasz

Abstract The daily care of patients who are mobilized in bed and their transfer from one device to another is a common and increasingly difficult problem for both people with disabilities and caregivers even if they are in a hospital or at home. There are many ideas, projects, and even products that come to their aid when it comes to the issue of patient transfer. Some solve this problem by providing transfer solutions, and others simply cancel the transfer with reconfiguration solutions. This article is a review of several types of reconfigurable wheelchairs, their functionalities, and the solutions used for the folding mechanisms with which they are reconfigured. The review covers mostly the reconfiguration part as well as the folding mechanisms used for it. In addition to the folding mechanisms and their functionality; the command-and-control system was also considered. Also, the limitations that must be overcome are addressed here, and a new solution is proposed for discussion.

Keywords Folding mechanisms · Reconfiguration · Wheelchairs · Patient transfer

3.1 Introduction

Many people suffer from diseases, and some of these diseases cause the person to have certain disabilities. Therefore, certain types of diseases require more care for the patient, such as paraplegics. The wheelchair is the most widespread mobile device

I.-E. Popescu (✉) · F. Florescu · C. Sticlaru · D. T. Mărgineanu · E.-C. Lovasz
Politehnica University of Timisoara, Timisoara, Romania
e-mail: ioan-emil.popescu@student.upt.ro

F. Florescu
e-mail: florin.florescu@student.upt.ro

C. Sticlaru
e-mail: carmen.sticlaru@upt.ro

D. T. Mărgineanu
e-mail: dan.margineanu@upt.ro

E.-C. Lovasz
e-mail: erwin.lovasz@upt.ro

I. Doroftei et al. (eds.), *Proceedings of SYROM 2022 & ROBOTICS 2022*, Mechanisms and Machine Science 127, https://doi.org/10.1007/978-3-031-25655-4_3

for people with disabilities around the world and a way to improve their accessibility. There is a huge potential to support the needs of care since robotic technology has evolved and developed a lot in recent years.

Our goal is to assess the current state of research and first find out where improvement is needed and then how we can do this as best and efficiently as possible to support these needs and contribute to the evolution of the technology used in this area. Following this evaluation of the current stage, we will make a proposal for improvement, bringing into the discussion a new concept that we intend to analyze, design, implement, verify, validate and finally get feedback from patients and caregivers in the following posts.

3.2 State-of-Art of Reconfigurable Wheelchairs

In this literature review study, we specifically looked at the following:

- wheelchair reconfiguration and functionality.
- folding mechanisms for reconfiguration.
- displacement (wheels used).
- command and control system.

In terms of reconfiguration and functionality, the most common and used wheelchairs are those that are reconfigured from a wheelchair to a stretcher and vice versa [1–17], but existing wheelchairs are beginning to incorporate more and more functionalities. The following features are some of them: wheelchair with reconfiguration function in bed (with docking) and vice versa [18, 19]; wheelchair with reconfiguration function in orthostatic position for recovery and vice versa [20–23]; tilt to one side [19–33]; function of amelioration and prevention of bedsores with pressure sensors [34, 35]; transfer function [18, 24–26, 36–40]; lifting function [41, 42]; toilet function [43–45]; other functionalities (e.g. seat adjusting [46]).

Regarding the folding mechanisms, there are many works in which different kinematic concepts have been presented that illustrate the structure of mechanical connections for reconfigurable wheelchairs. The most common mechanism used for actuation is the screw mechanism [27, 28] and as folding mechanisms used for multiple reconfigurations are slider-crank [1, 4]; four-bar linkage [31]; pantograph mechanism [47]; worm wheel mechanism [16]; rack and pinion mechanism [48]. Some of the identified mechanisms will be presented in more detail in the following chapter.

The most common approaches used for moving the wheelchair are those with swivel caster wheels [4–8, 29], but other wheel systems are also used such as mechanum wheels [32, 45], omni platform with 4 wheels [32], special 6-wheel omnidirectional mechanism [19], 2 wheels [49], wheels with suspension for rough terrain [50] and others. We also noticed that there is a high interest in wheel mechanisms that can climb the stairs on their own [30, 31].

Three major categories have been identified on the command-and-control side, as follows. Where the low budget for manufacturing but also the costs for maintenance was a priority, the operation is done manually by the patient. In cases where these were not the main purpose of the work, the operation of the mechanisms was done electrically either manually by the patient or caregiver when he presses a button; either automatically using artificial intelligence to read various gestures made by the patient with his head or hand (e.g., myoelectric signals from the forearm muscles [32]) or memorizing daily routes and routines.

3.3 Reconfigurable Mechanisms Used for Wheelchairs

The objective of this chapter is to present the main mechanisms for reconfiguring the backrest and footrest of wheelchairs. During the research stage, we noticed that the most frequently used mechanisms for raising and lowering the backrest and footrest are the inverted slider-crank mechanism, the four-bar linkage mechanism, and the slider-crank mechanism. Of course, besides these, other mechanisms are also used. Another contribution we made in this study besides reviewing literature and bringing solutions for improvement, was to make the drawings for all the mechanisms used in the presented works. Further are presented some of them.

Meng Ning and his team [19] developed and tested a reconfigurable system of wheelchair/bed. For the reconfiguration of the wheelchair, he used two inverted slider-crank mechanisms with linear actuation for reclining the backrest and footrest, which are operated by two linear actuators that work independently, Fig. 3.1a.

Jin Rui and Qiang Gao [17] used in their work two inverted slider-crank mechanisms with linear actuation for raising or lowering the backrest and the footrest. These two mechanisms work independently and do not interfere with each other in the process of reconfiguration. The height of the seat is fixed, and the mechanisms are operated by linear actuators, Fig. 3.1b. Also, they use an additional two-stage linkage mechanism for more adjusting of the backrest to the required angle of inclination.

Sreerag [4] has implemented in his wheelchair two inverted slider-crank mechanisms with linear actuation for adjusting the backrest and footrest. The mechanism for raising and lowering the backrest is operated independently of the mechanism for supporting the legs, and the seat height adjustment mechanism is dependent on those for the backrest and footrest. The seat has the possibility of height adjustment through a hydraulic mechanism, Fig. 3.1c.

Goher [21] uses an inverted slider-crank mechanism controlled by a linear actuator to adjust the backseat, while the footrest part is fixed and cannot be adjusted. An additional mechanism is used to allow the patient to be raised to a standing position, Fig. 3.1d.

Lingfeng Sang [45] operates two four-bar linkage mechanisms for reconfiguring the backrest and the footrest of his wheelchair project, Fig. 3.2a. The action of adjusting the backrest and footrest is dependent on each other. There is the possibility that the seat can be adjusted in height, and it contains a toilet function.

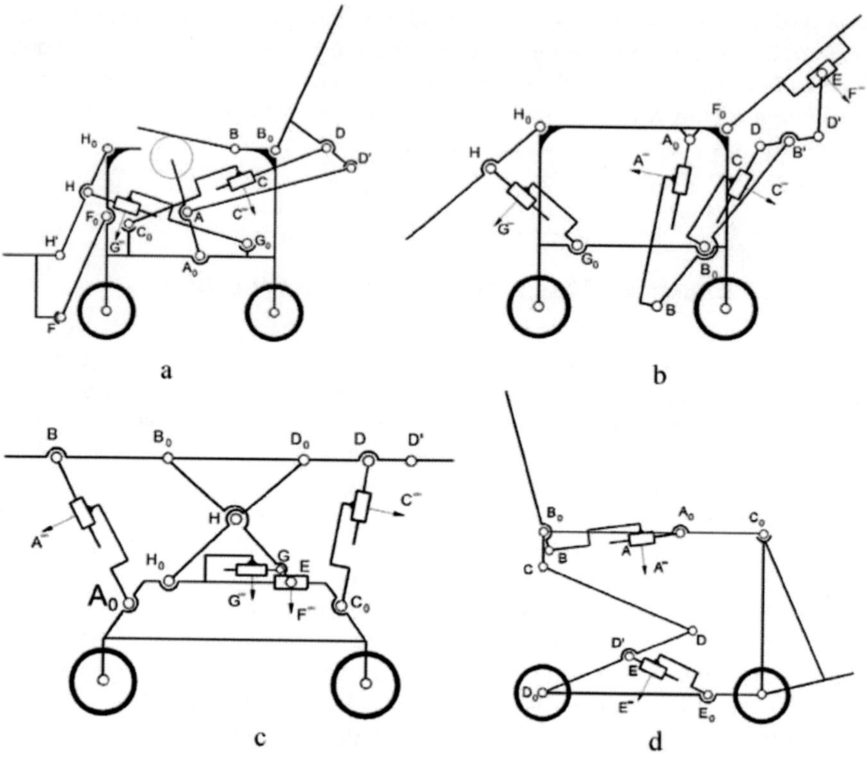

Fig. 3.1 Inverted slider-crank mechanisms used for reclining the backrest and footrest

Sandeep Kumar's work [28] presents a wheelchair that uses two four-bar linkage mechanisms for reclination of the backrest and the footrest, Fig. 3.2b. The mechanisms are dependent on each other and manually operated through a screw-dive mechanism, with the possibility of adding an electric motor. The seat is fixed, without height adjusting and standing capability.

Dr. Sukanta Roga [27] designed a wheelchair that operates two four-bar linkage mechanisms, one for adjusting the reclination of the backrest and the other one for adjusting the reclination of the footrest, Fig. 3.2c. The operation of the mechanisms is dependent on each other and with a linear movement. The mechanisms are operated by an electric DC motor, and the seat has no standing or height adjustment capabilities.

Rashid [3] has implemented a four-bar linkage mechanism with a dependent operation between the backrest and the footrest. The height of the system changes when it is reconfigured in the stretcher, Fig. 3.2d. The operation is linear and managed by pneumatic actuators.

Nirmal Mistry [20] developed a wheelchair that uses the slider crank mechanism driven by an electric motor for adjusting the backrest and the footrest. The actuation

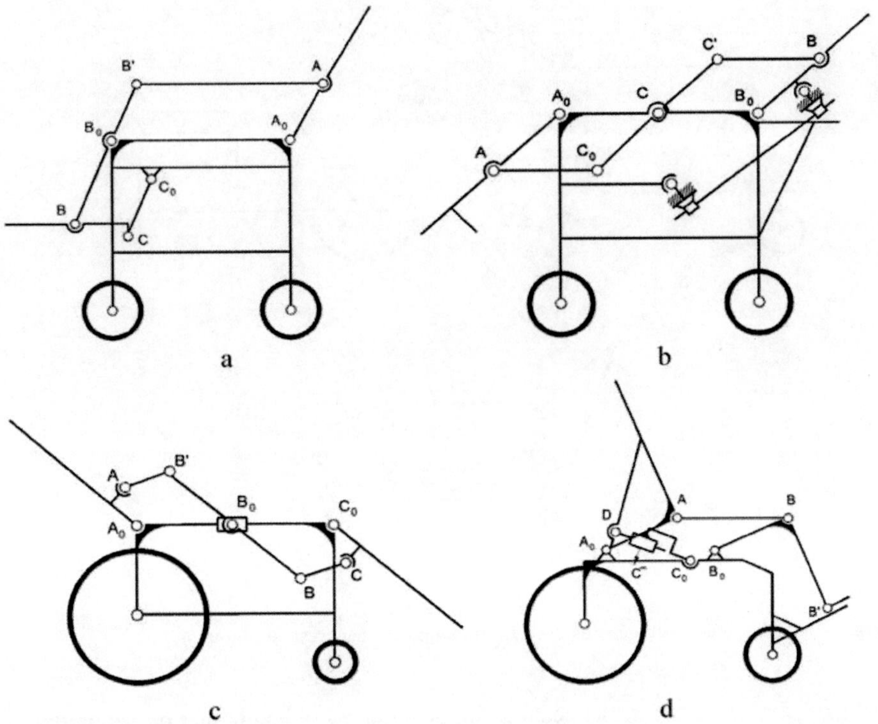

Fig. 3.2 Four-bar linkage mechanisms used for reclining the backrest and footrest

of the mechanisms is dependent on each other, and the seat that is driven by a linear actuator can rise for the standing position, Fig. 3.3a.

Meng-Hui Hsu [10] uses two different types of mechanisms for reclination of the backrest and footrest of his wheelchair. A slider-crank mechanism is used to adjust the backrest, while a four-bar linkage mechanism is used for the footrest. The actuation of the backrest is dependent on that of the footrest; the seat is fixed, without height adjustment, and with no standing function, Fig. 3.3b.

Smit Patel [48] designs a reconfigurable wheelchair that uses a slider crank mechanism to adjust the backrest and a scotch-yoke mechanism to recline the footrest. The seat is not height-adjustable, and the entire system has no standing function, Fig. 3.3c. The operation of the backrest is dependent on that of the footrest, the movement is linear, and the rack and pinion mechanism can be driven by an electric DC motor.

Sumit Desai [1] presented in his project a correlated inverted slider-crank mechanism for adjusting the backrest with a four-bar linkage mechanism for adjusting the footrest. Also, an additional four-bar linkage mechanism allows the patient to lift for the standing position, Fig. 3.4a. The whole mechanism is designed to use one rotary

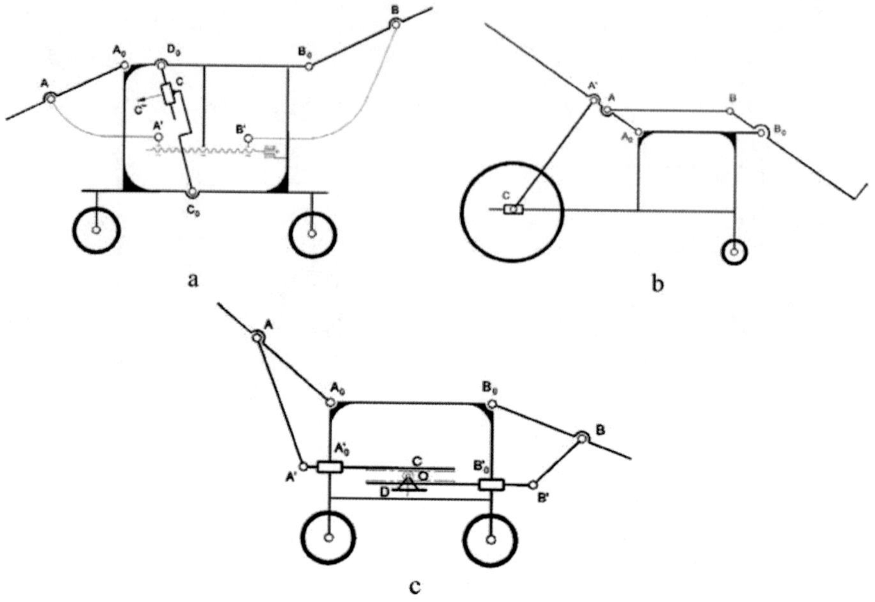

Fig. 3.3 Piston crank mechanisms used for reclining the backrest and footrest

actuator for all transfers; the seat cannot be adjusted in height and the actuation of the backrest and footrest is dependent on each other.

With the work of Gautam Dawar [23], a five-bar link mechanism (mechanism with a pivot and a fulcrum) with the manual operation was implemented to reconfigure the

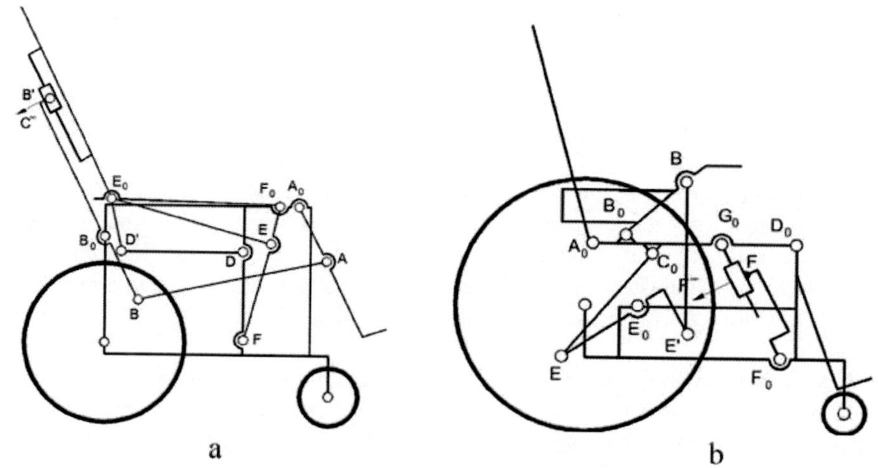

Fig. 3.4 Other mechanisms used for reclining the backrest and footrest

Table 3.1 Used types of mechanisms and their mode of operation

Paper Ref.	Backrest mechanism	Footrest mechanism	Backrest/Footrest actuation	Seat function	Actuation type
[19]	Inverted slider-crank	Inverted slider-crank	Independently	Fixed	Linear actuator
[17]	Inverted slider-crank	Inverted slider-crank	Independently	Fixed	Linear actuator
[4]	Inverted slider-crank	Inverted slider-crank	Independently	Height adjusting	Linear actuator
[21]	Inverted slider-crank	Fixed	Independently	Lift to standing	Linear actuator
[1]	Inverted slider-crank	Four-bar linkage	Dependent	Lift to standing	Rotary actuator
[45]	Four-bar linkage	Four-bar linkage	Dependent	Height adjusting	Linear actuator
[28]	Four-bar linkage	Four-bar linkage	Dependent	Fixed	Manual
[27]	Four-bar linkage	Four-bar linkage	Dependent	Fixed	DC Motor
[3]	Four-bar linkage	Four-bar linkage	Dependent	Height adjusting	Linear/Pneumatic
[20]	Slider crank	Slider crank	Dependent	Lift to standing	Linear/DC Motor
[10]	Slider crank	Four-bar linkage	Dependent	Fixed	Manual
[48]	Slider crank	Scotch-yoke	Dependent	Fixed	DC Motor

wheelchair. The seat is not adjusted in height but the whole system has the possibility to lift the patient into the standing position, Fig. 3.4b. The footrest is fixed, while the backrest is adjusted by actuating the rising mechanism.

Below is a Table 3.1 in which we took some of the reviewed literature works and compared them based on their mode of operation.

3.4 Summary and Conclusions

This paper was intended for a review on the folding mechanisms for the reconfiguration of wheelchairs, in order to find the needs in this field and bring solutions for improvement.

During the research stage, we noticed that the advantages of mechanisms with bars with 6 elements are not sufficiently exploited, respectively the seat, backrest and footrest do not represent the coupler of such mechanisms and deserve to be studied more in terms of reconfigurability.

References

1. Desai, S., Mantha, S., Phalle, V.: Design of a reconfigurable wheelchair withstand-sit-sleep capabilities for enhanced independence of long-term wheelchair users. Technol. Disabil. **30**, 135–151 (2018)
2. Kulkarni, S.B., Thakare, A.J., Tamann, S.H., Roman, G.S., Karankoti, S.V.: Design and fabrication of wheelchair-to-bed system using fluid power. IJSART **2**, 13–18 (2016)
3. Rashid, A.K., Safar, A.R., Shamil, S., Vishnu, P.K.V., Vishnu, C.R.: Design and fabrication of the pneumatically powered wheelchair-stretcher device. Int. J. Innov. Res. Sci. **4**, 10278–10289 (2015)
4. Sreerag, S., Gopinath, C., Mishra, M.R.: Design and development of conceptual wheelchair cum stretcher. SAS-TECH **10**, 78–86 (2011)
5. Kanisuru, M.: Adeyeri, Sesan PeterAyodeji, Abimbola Omotayo Orisawayi: development of a dual-purpose wheelchair for COVID-19 paraplegic patients using Nigerian anthropometry data. Scientific African **9**, 1–11 (2020)
6. Borkar, N.M., Apte, S.A., Deshmukh, T.N., Apte, S.M.: Mechanically operated wheelchair convertible stretcher. IJMET **7**, 261–265 (2016)
7. Arunkumar, S.M., Abhijith, P.K.: Haneepsab, A.K., Chetan, C: Design and fabrication of stretcher cum wheelchair. IJLERA **2**, 90–96 (2017)
8. Hirudkar, A., Bhusari, C., Khedkar, A., Parekar, A., Daf, S.P.: Automatic stretcher cum wheelchair. Int. Res. J. Eng. Technol. **4**, 1659–1663 (2017)
9. Jambhulkar, S., Ramteke, N., Prajapati, S., Nitnaware, A., Hargude, S.: Design and fabrication of wheelchair cum stretcher with suspension. Int. J. Scientific Res. Rev. **8**, 254–260 (2019)
10. Hsu, M.-H., Chen, H.-Y., Liu, J.-Y., Chen, C.-L.: Dual-purpose wheelchair mechanism designs. IMECS, 2 (2009)
11. Ugale, R.D., Virulkar, S.N., Bathe, A.T.: Design and development of Humanscale Wheelchair cum stretcher. IRJET **4**, 2510–2512 (2017)
12. Rajeev, V.R., Ramjith Krishnan, R., GopalakrishnaPillai, K.: Fabrication of Advanced wheelchair cum strecher. IJESTA **2**, 1–7 (2016)
13. Dr. Ramachandra, C.G., Shashank, S., Raghavendra, M.J., Kaushik Ranganath. T.G.: Wheelchair cum stretcher. IJERT **6**, 42–46 (2017)
14. Bobde, S., Borkar, N., Apte, S., Ghuguskar, S.: Automated wheelchair convertible stretcher. IRJET **4**, 889–892 (2017)
15. Gayatri Haritha, N., Choutele, A., Sawant, K.: Health monitoring wheelchair cum bed system. IJIRSET **9**, 1495–1498 (2020)
16. Paul, S., Shoyeb Raihan, A., Islam, F.: Automated Powered Wheelchair. Bangladesh University of Engineering & Technology, pp.1–83 (2020)
17. Rui, J., Gao, Q.: Design and analysis of a multifunctional wheelchair. IOP Conference Series 2019: Materials Science and Engineering, vol. 538, pp. 1–5 (2019)
18. Tian, Y., Wang, H., Zhang, Y., Su, B., Wang, L., Wang, X., Sang, L., Feng, Y., Niu, J.: Design and Evaluation of a Novel Person Transfer Assist System. IEEE, 1–9 (2021)
19. Ning, M., Ren, M., Fan, Q., Zhang, L.: Mechanism design of a robotic chair/bed system for bedridden aged. Adv. Mech. Eng. **9**, 1–8 (2017)
20. Mistry, N., Panchal, M., Mangroliya, B., Labana, A.K.: Design and fabrication of convertible wheelchair. IJARIIT **4**, 485–492 (2018)
21. Khaled, M.: Goher: a reconfigurable wheelchair for mobility and rehabilitation: design and development. Cogent Eng. **3**, 1–23 (2016)
22. Abdul Ghani, N.M., Tokhi, M.O.: Sit-to-stand and stand-to-sit control mechanisms of two-wheeled wheelchair. J. Biomech. Eng. **138**, 041007–1 to 041007–12 (2016)
23. Dawar, G., Kejariwal, A., Kumar, D.: Design of a modular wheelchair with posture transformation capabilities from sitting to standing. Disabil. Rehabil. Assist. Technol. **15**, 1–14 (2019)

24. Wei, C.-H., Tung, T.-C., Hsiao, S.-C., Chen, W.-C., Chiu, Y.-M., Tu, K.-T., Yeh, C.-W., Chen, K.-Y.: Hospital bed with auxiliary functions of lateral positioning and transferring for immobilized patients. IEEE Industrial Electronics Society (IECON), pp. 2991–2995 (2007)
25. Suryawanshi, S.J., Janardhan Reddy, K.: Conceptual product development of wheelchair for people disabled in legs. Int. J. Res. Mech. Eng. **1**, 1–10 (2013)
26. Mukai, T., Hirano, S., Nakashima, H., Sakaida, Y., Guo, S.: Realization and safety measures of patient transfer by nursing-care assistant robot RIBA with tactile sensor. J. Robot. Mechatron. **23**, 360–369 (2011)
27. Roga, S., AKumar, A., Aman, Singh, A., Kumar, B.: Design and fabrication of wheelchair cum stretcher with multi fold. IJAIEM **6**, 167–175 (2017)
28. Kumar, S., Sahni, G., Kaintura, G., Thapliyal, A., Bisht, N., Bijalwan, V.-K., Singh, G.: Fabrication of collapsible chair cum bed model-an engineering approach. Int. J. Sci. Eng. Res. **9**, 50–54 (2018)
29. Ujawane, R., Mahalle, A.K.: Mechanically operated-wheelchair convertible stretcher and walker. IJARIIE **3**, 484–490 (2017)
30. Ghani, N.M.A., Nasir, A.N.K., Hassan, M.A.H., Tokhi, M.O.: PD-fuzzy control of a stair climbing wheelchair. AASRI Procedia **4**, 18–25 (2013)
31. Morales, R., Gonzalez, A., Feliu, V.: Mechanical and kinematics design methodology of a new wheelchair with additional capabilities. InTech—Bioinspiration and Robotics Walking and Climbing Robots, Habib, M.K. (Ed.), pp. 463–482 (2007)
32. Kundu, A.S., Mazumder, O., Lenka, P.K., Bhaumik, S.: Omnidirectional assistive wheelchair: design and control with isometric myoelectric based intention classification. IEEE IRIS—Procedia Comput. Sci. **105**, 68–74 (2017)
33. Sivadas, A., Philip, E., Jacob, C.J., Varghese, F.: An evaluation of wheelchair cum bed mechanism with side panel movement for bed. IJIRST **2**, 287–292 (2016)
34. Park, K.-H., Bien, Z., Lee, J.-J., Kim, B.K., Lim, J.-T., Kim, J.-O., Lee, H., Stefanov, D.H., Kim, D.-J., Jung, J.-W., Do, J.-H., Seo, K.-H., Kim, C.H., Song, W.-G., Lee, W.-J.: Robotic smart house to assist people with movement disabilities. Autonom. Robots **22**, 183–198 (2007)
35. Seo, K.-H., Oh, C., Choi, T.-Y., Lee, J.-J.: Bed-type robotic system for the bedridden. International Conference on Advanced Intelligent Mechatronics 2005, IEEE/ASME, pp. 1170–1175 (2005)
36. Wang, H., Tsai, C.-Y., Jeannis, H., Chung, C.-S., Kelleher, A., Grindle, G.G., Cooper, R.A.: Stability analysis of electrically powered wheelchair-mounted robotic-assisted transfer device. JARD **51**, 761–774 (2014)
37. Burkman, J., Grindle, G., Wang, H., Kelleher, A.: Rory a cooper: further development of a robotic-assisted transfer device. Top Spinal Cord Inj. Rehabil. **23**, 140–146 (2017)
38. Grindle, G.G., Wang, H., Jeannis, H., Teodorski, E., Rory, A.: Cooper: Design and User Evaluation of a Wheelchair Mounted Robotic Assisted Transfer Device. Hindawi Publishing Corporation BioMed Research International, 1–9 (2015)
39. Wang, H., Xie, L., Zhao, X., Niu, J., Liu, Z., Vladareanu, L., Mihai, R.-V.: Mechatronics system design and experiment research for a novel patient transfer apparatus. 3rd ICMRA, pp. 1140–1143 (2015)
40. Krishnan, R.H. and Pugazhenthi, S.: Concept development and design of self-transfer devices for wheelchair users. Int. J. Autom. Smart Technol. **9**, 1–11 (2019)
41. Mori, Y., Sakai, N., Katsumura, K.: Development of a wheelchair with a lifting function. Hindawi Publishing Corporation - Advances in Mechanical Engineering, pp. 1–9 (2012)
42. Bostelman, R., Albus, J.: A multipurpose robotic wheelchair and rehabilitation device for the home. IEEE/RSJ International Conference on Intelligent Robots and Systems 2007, pp. 1–6 (2007)
43. Krishnan, R.H., Pugazhenthi, S.: Mobility assistive devices and self-transfer robotic systems for elderly, a review. Intel Serv. Robot. **7**, 37–49 (2014)
44. Mohan Kumar, R., Lohit, H.S., Mishra, M.R., Basheer Ahamed, Md.: Design of multipurpose wheelchair for physically challenged and elder people. SASTECH **11**, 107–117 (2012)

45. Sang, L., Yamamur, M., Dong, F., Gan, Z., Fu, J., Wang, H., Tian, Y.: Analysis, design, and experimental research of a novel wheelchair-stretcher assistive robot. Appl. Sci. **10**, 264 (2020)
46. Hsu, P.E., Hsu, Y.L., Lu, J.M. and Chang, C.H.: Seat adjustment design of an intelligent robotic wheelchair based on the stewart platform. INTECH (Int. J. Adv. Robot. Syst.) **10**, 1–11 (2013)
47. Yukawa, T., Kuramochi, Y., Takahashi, T., Takahashi, K.: Nursing-care system for bedridden patients with electric wheelchair, lift, portable bath. mobile robot portable toilet. IJEIT **2**, 100–108 (2012)
48. Patel, S., Koli, Y., Talekar, B., Rathod, V., Baria, J., Patel, P.: Modification and fabrication of wheelchair cum stretcher. IRJET **7**, 1900–1903 (2020)
49. Nikpour, M., Huang, L., Al-Jumaily, A.M.: Stability and direction control of a two-wheeled robotic wheelchair through a movable mechanism. IEEE **8**, 45221–45230 (2020)
50. Madanhire, I., Gudukeya, L., Mushonga, R.: Development of a Rough Terrain Wheelchair Design. Public Health in Developing Countries—Challenges and Opportunities, 1–16 (2020)

Chapter 4
Type Synthesis of the Mechanisms with Linear Actuation Useful for Orientation Modules

F. Florescu, I.-E. Popescu, C. E. Moldovan, and E.-C. Lovasz

Abstract The paper shows a study of the mechanism structures which have a linear actuation as drive element and an oscillating element as driven element. The systematic type synthesis of all the possible structures takes in account the elementary kinematic chains of the mechanisms using linkages and/or gears with three, four, five and six elements. The mechanism structures outcome from the type synthesis can be useful for orientation modules with linear actuation in serial robotics. These mechanisms will be further kinematical and kinetostatical characterized in order to optimize their performances.

Keywords Kinematic chains · Mechanisms structures · Orientation mechanism module · Linear actuator

4.1 Introduction

The paper studies the structural design of the mechanism with linear actuators and oscillating output elements. The type synthesis shows a systematic development of the mechanism structures starting from the elementary kinematic chains and eliminating the structures were: not all of the elements are involved in the motion transmission, respectively the linear actuator is not adjacent to the frame. This study is an important issue in the process of fixing all the possible structures able to ensure a large

F. Florescu · I.-E. Popescu · C. E. Moldovan · E.-C. Lovasz (✉)
Politehnica University of Timisoara, Timisoara, Romania
e-mail: erwin.lovasz@upt.ro

F. Florescu
e-mail: florin.florescu@student.upt.ro

I.-E. Popescu
e-mail: ioan-emil.popescu@student.upt.ro

C. E. Moldovan
e-mail: cristian.moldovan@student.upt.ro

I. Doroftei et al. (eds.), *Proceedings of SYROM 2022 & ROBOTICS 2022*, Mechanisms and Machine Science 127, https://doi.org/10.1007/978-3-031-25655-4_4

oscillating angle without singularities and with a minimum number of elements. The further kinematic and kinetostatic analysis of the useful mechanism structures with linear actuators and oscillating output elements for robotic orientation mechanisms allows the optimization of their performances.

Applications of simple and complex closed kinematic chains used for orientation mechanisms are numerous. Huang et all designed a 2-DOF parallel robot for fast sorting, studied a hybrid method that enables the optimal kinematic design to be implemented and to maximize the structural length index. So, a modified global index was proposed [1].

An 2-DOF-oriented mechanism was developed for robots designed for welding, painting or servicing in which the two movement of orientation and their decoupling, a differential planetary unit with toothed wheels, are made with two linear motors [2].

Several industrial robots are gifted with a bi-mobile orientation mechanism (MOr-2R), with two exit rotations, or trimobile orientation mechanism (MOr-3R), with three exit rotations [3]. Kinematic schemes of the planetary mechanism with cylindrical gears and conical gears are homomorphic structures [4].

The design and model of a reconfigurable 6-DOF robot manipulator, equipped with joints with no actuator or sensor, on-board a controlled spacecraft for on-orbit servicing was proposed by Aghili [5].

The problem of singular configurations of different kinds appear in closed-loop kinematic chains, defined as those in which the Jacobian matrices are involved. The Watt and Stephenson linkages evidently constitute planar 6-link complex kinematic chains. Performing a precise repetitive task it is nowadays possible since the advances in computer-aided synthesis of linkages so complex architectures enhance both the accuracy and the load-carrying capacity of the systems at hand [6].

4.2 Type Synthesis of Elementary Kinematic Chains

The current type synthesis takes in account all the elementary kinematic chains of the mechanisms using linkages and/or gears (limited to one gear kinematic pair) with three, four, five and six elements useful for a robotic orientation module with linear actuation.

All the considered kinematic chains of the mechanism structures have one prismatic kinematic pair, which generate the linear motion of the drive element. The driven (output) element.

The condition to have a constrain motion of a mechanism structure, according to Alt [7], is:

$$2 \cdot (e_1 + e_2/2) - 3 \cdot n + 3 + M = 0, \tag{4.1}$$

where: e_1—number of the kinematic pairs with DoF = 1,

e_2—number of the kinematic pairs with DoF = 2,

n—number of elements,

M—mechanism degree of freedom (mobility).

4.2.1 Type Synthesis of the 4-bar and 6-bar Linkages

By considering all the elements of the kinematic chain bars, the input conditions for the kinematic chains are: to contain only kinematic pairs with DoF = 1 ($e_2 = 0$), one of them should be a prismatic kinematic pair and to have the mobility of the mechanism $M = 1$. That means, the constrain motion condition (4.1) with the previous input conditions will give the correlation between e_1 and n:

$$e_1 = \frac{3 \cdot n - 4}{2} \in N. \tag{4.2}$$

The first solution of the Eq. 4.2 follows for:

$$n = 2 \rightarrow e_1 = 1 \in N, \tag{4.3}$$

which is not convenient because we have only a drive prismatic kinematic pair.

The second solution of the Eq. 4.2 is:

$$n = 4 \rightarrow e_1 = 4 \in N, \tag{4.4}$$

The numbers of the elements of different ranks satisfy the diophantine equations system:

$$\begin{cases} n = \mathrm{n}_2 + \mathrm{n}_3 \\ 2 \cdot \sum\limits_{i=1}^{2} \mathrm{e}_i = 2 \cdot \mathrm{n}_2 + 3 \cdot \mathrm{n}_3 \end{cases} \quad \begin{cases} 4 = \mathrm{n}_2 + \mathrm{n}_3 \\ 8 = 2 \cdot \mathrm{n}_2 + 3 \cdot \mathrm{n}_3 \end{cases} \tag{4.5}$$

where n_2, n_3 are the number of binary and ternary elements (higher ranks elements are neglected). The computed number of binary and ternary elements will be:

$$n_2 = 4, n_3 = 0. \tag{4.6}$$

The kinematic chain for this case and the developed four bar linkages applying the Reuleaux method are shown in Fig. 4.1. The notation used in the developed mechanisms structures started with the frame joint followed successive from the next joints of the frame closed loop, respectively in brackets () is indicated the parallel connected kinematic chain, where R—revolute joint, P—prismatic joint and G—gear pair are. The actuated joint is underline indicated. In the considered cases

Four bars kinematic chain		Frame a	Frame b	Frame c	Frame d
1	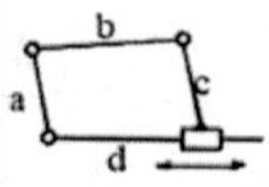	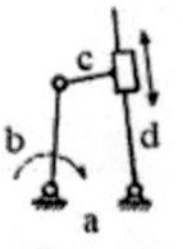 RPRR(a)		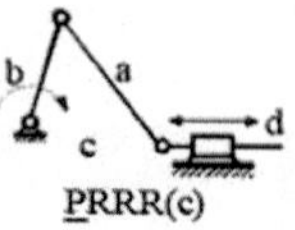 PRRR(c)	

Fig. 4.1 Development by Reuleaux method four bar linkages with linear actuation

the actuated joint is a prismatic joint P. The last bracket shows the considered frame element of the kinematic chain. Identical structures are no more represented.

The resulted mechanism structures are the inverted slider crank RPRR(a) and the slider crank PRRR(c).

The third solution of the Eq. 4.2 is:

$$n = 6 \rightarrow e_1 = 7 \in N, \tag{4.7}$$

and from the diophantine equation system (4.5):

$$\begin{cases} 6 = \mathrm{n}_2 + \mathrm{n}_3 \\ 14 = 2 \cdot \mathrm{n}_2 + 3 \cdot \mathrm{n}_3 \end{cases} \tag{4.8}$$

results the number of the of the binary and ternary elements:

$$n_2 = 4, n_3 = 2. \tag{4.9}$$

The chains in this case are the known Watt and Stephenson six elements chains. Using the same Reuleaux method results the mechanism structure shown in Figs. 4.2 and 4.3.

The resulted mechanism structures shown in Figs. 4.2 and 4.3, where the useful six bar linkages have the linear actuator close to the frame and they are drawn with black. Also, the mechanism structures drawn with red have as main transmission loop a four-bar linkage and the parallel connected loop don't influence the motion of the driven element.

4.2.2 *Type Synthesis of Geared Linkages*

By considering the input conditions for the kinematic chains to contain one higher kinematic pairs with DoF $= 2$ ($e_2 = 1$), one of the lower kinematic pair should be a prismatic pair and to have the mobility of the mechanism $M = 1$, follows the

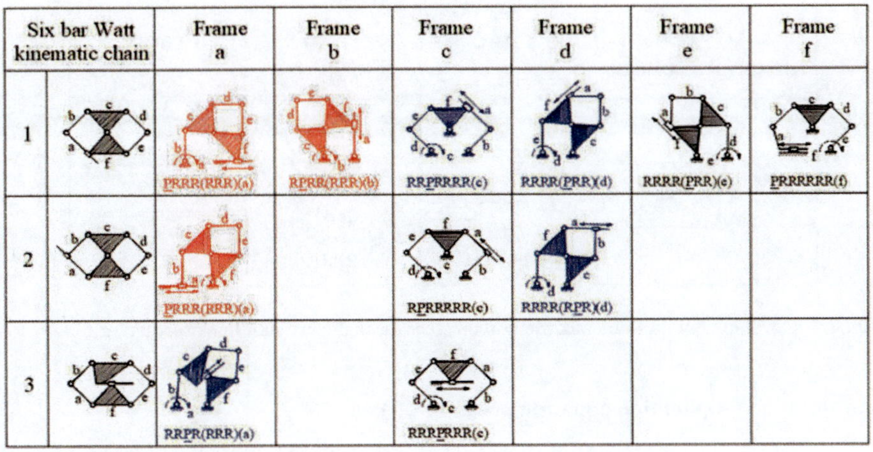

Fig. 4.2 Development by Reuleaux method the six bar Watt linkages with linear actuation

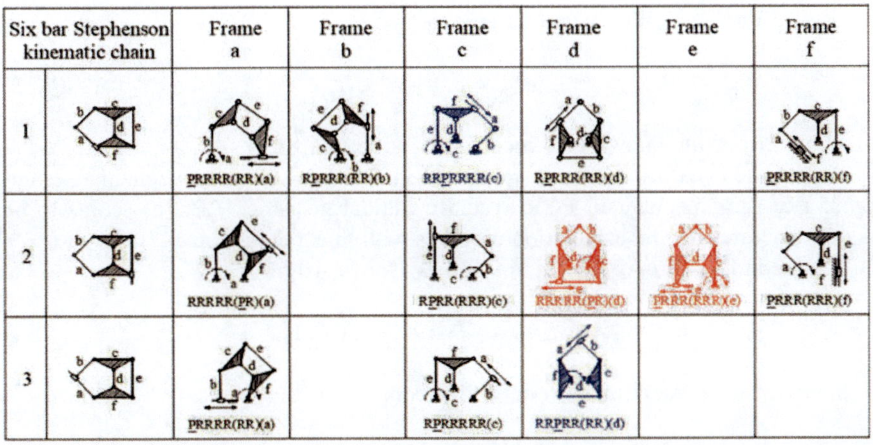

Fig. 4.3 Development by Reuleaux method the six bar Stephenson linkages with linear actuation

geared linkages. The correlation between e_1 and n considering the constrain motion condition (4.1) is:

$$e_1 = \frac{3 \cdot n - 5}{2} \in N. \tag{4.10}$$

The first solution of the Eq. 4.9 follows for:

$$n = 3 \rightarrow e_1 = 2 \in N, e_2 = 1, \tag{4.11}$$

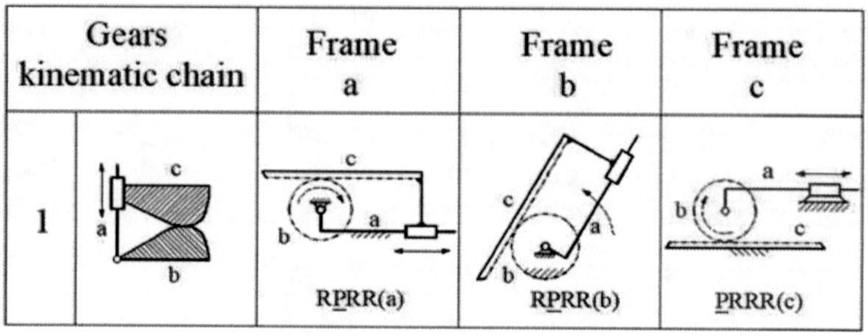

Fig. 4.4 Development by Reuleaux method the gear mechanisms with linear actuation

and from the diophantine equation system (4.5):

$$\begin{cases} 3 = \mathrm{n}_2 + \mathrm{n}_3 \\ 6 = 2 \cdot \mathrm{n}_2 + 3 \cdot \mathrm{n}_3 \end{cases} \tag{4.12}$$

results the number of binary and ternary elements:

$$n_2 = 3, n_3 = 0, \tag{4.13}$$

which described the known gear mechanism structure [8].

The resulted gear mechanisms are shown in Fig. 4.4. Two gear mechanism structures: gear-rack mechanism PGR(a) and cycloidal gear-rack RPG(b) seems to be useful structures for the orientation modules with linear actuation.

The second solution of the Eq. 4.9 follows for [9, 10]:

$$n = 5 \to e_1 = 5 \in N, e_2 = 1, \tag{4.14}$$

In this case the equation system (4.5) becomes:

$$\begin{cases} 5 = \mathrm{n}_2 + \mathrm{n}_3 \\ 12 = 2 \cdot \mathrm{n}_2 + 3 \cdot \mathrm{n}_3 \end{cases} \tag{4.15}$$

where the computed number of binary and ternary elements are:

$$n_2 = 3, n_3 = 2 \tag{4.16}$$

and the number of closed loops for the kinematic chain follows with:

$$N = \sum_{i=1}^{2} e_i - n + 1 = 2. \tag{4.17}$$

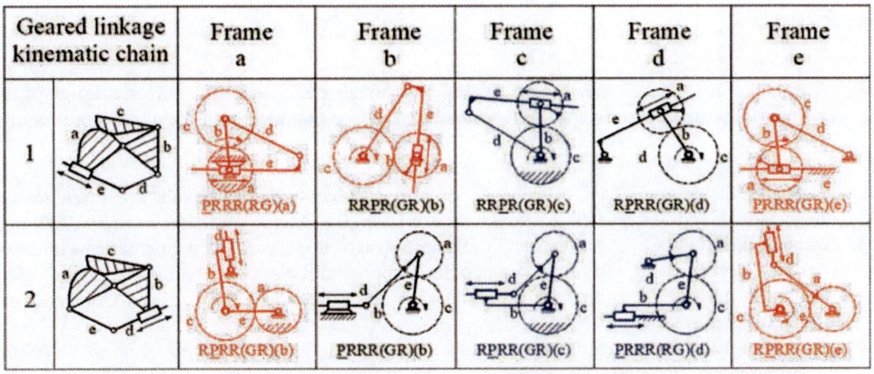

Fig. 4.5 Development by Reuleaux method the geared linkages with linear actuation

The systematic development of all possible connections between the elements with kinematic pairs give 11 possible kinematic chains. Through the additional conditions to have the loops with positive degree of freedom and to fulfil the gear contact condition remain available 6 kinematic chains [8].

The resulted mechanism structures useful for the orientation modules are shown in Fig. 4.5 (drawn with black). The useful geared structures are inverted slider-crank RPRR(GR)(d) and slider-crank PRRR(GR)(b). The other shown structures have not all the elements involved in the motion transmission or have the prismatic pair not close to the frame.

4.3 Conclusions

The common solutions for actuating the orientation modules of the serial robots used rotational servomotors, which actuate through an ordinar or cycloidal gear train the rotation joint. Some solutions shows correlated motions of two orientation motions by using an cycloidal gear train with 2 DoF actuated by two rotational motors.

Because several advantages of the linear servoactuators consisting in higher accuracy and repetability, lower noise, higer velocity respectively self-locking condition ensured without using a break. The study shows a systematic type synthesis of mechanism structures using linkages and/or gears with three, four, five and six elements.

The resulted mechanism structures will be further kinematical and kinetostatical analyzed and compared, in order to find out the optimal solution or solutions for the orientation module of a serial robot. The additional conditions need to ensure that the mechanism structure allow a large oscillation (orientation) angle, a favorable transmission angle, avoid the singularities, has a compact design and a linear or approximately linear transmission function.

References

1. Huang, T., Li, M., Li, Z., Chetwynd, D.G., Whitehouse, D.J.: Optimal kinematic design of 2-DOF parallel manipulators with well-shaped workspace bounded by a specified conditioning index. IEEE Trans. Robot. Automat. **20**, 538–543 (2004). https://doi.org/10.1109/TRA.2004.824690
2. Dudiţă, F., Diaconescu, D.V., Enache, V., Stareţu, I., Macovei, O.: Mecanism de orientare cu două grade de mobilitate. Oficiul de Stat pentru Invenţii şi Mărci—www.osim.ro 3 (1990)
3. Antonescu, P., Grecu, B., Antonescu, O.: Sinteza structural-geometrică şi cinematică a mecanismelor de orientare cu roţi dinţate conice. In: Simpozionul Naţional Mecanisme, Tribologie, Organe de maşini, Transmisii mecanice, Design de produs (7, 2002, Braşov). Universitatea Transilvania din Braşov 6 (2002)
4. Antonescu, P.: Contribuţii la sinteza mecanismelor de orientare din structura roboţilor. SYROM 93. Bucureşti **II**, 9–16 (1993)
5. Aghili, F., Su, C.: Reconfigurable space manipulators for in-orbit servicing and space exploration. In: International Conference on Intelligent Robotics and Applications, 3–5 October 2012, Montreal, QC, Canada (2012)
6. Gosselin, C., Angeles, J.: Singularity analysis of closed-loop kinematic chains. IEEE Trans. Robot. Automat. **6**, 281–290 (1990). https://doi.org/10.1109/70.56660
7. Luck, K., Modler, K.-H.: Getriebetechnik. Akademie-Verlag, Berlin, Analyse—Synthese—Optimierung (1990)
8. Lovasz, E.-C., Cărăbaş, I.: Principii de sinteză a mecanismelor cu roţi dinţate şi came, Editura Politehnica, Timişoara (2006)
9. Lovasz, E.-C., Modler, K.-H., Pop, C., Pop, F., Mărgineanu, D.T., Maniu, I.: Type synthesis and analysis of geared linkages with linear actuation. Mechanika **24**(1), 108–114 (2018)
10. Modler, K.-H., Lovasz, E.-C., Perju, D., Hollmann, Ch.: Geared linkages with linear displacement actuator used as function generating mechanisms. 11th World Congress in Mechanism and Machine Science, Tianjin, vol. 3, 1254–1259 (2004)

Chapter 5
Forces Transmission Index of the Structural Group 0/3/3

C. D. Duca and F. Buium

Abstract In this work the structural groups procedure is applied, based by authors in a previous series of papers, referring to the structural group 0/3/3. This group is the most complex between usual groups because it has the configuration DoF equals three, whereas the others usual structural groups have DoF equals one. In the first section of the paper, the calculation procedure was shown of forces transmission index, noted T, as function of configuration parameters. It ends up therefore with an algorithm, usable in the ulterior applications. In the second section of the paper the connection is shown between structural group 0/3/3 and the mechanism to which it belongs. It results therefore that primary configuration parameters are distances between exterior joints. Its expressions allow writing of the equations, leading to the necessary parameters for index T calculation. For example, a mechanism of a parallel robot is taken into consideration, and two procedures was shown of forces transmission quality analysis, combining the analytical relationships with computer aided graphical representations.

Keywords Structural group 0/3/3 · Forces transmission index · Group's configuration

5.1 Introduction

In a series of preliminary papers [1–4], a procedure was performed in order to appreciate forces transmission in mechanisms, entitled Procedure of Structural Groups. Until now, the common accepted criterion for this purpose has been transmission angle (or pressure angle), which has the inconvenience of being applicable only to the simple mechanisms (four bar and slider crank mechanisms). The essential characteristic of our proposed procedure is that forces transmission index is associate to structural groups and not to mechanisms as wholes. This index, noted T defines as [3–9]:

C. D. Duca · F. Buium (✉)
"Gheorghe Asachi" Technical University of Iasi, Iasi, Romania
e-mail: fbuium@gmail.com; florentin.buium@academic.tuiasi.ro

I. Doroftei et al. (eds.), *Proceedings of SYROM 2022 & ROBOTICS 2022*, Mechanisms and Machine Science 127, https://doi.org/10.1007/978-3-031-25655-4_5

$$T = \frac{D}{D_r} \tag{5.1}$$

where, $D = |D_k|$, and D_k is the coefficients determinant in the linear system which solves the group's k statical analysis. D_r is a referential value equals maximum value of D, if this is finite. In this way, a mechanism can have more T indices, after how many structural groups it contains. For a proper operation, each index must respect the condition $T \geq T_a$, T_a being an admissible value for T [4]. The justification to adopt this procedure and its practical advantages were widely exposed in [4].

From those shown it results that to analyze a mechanism, its component groups must put in highlight, in a first step. Then, for each group applies the specific calculus procedure of T index, starting of definition formula (5.1). The structural group 0/3/3, approached in this paper, is the most complex between usual structural groups. At the same time, we emphasize that it is long enough used, especially to parallel robots with 2 or 3 DoF. We mention that problem of this group has been treated in the papers [1, 2, 4] too, but now we perform an extended analysis and clarify some aspects, which were not enough outlined in beforementioned papers.

5.2 Determination of T—Index

As was shown in [4], the group 0/3/3 has the configuration degree of freedom $L_c = 3$. This shows that D_k, D and T depend on three independent parameters, therefore these are functions of 3 variables. Let adopt as first option, parameters $(\alpha_E, \alpha_F, \alpha_G)$-position angles of bars AE, BF and CG reported to an axes system fixed to link 4 (Fig. 5.1).

In order to build determinant D_k, the equations system of the group's statical analysis will be written, referring the notation of Fig. 5.1 and $e = AE$, $f = BF$, $g = CG$. Considering, also, that a link of n-order is load with a resultant torsor of applied forces of the form:

$$\left(\overline{F}_n\left(F_{nx}, F_{ny}, F_{nz}\right), \overline{M}_n\right).$$

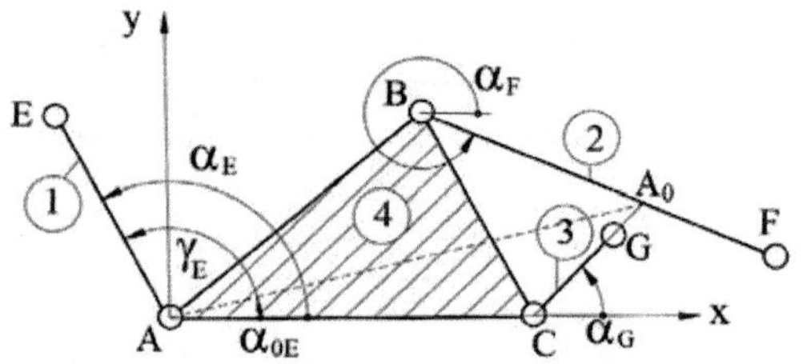

Fig. 5.1 Structural group 0/3/3 defining of parameter α_E

$$\begin{cases} \sum M_E(1) = 0 : R_{41x} \cdot e \cdot \sin\alpha_E - R_{41y} \cdot e \cdot \cos\alpha_E = -M_1 - M_E(\overline{F}_1) \\ \sum M_F(2) = 0 : R_{42x} \cdot f \cdot \sin\alpha_F - R_{42y} \cdot f \cdot \cos\alpha_F = -M_2 - M_F(\overline{F}_2) \\ \sum M_G(3) = 0 : R_{43x} \cdot g \cdot \sin\alpha_G - R_{43y} \cdot g \cdot \cos\alpha_G = -M_3 - M_G(\overline{F}_3) \\ \sum M_A(4) = 0 : -R_{42x} \cdot (y_A - y_B) + R_{42y} \cdot (x_A - x_B) - \\ \qquad -R_{43x} \cdot (y_A - y_C) + R_{43y} \cdot (x_A - x_C) = -M_4 - M_A(\overline{F}_4) \\ \sum F_x(4) = 0 : R_{41x} + R_{42x} + R_{43x} = -F_{4x} \\ \sum F_y(4) = 0 : R_{41y} + R_{42y} + R_{43y} = -F_{4y} \end{cases} \tag{5.2}$$

System (5.2) contains equilibrium kineto-static equations of the groups component elements, M—represents resultant moments loading elements, reported to the points marked with letters in subscript indices, F—represents forces loading elements, projected on axis x, y of the referential system (Fig. 5.1). The numbers between brackets represent the element's order.

The determinant $D(\alpha_E, \alpha_F, \alpha_G)$ of systems coefficients is:

$$D_k = \begin{vmatrix} e \cdot \sin\alpha_E & -e \cdot \cos\alpha_E & 0 & 0 & 0 & 0 \\ 0 & 0 & f \cdot \sin\alpha_E & -f \cdot \cos\alpha_E & 0 & 0 \\ 0 & 0 & 0 & 0 & g \cdot \sin\alpha_G & -g \cdot \cos\alpha_G \\ 0 & 0 & -(y_A - y_B) & x_A - x_B & -(y_A - y_C) & x_A - x_C \\ 1 & 0 & 1 & 0 & 1 & 0 \\ 0 & 1 & 0 & 1 & 0 & 1 \end{vmatrix} \tag{5.3}$$

The determinant $D_r = D_M$—maximum of D is obtained when bars AE, BF and CG are perpendicular on the bisectors of triangle ABC [10]. In this configuration, the parameters $(\alpha_E, \alpha_F, \alpha_G)$ acquire the values:

$$\alpha_{EM} = \frac{\pi}{2} + \frac{\widehat{A}}{2}, \ \alpha_{FM} = \frac{\pi}{2} - \left(\widehat{C} + \frac{\widehat{B}}{2}\right), \ \alpha_{GM} = \frac{\pi}{2} - \frac{\widehat{C}}{2}, \tag{5.4}$$

and

$$D_M = |D_k(\alpha_{EM}, \alpha_{FM}, \alpha_{GM})|. \tag{5.5}$$

Based on this data, index T calculates consequently to a configuration characterized by parameters $(\alpha_E, \alpha_F, \alpha_G)$:

$$T = \frac{|D_k(\alpha_E, \alpha_F, \alpha_G)|}{|D_k(\alpha_{EM}, \alpha_{FM}, \alpha_{GM})|}. \tag{5.6}$$

In some analysis or synthesis problems it is useful to know another expression of index T, as well. Expanding the determinant D_k along first line and then along first column, it obtains:

$$D_k = D_{k1} \sin \alpha_E + D_{k2} \cos \alpha_E, \tag{5.7}$$

where,

$$D_{k1}(\alpha_F, \alpha_G) = e \cdot \begin{vmatrix} f \sin \alpha_F & -f \cos \alpha_F & 0 & 0 \\ 0 & 0 & g \sin \alpha_G & -g \cos \alpha_G \\ -(y_A - y_B) & x_A - x_B & -(y_A - y_C) & x_A - x_C \\ 1 & 0 & 1 & 0 \end{vmatrix},$$

$$D_{k2}(\alpha_F, \alpha_G) = -e \cdot \begin{vmatrix} f \sin \alpha_F & -f \cos \alpha_F & 0 & 0 \\ 0 & 0 & g \sin \alpha_G & -g \cos \alpha_G \\ -(y_A - y_B) & x_A - x_B & -(y_A - y_C) & x_A - x_C \\ 0 & 1 & 0 & 1 \end{vmatrix}. \tag{5.8}$$

Expression (5.7) has been transformed using harmonic analysis thus:

$$D_k = D_{me} \sin(\alpha_E - \alpha_{0E}), \tag{5.9}$$

where,

$$D_{mE} = \operatorname{sgn}(D_{k1})\sqrt{D_{k1}^2 + D_{k2}^2}, \quad \alpha_{0E} = \tan^{-1}\left(-\frac{D_{k2}}{D_{k1}}\right). \tag{5.10}$$

Has been observed that D_{mE} and α_{0E} depend only on parameters α_F and α_G because D_{k1}, D_{k2} depend on these parameters only. The formula (5.9) and (5.10) allow an additional interpretation for D_{mE} and α_{0E}. By extension of determinants D_{k1}, D_{k2} becomes to conclusions that angle α_{0E} is formed by line AA_0 (A_0-intersection point of lines BF and CG) with axis Ox. Consequently, results this angle can be calculated from geometrical considerations as well, based on Fig. 5.1:

$$x_{A0} = \frac{x_B \tan \alpha_F - x_C \tan \alpha_G - (y_B - y_C)}{\tan \alpha_F - \tan \alpha_G}, y_{A0} = y_B + (x_{A0} - x_B) \tan \alpha_F,$$

$$\alpha_{0E} = \tan^{-1}\left(\frac{y_{A0}}{x_{A0}}\right) \tag{5.11}$$

From (5.9) can be deducted that D_{mE} is the amplitude of sinusoidal function D_k in conditions when α_F and α_G are constant and α_E is variable. It results maximum value of D_k is obtained when $\alpha_E - \alpha_{0E} = \frac{\pi}{2}$, and $\alpha_E = \frac{\pi}{2} + \alpha_{0E}$. From here results

$$D_{mE} = D_k\left(\frac{\pi}{2} + \alpha_{0E}, \alpha_F, \alpha_G\right), \tag{5.12}$$

and another expression of T:

$$T = \frac{|D_{mE}| \cdot |\sin(\alpha_E - \alpha_{E0})|}{D_M}. \tag{5.13}$$

Formula (5.13) can be written analog, taking into consideration as variable parameters, angles α_F and α_G. Of course, in these conditions, the value of T remains the same.

Another mod to adopt the configuration parameters, refers to bar's lengths EF, EG and FG. Its relationships with angles $(\alpha_E, \alpha_F, \alpha_G)$ are:

$$x_E = x_A + e \cdot \cos\alpha_E,\ y_E = y_A + e \cdot \sin\alpha_E,\ x_F = x_B + f \cdot \cos\alpha_F$$
$$y_F = y_B + f \cdot \sin\alpha_F,\ x_G = x_C + g \cdot \cos\alpha_G,\ y_G = y_C + g \cdot \sin\alpha_G. \tag{5.14}$$

$$EF = \sqrt{(x_E - x_F)^2 + (y_E - y_F)^2},\ EG = \sqrt{(x_E - x_G)^2 + (y_E - y_G)^2},$$
$$FG = \sqrt{(x_F - x_G)^2 + (y_F - y_G)^2}. \tag{5.15}$$

5.3 Connection Structural Group—Mechanism

The before found expressions for T—index, refers strictly to group 0/3/3 and are functions of the triplet $(\alpha_E, \alpha_F, \alpha_G)$. Connection between structural group and mechanism from which the group takes part, is done using basic parameters $(EF,\ EG,\ FG)$. Its values represent the result of positional analysis of the mechanism to which the group is attached. In the same time, the triplet $(EF,\ EG,\ FG)$ determines the value of index T (5.6) through parameters $(\alpha_E, \alpha_F, \alpha_G)$, which was calculated aided equations system (5.15). This system represents the essential of relation structural group—mechanism. Further two procedures is presented, which combine numerical calculations and graphical representations, regarding mechanisms behavior in terms of forces transmission.

(a) For example, the mechanism with DoF = 3, (Fig. 5.2), frequently used as parallel robot, is taken into consideration. The mechanism is formed by driving links 5, 6, 7 which rotates around points P, Q, R to which group 0/3/3 is connected by joints E, F, G. Since it is one group only, index T characterizes the structural group and the whole mechanism as well.

It is proposed a calculus procedure for T, in order to establish if condition $T > T_a$ will be satisfied. Although the mechanism has 3 DoF, the angles of driving links can be expressed as function of one parameter only $(t)(\varphi_5(t), \varphi_6(t), \varphi_7(t))$. This possibility is explained by fact that end-effector's (5.4) movement is coordinated by a unique

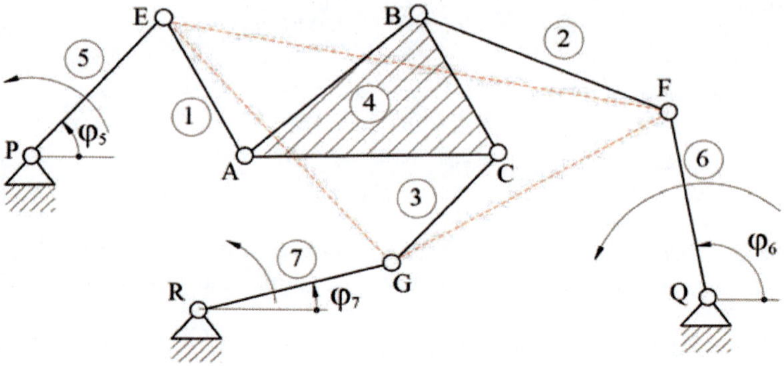

Fig. 5.2 The mechanism with DoF = 3

parameter (time for example). This way, coordinates of E, F, G are expressed as function of t. The flowchart of this procedure has been shown in Fig. 5.3, where calculus formulae are indicated with notations from previous chapter. Calculus of angles ($\varphi_E(t), \varphi_F(t), \varphi_G(t)$) assumes to solve non-linear system (5.15). It comes this way to function $T(t)$ (5.6) which is graphically represented (Fig. 5.5) and it compares to line $T = const$. This analysis is useful in design stage, in order to see that obtained in kinematic synthesis mechanism, is efficient in terms of forces transmission, without requiring to perform kinematic or dynamic analysis (Figs. 5.3 and 5.4).

(b) The presented procedure is presented in flowchart shown in Fig. 5.6, which has two branches: branch M, that refers to initial mechanism and branch G, associated

Fig. 5.3 The flowchart of the procedure

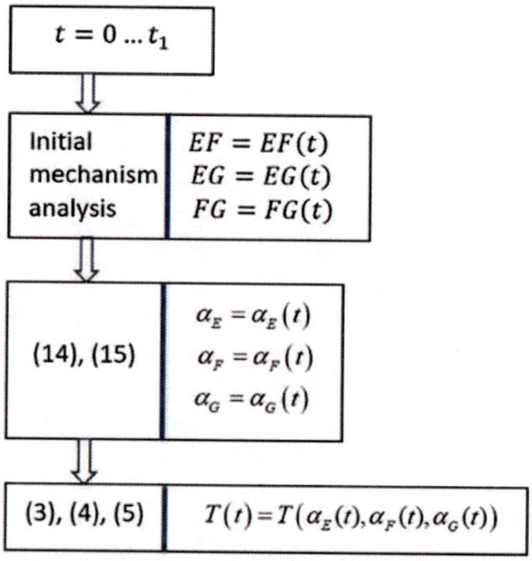

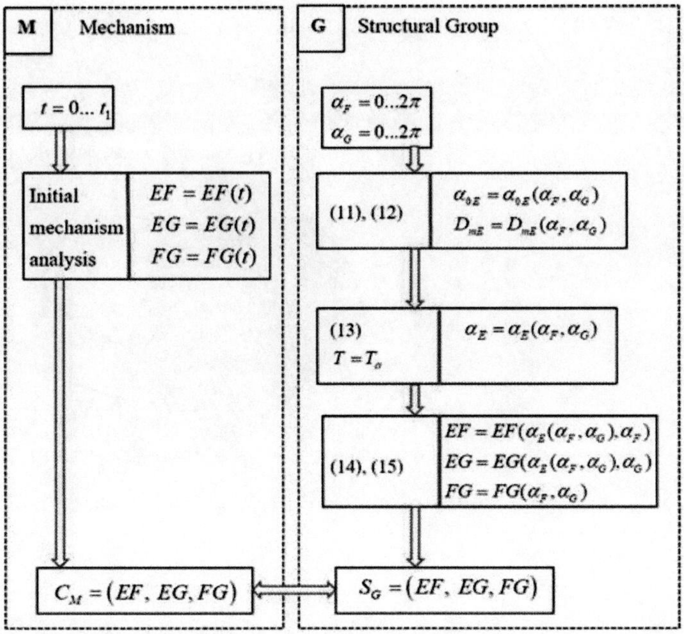

Fig. 5.4 The flowchart with two branches

Fig. 5.5 Diagram $T(t)$

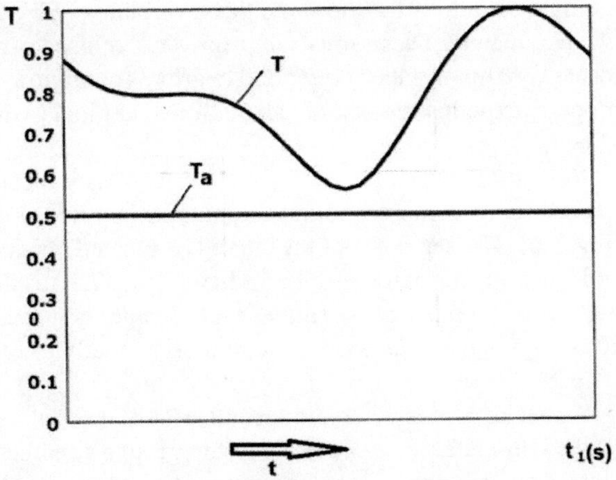

to the structural group. Branch M leads to three parametric equations for variables (EF, EG, FG). These equations represent a spatial curve C_M (Fig. 5.6) in the referential system (EF, EG, FG). The points located on the curve C_M depict evolution of real distances (EF, EG, FG), depending on parameter t. The branch G starts

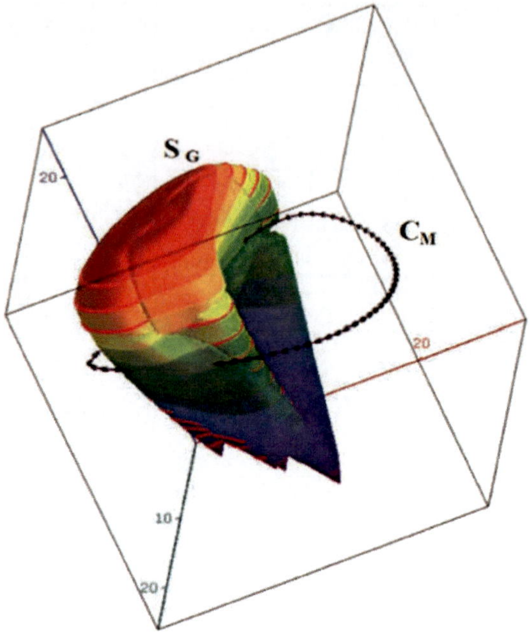

Fig. 5.6 Curve C_M and surface S_G

from initial parameters (φ_F, φ_G) and comes by successive calculations to three parametrical equations for the same three variables (EF, EG, FG), but as function of two parameters. These equations represent a closed surface S_G (Fig. 5.6). The points located on this surface correspond to groups configurations which have $T = T_a$. This property explains because at calculation of angle α_E with formula (5.13), it considers $T = T_a$.

Relation between curve C_M and surface S_G indicates the mechanisms behavior. Therefore, in common points the mechanism has as the structural group $T = T_a$ (Fig. 5.6). The segments of the curve C_M located inside the surface S_G, have $T > T_a$, and those outside of the surface S_G have $T < T_a$. It results that for a favorable running, the curve C_M must be located entirely inside of surface S_G. This procedure has the advantage that it does not require numerical methods of solving, being thereby easier for use.

A special case is where, in formula (5.13) it adopts $T = 0$. This means that points located on the surface S_G leads to the groups singular configurations and for those located outside this surface, the group cannot be built, because positional analysis has complex solutions and index T is senseless. It is important to notice that to $T = 0$, $\alpha_E = \alpha_{0E}$ (5.13) showing that lines AE, BF and CG (Fig. 5.1) are concurrent. This fact confirms the known property [4] referring to singularities of group 0/3/3. The relation of the surface S_G (to $T = 0$) with the curve C_M allows the formulation of a new singularities' classification, as is shown in [4].

5.4 Conclusions

1. The paper solves the problem of forces favorable transmission in the case of the structural group 0/3/3, using the structural group-based procedure, previously approached. On the authors opinion this problem has not been treated properly until now.
2. In the paper, the calculus algorithm has been indicated for the force's transmission index, depending on the characteristic structural group's parameters. It formulates also the equations allowing calculation of these parameters.
3. In order to show the proposed procedure's utility, this has been applied into a practical case of a parallel robot mechanism. Two modes to use this approach has been shown, which allow forces transmission quality appreciating.

References

1. Duca, C., Buium, F.: Forces transmission in the 0/3/3 structural group. Bul. Inst. Pol. Iași, Tom. LIX(LXIII), Fasc. **1**, 17–22 (2013)
2. Duca, C., Buium, F.: Transmission indices adoptation for 6R structural group. Appl. Mech. Mater. **658**, 55–58 (2014). Trans Tech Publications, Switzerland. https://doi.org/10.4028/www.scientific.net/AMM.658.55
3. Duca, C., Buium, F.: The singularities, forces transmission index and mechanism selfblocking, ACME 2016. IOP Conference Series: Materials Science and Engineering, vol. 147, 012073 (2016). https://doi.org/10.1088/1757-899X/147/1/012072
4. Duca, C., Buium, F.: The structural groups approach and transmission angles boundaries, ACME 2022. The 10th International Conference on Advanced Concepts in Mechanical Engineering IOP Publishing
5. Erdman, A.G., Sandor, G.N., Kota, S.: Mechanism design. Analysis and Synthesis, Vol.1, Fourth Edition (2001)
6. Shrinivas, S.: Balli, satish chand, transmission angle in mechanisms (Triangle in mech). Mech. Mach. Theory **37**, 175–195 (2002)
7. Lin, C.-C., Chang, W.-T.: The force transmissivity index of planar linkage mechanisms. Mechanism Nachine Theory **37**, 1465–1485 (2002)
8. Chen, C., Angeles, J.: Generalized transmission index and transmission quality for spatial linkages. Mechanism Nachine Theory **42**, 1225–2123 (2007)
9. Jensen, W.: Preben. Mankato State University, Minesota, Classical and modern mechanisms for engineers and inventors (1991)
10. Arobolevski, 1.1.: Terina mecanismelor și a mașinilor, Ed. Tehnică, București (1955)

Chapter 6
Design and Performance Analysis of a Compliant Finger Mechanism

F. J. Espinosa-Garcia, M. Ceccarelli, and M. Arias-Montiel

Abstract A new robotic finger mechanism design based on a combination of rigid and compliant elements is presented. Aspects like anthropometric, anthropomorphic, underactuated, and easy operation were considered for a feasible solution. For finger mechanism, a problem optimization based on flexion–extension movements (F-E) is proposed. As a solution method, the Grey Wolf optimizer (GWO) algorithm is implemented. As a result, the error between the proposed and obtained trajectory is 1.55×10^{-4} mm. The CAD design and a short description of the selected flexible elements are presented.

Keywords Compliant finger mechanisms · Underactuated mechanisms · Optimization

6.1 Introduction

Robotic hand design continues to be an interesting topic for researchers, since one of the main challenges in this field is to replicate human hand capabilities (grasp and manipulate, hold, surface recognition, feeling of pressure and force) [1]. Currently, this is a complex task so that the efforts are focused on the development of improved prototypes to do a specific task. Basically, the existing models can be divided in two types: anthropomorphic models (based on anatomy of human hand) and non-anthropomorphic (grippers). It depends on the application area, these can be used in assistive robotics, supervised manipulation, prosthesis, and service robotics [2].

F. J. Espinosa-Garcia (✉)
Postgraduate Division, Technological University of the Mixteca, León, Mexico
e-mail: fjeg_1234@hotmail.com

F. J. Espinosa-Garcia · M. Ceccarelli
Laboratory of Robots and Mechatronics (LARM2), University of Rome Tor Vergata, Rome, Italy
e-mail: marco.ceccarelli@uniroma2.it

M. Arias-Montiel
Institute of Electronics and Mechatronics, Technological University of the Mixteca, León, Mexico
e-mail: mam@mixteco.utm.mx

I. Doroftei et al. (eds.), *Proceedings of SYROM 2022 & ROBOTICS 2022*, Mechanisms and Machine Science 127, https://doi.org/10.1007/978-3-031-25655-4_6

Regardless of the application type, there is an important feature that all designers consider, which is "grasping capability". Based on it, the purpose is to generate prototypes than can achieve the greatest number of grasping configurations according to a specific taxonomy. Since a finger is the element that present more contact with the object's surface, its design is the "key" that defines what kind of objects can be grasped. Several designs have been developed in the past, for example in [3] the finger design has four degrees of freedom, and it is actuated by a servomotor. This design is capable to perform cylindrical and spherical grasping. In [4] Dechev present a mechanism composed of 6 links with one degree of freedom and it is made of aluminum. Most of the available designs have a high number of degrees of freedom but they present problems like complex actuation, high cost, and difficult to manufacture and maintain [5]. To solve these problems, the implementation of a underactuated mechanism can be a solution. Prototypes designs under this approach has reduced weight, volume, and cost as in the case in [6]. In addition, in [7] authors present a underactuated mechanism as solution to solve the problems of weight, cost, and complex manufacture. Additionally, the finger proposed present an easy operation.

Although underactuated fingers present advantages under the traditional solutions, there is a common problem in these kinds of designs, which is "adaptability". The advantages of the compliant fingers are that they can adapt to the shape of different objects in a better way that underactuated or classical finger structures. Furthermore, they present a larger motion range and better flexibility [8]. Notwithstanding, this design approach present issues like the prototype's appearance is not anthropomorphic, they mostly need an external air source, the force applied is less compared with underactuated and classical solution, also their application scope is limited [9].

This paper presents a new proposal finger design based on a combination of compliant and rigid elements with underactuated characteristics. Also, aspects like anthropomorphism, anthropometric, low cost, easy operation and low level of manufacture are considered.

6.2 Finger Design

Based on the literature review and considering advantages of the underactuated compliant mechanisms, the requirements for the present solution are:

1. Anthropomorphic and anthropometric finger design
2. Lightweight mechanism design
3. Flexible elements to get an adaptability of regular and irregular objects
4. Underactuated motion
5. Flexion-extension motion capability like a human finger
6. Easy manufacture and assembly.

To fulfill requirements 1 and 5, the length of each phalange of the human finger is considered based on measurements shown in [10]. To obtain a prototype capable to replicate the proposed trajectory (F–E), a path generation synthesis is proposed. Through this process the optimal length for the links is figured out. The desired points (Dp) for the trajectory are shown in Table 6.1.

Attending the requirements 4 and 6 (easy manufacture and assembly) the mechanism proposed present a simple configuration and all the phalanges present movement. The kinematic scheme is shown in Fig. 6.1.

The mechanism is composed of 7 links, 6 binary and 1 ternary. For the actuation of the proximal phalanx a cross four bar mechanism is considered. The middle phalange is represented by two links (Mp and R6). The distal phalanx is represented by a ternary link. Definition of the input data and design variables ranges for the synthesis process are shown in Table 6.2.

The values of the phalanges are obtained from [10]. The ranges of the design variables are proposed considering not to exceed the length of a human finger (Table 6.2).

Generated point (Gp) by the proposed mechanism is obtained using the vector loop equation for the four bar linkages with respect to coordinate system O2XoYo:

Table 6.1 Desired point obtained from [10]

Point position	Coordinate (x, y)
Point 1	(0.00, 43.76)
Point 2	(−17.52, 39.40)
Point 3	(−37.39, 21.67)
Point 4	(−38.95, 7.36)
Point 5	(34.77, −12.43)

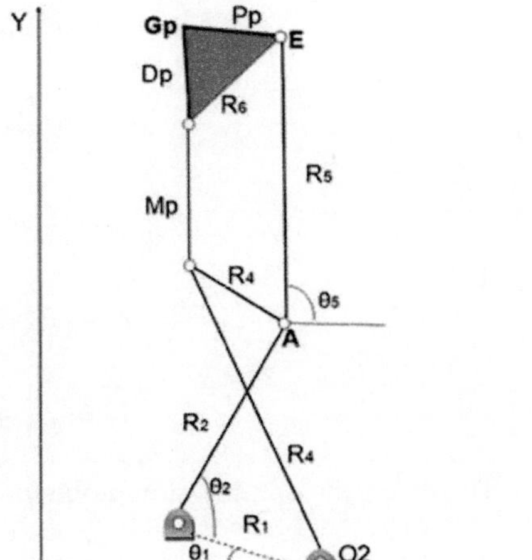

Fig. 6.1 Kinematic scheme of the proposed mechanism

Table 6.2 Input data and design variables for the synthesis process

Input data	Design variables
R4: Proximal phalanx length (35 mm)	R1: Fixed link length [5–22] mm
Dp: Distal phalanx length (16.75 mm)	R2: Link 2 length [20–40] mm
Mp: Middle phalanx length (27.01 mm)	R3: Link 3 length [15–20] mm
R6: Distal width (20 mm)	R5: Link 5 length [35–40] mm
PP. Dp complement (15 mm)	Xo, Yo: Coordinate point O2 [0–30] mm
	Θ1: R1 angular position [0–2pi] rad Θ2: Input angular position [0–2pi] rad Θ5: R5 angular position [0–2pi] rad

$$R_1 \sin\theta_1 + R_4 \sin\theta_4 = R_2 \sin\theta_2 + R_3 \sin\theta_3$$

$$R_1 \cos\theta_1 + R_4 \cos\theta_4 = R_2 \cos\theta_2 + R_3 \cos\theta_3 \tag{6.1}$$

Once Eq. (6.1) has been solved, the position of points A, E and Gp can be obtained by computing

$$A_x = R_1 \cos\theta_1 + R_2 \cos\theta_2$$

$$A_y = R_1 \sin\theta_1 + R_2 \sin\theta_2 \tag{6.2}$$

$$E_x = A_x + R_5 \cos\theta_5$$

$$E_y = A_y + R_5 \sin\theta_5 \tag{6.3}$$

$$G_x = E_x + Pp\cos(30°)$$

$$G_y = E_y + Pp\sin(30°) \tag{6.4}$$

Therefore, the optimization problem is defined as:

$$min \sum_{i=1}^{n} \left[\left(Pd_x^i - G_x^i\right)^2 + \left(Pd_y^i - G_y^i\right)^2 \right] + (M_1 * h_1(x)) + (M_2 * h_2(x)) \tag{6.5}$$

subject to:

$$(a)\,(R_1 + R_2 < R_3 + R_4)\,where\,(R_2 < R_3 < R_4 < R_1)$$

$$(b)\,\theta_2^i > \theta_2^{i+1} > \ldots > \theta_2^{i+n}\,where\,\mathrm{i} = 1 \text{ and } \mathrm{n} = 5$$

$$(c)\,x_i \in [li_i, ls_i]\forall\, x_i \in X$$

$$where\,X = [R_1, R_2, R_3, R_5, x_0, y_0, \theta_1, \theta_2^1, \theta_2^2, \ldots \theta_2^5, \theta_5^1, \theta_5^2, \ldots \theta_5^5] \tag{6.6}$$

where n is the number of the points (Table 6.1). $Pd_x{}^i$, $Pd_y{}^i$ are the coordinates of the trajectory points and G_x, G_y the coordinates of the point generated in the synthesis. $h_1(x)$ evaluates the sequence condition for the input angles $\theta_2{}^i > \theta_2{}^{i+1} > \ldots > \theta_2{}^{i+n}$. Also, $h_2(x)$ is a constraint to evaluate the Grashof condition. M_1 and M_2 are constants of a very high value to penalize the goal function Eq, (6.5) when associate constraint fails.

6.3 Grey Wolf Optimizer (GWO)

This metaheuristic algorithm is based on the population that simulate the social hierarchy and hunting mechanism of gray wolves in the wild [11]. The main phases of grey wolf hunting are as follows:

- Tracking, chasing, and approaching the prey
- Pursuing, encircling, and harassing the prey until it stops moving
- Attack towards the prey.

In this algorithm the solutions are classified based on a social hierarchy. Fittest solution is considered as the alpha (α). The second and third best solutions are beta (β) and delta (δ) respectively. Rest of the solutions are named as omega (ω). To encircling behaviour the Eq. (6.7) is used.

$$\vec{D} = [\vec{D} \cdot \vec{X}_p(t) - \vec{X}(t)]$$

$$\vec{X}(t+1) = \vec{X}_p(t) - \vec{A} \cdot \vec{D} \tag{6.7}$$

where t represents the current iteration, $\vec{X}_p$ is the position vector of the prey and $\vec{X}$ indicates the position vector of grey wolf. The vectors $\vec{A}$ and $\vec{C}$ are coefficient calculated as follows:

$$\vec{A} = 2\vec{a} \cdot \vec{r}_1 \cdot \vec{a}$$

$$\vec{C} = 2 \cdot \vec{r}_2 \tag{6.8}$$

Components of $\vec{a}$ are linearly decreased from 2 to 0 and $\vec{r}_1$, $\vec{r}_2$ are random vectors in [0, 1]. For hunting phase, ω solutions update their positions based on α, β and δ position according to following equations.

$$\vec{D}_\alpha = \left|\vec{C}_1 \cdot \vec{X}_\alpha - \vec{X}\right|, \vec{D}_\beta = \left|\vec{C}_2 \cdot \vec{X}_\beta - \vec{X}\right|, \vec{D}_\delta = \left|\vec{C}_3 \cdot \vec{X}_\delta - \vec{X}\right|$$

$$\vec{X}_1 = \vec{X}_\alpha - \vec{A}_1 \cdot \left(\vec{D}_\alpha\right), \vec{X}_2 = \vec{X}_\beta - \vec{A}_2 \cdot \left(\vec{D}_\beta\right), \vec{X}_3 = \vec{X}_\delta - \vec{A}_3 \cdot \left(\vec{D}_\delta\right)$$

$$\vec{X}(t+1) = (\vec{X}_1 + \vec{X}_2 + \vec{X}_3)/3 \tag{6.9}$$

The last phase is when wolves finish hunting the prey. This behavior is defined by the A value. If |A| < 1 the algorithm finish, otherwise, the algorithm continues searching for optimum. The GWO process is shown in Fig. 6.2.

```
Input: Problem Size, Population size
Output: Pg_best
Start
    Initialization of the population of grey wolves Xi (i = 1, 2, ... , n)
    Initialization of a, A, and C
    Calculation of the fitness values of search agents and grading of agents.
    (Xα = the best solution in the search agent, Xβ = the second best solution
    in the search agent, and Xδ = the third best solution in the search agent)
    t= 0
    While (t < Maximum number of iterations)
        For each search agent
            Updating the position of the current search agent by Equation
        End for
    Updating of a, A, and C
    Calculation of the fitness values of all search agents and grading them
    Updating the positions of Xα, Xβ, and Xδ
    t= t+1
    End while
End
```

Fig. 6.2 GWO Pseudo algorithm

6.4 Implementation and Results

The numerical experiments were carried out on a laptop with core i7 9th gen processor, 16 GB RAM, and the clock speed of 2.6 GHz with turbo boost of up to 4.5 GHz. The parameters of the algorithm are population = 600, problem dimension = 16 and interaction max = 200. To evaluate the quality of solutions the algorithm was iterated 12 times. As a result, 75% of the solutions are in the 1 × 10–3 mm range. The worst solution was 7.3379×10^{-1} and the best obtained solutions was 1.55×10^{-4} mm. Figure 6.3 shows a comparison between desired trajectory (F-E) and obtained trajectory. Table 6.3 shows the vector solution for the mechanism.

The CAD model is shown in Fig. 6.4. Two materials have been selected for the prototype design: ABS (Acrylonitrile Butadiene Styrene) for rigid pieces (R2 and R3) and TPU (Thermoplastic Polyurethane) for flexible elements (Dp, Mp, R1 and R6).

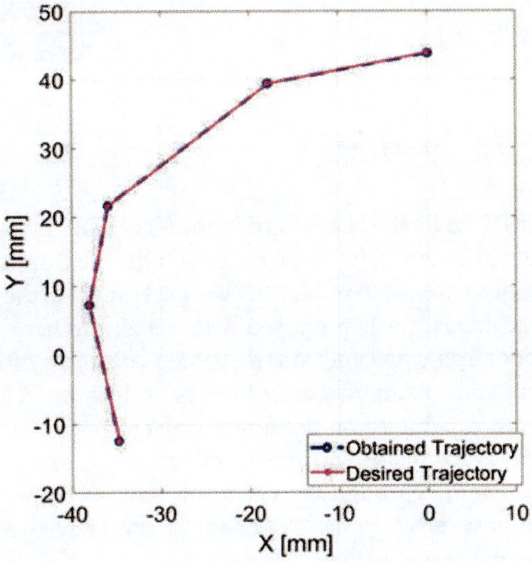

Fig. 6.3 Comparison between F-E trajectory generated for the finger and for the proposed design

Table 6.3 Results obtained from the algorithm

Design variable	Value obtained by GWO
R1	6.77 mm
R2	32.44 mm
R3	17.53 mm
R5	40 mm
Xo	3.20 mm
Yo	28.48 mm
Θ1	2.87 rad

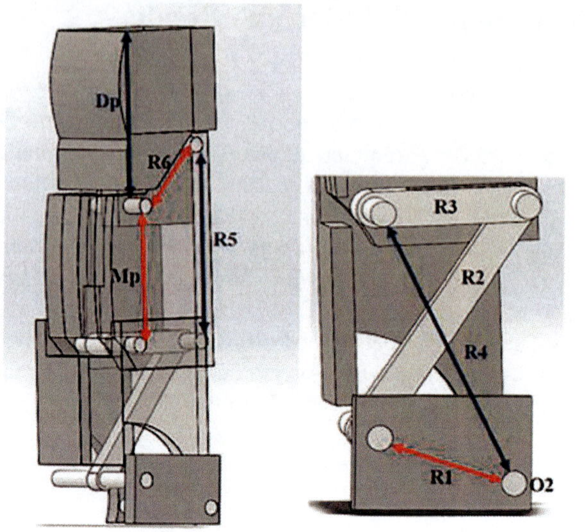

Fig. 6.4 CAD design

6.5 Conclusion

Inspired by the concepts of compliant and underactuated mechanism this paper introduces a new robotic finger which has flexible and rigid elements. According to the requirements, to reduce the weight respect to the fully actuated finger a underactuated mechanism was proposed with the aim to use one actuator. Referring to the anthropomorphic and anthropometric aspects in a synthesis process the values of phalanges are taken from the real measures of the human finger. Moreover, to generate a prototype capable to do the finger trajectory as a human finger does, a relation between the mechanism and points of the trajectory was used to propose an optimization problem. Through the GWO algorithm the problems were solved, and the error is almost zero (1×10^{-4}). For easy manufacture, assembly, and low cost the design can be generated using a 3D printer.

Acknowledgements The principal author is grateful to the Double PhD. program at Technological University of the Mixteca and University of Rome Tor Vergata, moreover, to the National Council of Science and Technology (CONACYT) for supporting the research presented in this paper.

References

1. Ceccarelli, M.: Notes for a history of grasping devices. Mechanisms Mach. Sci. **10**, 3–16 (2012)
2. Andrychowicz, O.M., Baker, B., Chociej, M., Jozefowicz, R., McGrew, B., ächocki, J., Zaremba, W.: Learning dexterous in-hand manipulation. Int. J. Robot. Res. **39**(1), 3–20 (2020)
3. Zhang, Y., Han, Z., Shang, H., Wang, T., Guo, W.: Design and control of the BUAA four-fingered hand. In: IEEE Transaction on Robotics and Automation, pp. 2517–2522 (2001)

4. Dechev, N., Cleghorn, W., Naumann, S.: Multiple finger, passive adaptive grasp prosthetic hand. Mech. Mach. Theory **36**(10), 1157–1173 (2001)
5. Licheng, W., Kong, Y., Li, X.: Review and research issues on underactuated finger mechanism. In: Proceedings of the Chinese Intelligent Automation Conference, Berlin, pp. 171–180 (2015)
6. Espinosa-Garcia, F.J., Tapia-Herrera, R., Lugo-Gonzalez, E., Arias-Montiel, M.: Development of a robotic hand based on a palm with a metamorphic mechanism for extending the thumb's functionality. J. Braz. Soc. Mech. Sci. Eng. **43**(8), 1–16 (2021)
7. Ceccarelli, M., Zottola, M.: Design and simulation of an underactuated finger mechanism for LARM hand. Robotica **35**(3), 483–497 (2015)
8. Odhner, L.U., Jentoft, L.P., Clafee, M.R., Corson, N., Tenzer, Y., Ma, R.R., Buehler, M., Kohout, R., Howe, R.D., Dollar, A.M.: A compliant, underactuated hand for robust manipulation. Int. J. Robot. Res. **33**(5), 736–735 (2014)
9. Bai, G., Rojas, N.: Self-adaptive monolithic anthropomorphic finger with teeth-guided compliant cross-four-bar joints for underactuated hands. In: IEEE-RAS 18th International Conference on Humanoid Robots (Humanoids), pp. 145–152 (2018)
10. Espinosa-Garcia, F.J., Tapia-Herrera, R., Lugo-González, E., Arias-Montiel, M.: Design and simulation of a robotic hand with foldable palm based on mechanisms with variable. In: Proceedings of XXI Congreso Mexicano de Robótica, pp. 300–305 (2019)
11. Mirjalili, S., Mirjalili, S.M., Lewis, A.: Grey wolf optimizer. Adv. Eng. Softw. **69**, 46–61 (2014)

Chapter 7
Fixed Sequence Agricultural Robot Functional Characteristics

A. Comanescu, D. Boblea, I. Dugaesescu, and G. Soare

Abstract In this paper, a specific fixed sequence robot is developed and analyzed. It may have various applications in agricultural operations, such as seeding and planting. Its mono-mobile mechanism belongs to an important linkages class ***, which gives the possibility to create new solutions for robots. The main structural, kineto-dynamic characteristics highlighted by relevant graphic representations and their general positional-kinematic and kineto-static and dynamic algorithms are presented. The mechanism may be integrated in various manipulators and other equipment.

Keywords Agricultural robot · Fixed sequence · Mechanism · Structural model · Modular groups' connection · Kineto-dynamic algorithm · Constructive model

7.1 Introduction

The main aim of any designer or researcher is to create new adequate constructive models. This paper demonstrates how the theoretical knowledge existing in mechanisms and robots' theory may be applied to create and to verify the designed models. The authors selected the mechanism for the fixed sequence robot from specialized literature [1]. Any fixed sequence robot solution with one degree of mobility may be found between 16 mono-mobile linkages (Table 7.1) mentioned in the classical literature [2, 3], all representing an important basis for the structural modelling of new robots' solutions. By using the fundamental theorems for structural modelling [4] and starting from any linkage, the designer may only select the basis and the effector for the robot placing the active pairs in suitable positions in the linkages. Based on the achieved structural model, one may create the constructive model of the robot. A technological process adequate device (i.e. gripper) may be placed at the extreme point of the effector.

A. Comanescu (✉) · D. Boblea · I. Dugaesescu · G. Soare
University Politehnica of Bucharest, Bucharest, Romania
e-mail: adrianacomanescu@yahoo.com

I. Doroftei et al. (eds.), *Proceedings of SYROM 2022 & ROBOTICS 2022*, Mechanisms and Machine Science 127, https://doi.org/10.1007/978-3-031-25655-4_7

Table 7.1 Mono-mobile planar linkages with three contours

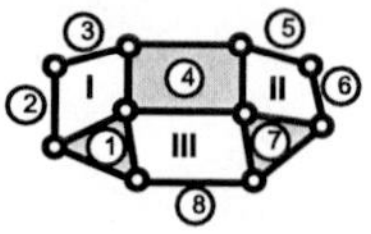 L1	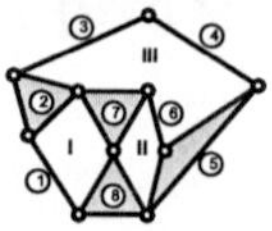 L2	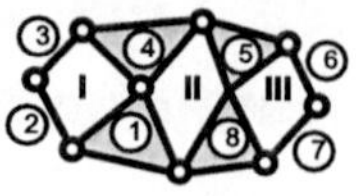 L3	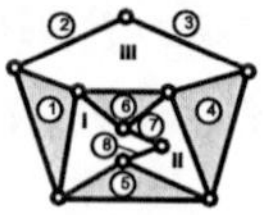 L4
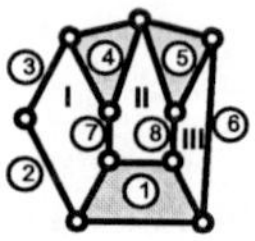 L5	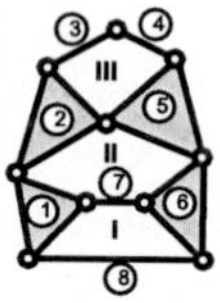 L6	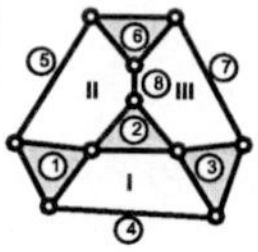 L7	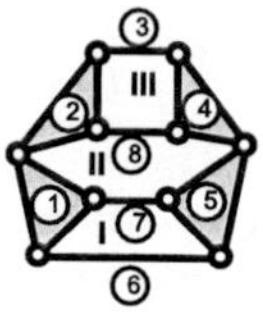 L8
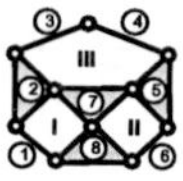 L9	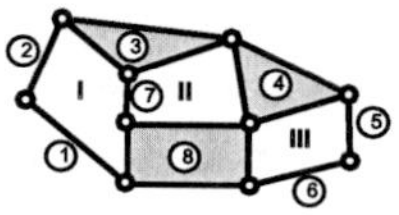 L10	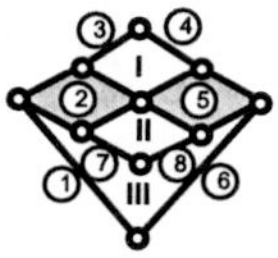 L11	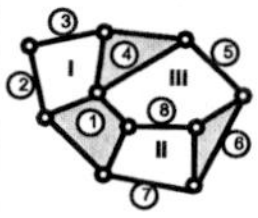 L12
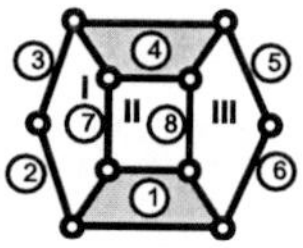 L13	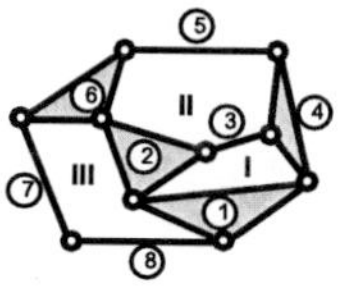 L14	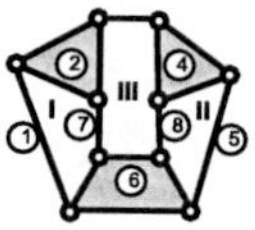 L15	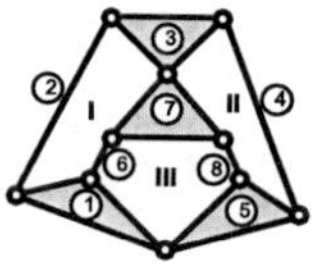 L16

The paper highlights the structural characteristics of the system, such as the modular structure for establishing the kineto-dynamic calculation algorithm. These algorithms, using calculation models specific to various active and passive modular groups [4], are verified by illustrations and graphical animations.

7.2 Structural Considerations

The manipulator [1] for seeding/planting operations (Fig. 7.1) is a plan system with one degree of mobility determined by the independent parameter $\varphi 1$ and three independent contours of the structural model (Fig. 7.2). This is obtained from the fundamental Watt chain by amplification with a dyad [2–6]. The mechanism dependent parameters are mentioned in Fig. 7.3 and Table 7.1.

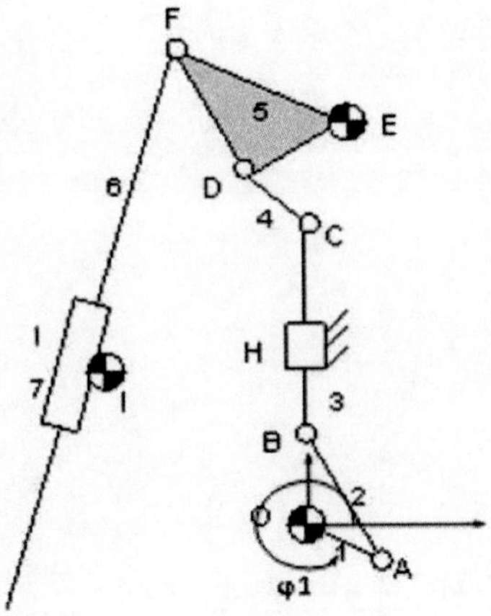

Fig. 7.1 Constructive model

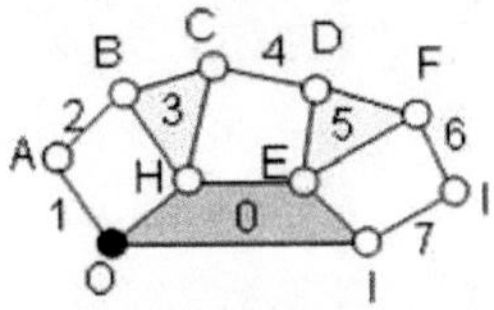

Fig. 7.2 Structural model

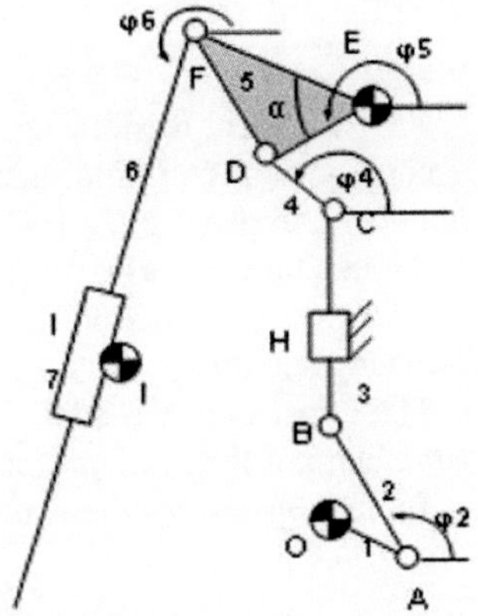

Fig. 7.3 Mechanism's dependent parameters

The structural model (Fig. 7.2) has 7 mobile kinematic elements and 10 inferior ki-nematic pairs (rotation and translation) and hence establishes the connection of the modular groups (Fig. 7.4). Figure 7.4 illustrates: one active modular group GMAI (O,1) and three dyad-type groups, respectively RRT, RRR and RTR [2–6]. This connection represents the basis of the kineto-dynamic algorithm developed in the following sections.

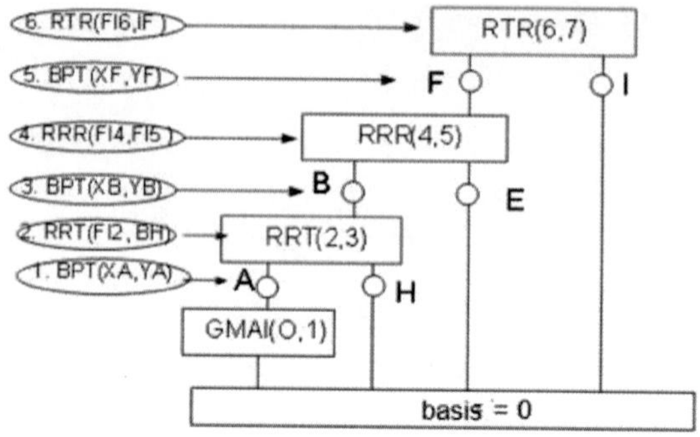

Fig. 7.4 Modular groups 'connection'

Table 7.2 Geometric parameters

Constant geometric parameters [mm]						
$XO = 0$	$XE = 24$	$YE = 379$	$XI = -177$	$YI = 164$		
$OA = 53$	$AB = 137$	$BC = 185$	$CD = 83$	$ED = 59$	$EF = 173$	$FT = 543$
$\alpha = 1.39$	$\alpha 0 = \alpha \cdot \frac{180}{\pi}$	α ≡ angle (DEF)–Fig. 7.3				

Table 7.3 Independent parameters

Independent parameters		
$k = 0.36$	$\phi 1_k = k \cdot \frac{\pi}{18}$	$\omega 1 = 1$

The structural model (Fig. 7.2) has 7 mobile kinematic elements and 10 inferior ki-nematic pairs (rotation and translation) and hence establishes the connection of the modular groups (Fig. 7.4). Figure 7.4 illustrates: one active modular group GMAI (O,1) and three dyad-type groups, respectively RRT, RRR and RTR [2–6]. This connection represents the basis of the kineto-dynamic algorithm developed in the following sections.

The constructive model (Figs. 7.1 and 7.3) establishes the constant geometric parameters of the system (Table 7.2).

Table 7.3 presents the values used for independent parameters of the mechanism (Fig. 7.1).

7.3 Mechanism Positional-Kinematic Modelling

Modular groups' connection presented in Fig. 7.4 along with the kinematic calculation models lead to the calculation algorithm in Table 7.4, developed in Mathcad.

Table 7.4 Dependent parameters

Modulus	Dependent parameters		
BPT(A)	$XA_k = 0 + OA \cdot \cos(\phi 1_k)$	$YA_k = 0 + OA \cdot \sin(\phi 1_k)$	
	$X1A_k = 0 - OA \cdot \sin(\phi 1_k)$	$Y1A_k = 0 + OA \cdot \cos(\phi 1_k)$	
RRT(2,3)			
	$XA_k + AB \cdot \cos(\phi 2) = 0$	$YA_k + AB \cdot \sin(\phi 2) - s = 0$	
	$sol_k = Find(\phi 2, s)$	$\begin{pmatrix} \phi 2_k \\ s_k \end{pmatrix} = sol_k$	$\phi 20_k = \phi 2_k \cdot \frac{180}{\pi}$
	$A_k = \begin{pmatrix} -AB \cdot \sin(\phi 2_k) & 0 \\ AB \cdot \cos(\phi 2_k) & -1 \end{pmatrix}$	$B_k = \begin{bmatrix} -(X1A_k - 0) \\ -(Y1A_k - 0) \end{bmatrix}$	
	$solv_k = lsolve(A_k, B_k)$	$\begin{pmatrix} \omega 2_k \\ s1_k \end{pmatrix} = solv_k$	
BPT(B)	$XB_k = XA_k + AB \cdot \cos(\phi 2_k)$	$X1B_k = X1A_k - AB \cdot \sin(\phi 2_k) \cdot \omega 2_k$	
	$YB_k = YA_k + AB \cdot \sin(\phi 2_k)$	$Y1B_k = Y1A_k + AB \cdot \cos(\phi 2_k) \cdot \omega 2_k$	
RRR(4,5)	$XC_k + CD \cdot \cos(\phi 4) - XE - ED \cdot \cos(\phi 5) = 0$		
	$YC_k + CD \cdot \sin(\phi 4) - YE - ED \cdot \sin(\phi 5) = 0$		
	$sol_k = Find(\phi 4, \phi 5)$	$\begin{pmatrix} \phi 4_k \\ \phi 5_k \end{pmatrix} = sol_k$	$\begin{pmatrix} \phi 40_k \\ \phi 50_k \end{pmatrix} = sol_k \cdot \frac{180}{\pi}$
	$A_k = \begin{pmatrix} -CD \cdot \sin(\phi 4_k) & ED \cdot \sin(\phi 5_k) \\ CD \cdot \cos(\phi 4_k) & -ED \cdot \cos(\phi 5_k) \end{pmatrix}$	$B_k = \begin{bmatrix} -(X1C_k - 0) \\ -(Y1C_k - 0) \end{bmatrix}$	
	$solv_k = lsolve(A_k, B_k)$	$\begin{pmatrix} \omega 4_k \\ \omega 5_k \end{pmatrix} = solv_k$	
BPT(F)	$XF_k = XE + EF \cdot \cos(\phi 5_k - \propto)$	$X1F_k = 0 - EF \cdot \sin(\phi 5_k - \propto) \cdot \omega 5_k$	
	$YF_k = YE + EF \cdot \sin(\phi 5_k - \propto)$	$Y1F_k = 0 + EF \cdot cos(\phi 5_k - \propto) \cdot \omega 5_k$	

(continued)

Table 7.4 (continued)

Modulus	Dependent parameters		
RTR(6,7)	$XF_k + FI \cdot \cos(\phi 6) - XI = 0$	$YF_k + FI \cdot \sin(\phi 6) - YI = 0$	
	$sol_k = Find(FI, \phi 6)$	$\begin{pmatrix} FI_k \\ \phi 6_k \end{pmatrix} = sol_k$	$\phi 60_k = \phi 6_k \cdot \frac{180}{\pi}$
	$A_k = \begin{pmatrix} \cos(\phi 6_k) & -FI_k \cdot \sin(\phi 6_k) \\ \sin(\phi 6_k) & FI_k \cdot \cos(\phi 6_k) \end{pmatrix}$	$B_k = \begin{pmatrix} -X1F_k \\ -Y1F_k \end{pmatrix}$	
	$solv_k = lsolve(A_k, B_k)$	$\begin{pmatrix} FI_k \\ \omega 6_k \end{pmatrix} = solv_k$	
BPT(T)	$XT_k = XF_k + FT \cdot \cos(\phi 6_k)$	$X1T_k = X1F_k - FT \cdot \sin(\phi 6_k) \cdot \omega 6_k$	
	$YT_k = YF_k + FT \cdot \sin(\phi 6_k)$	$Y1T_k = Y1F_k + FT \cdot cos(\phi 6_k) \cdot \omega 6_k$	

Kinematic parameters of point T—the end of the execution element 6 is given in Fig. 7.5. Its trajectory must satisfy the technological process necessities and an adequate tool will be attached. In k successive discrete positions of the mechanism, the parameters of P point placed to any link are denoted by XP_k, YP_k and its velocity components are respectively $X1P_k$, $Y1P_k$ (Table 7.4). All parameters may be found on Figs. 7.1 and 7.3.

The trajectory of the T extreme point (Fig. 7.5) of the effector demonstrates the optimal selection of all geometrical parameters and the aim for which the constructive model was developed.

7.4 Equilibrium Moment Determination

The virtual powers method allows the determination of the equilibrium moment, ensuring the system is in equilibrium under the action of the exterior and inertia system of forces.

This work highlights the equilibrium moment required to be applied in the active coupling considering exclusively the exterior force RT, applied at the end T of the execution element, all other force categories being disregarded. Prior to this, the velocity parameters ($X1T_k$, $Y1T_k$) of point T are established, such that the value of the equilibrium moment, marked with CM_k, and can be calculated from Table 7.5. Variation curve of the equilibrium moment is presented in Fig. 7.6.

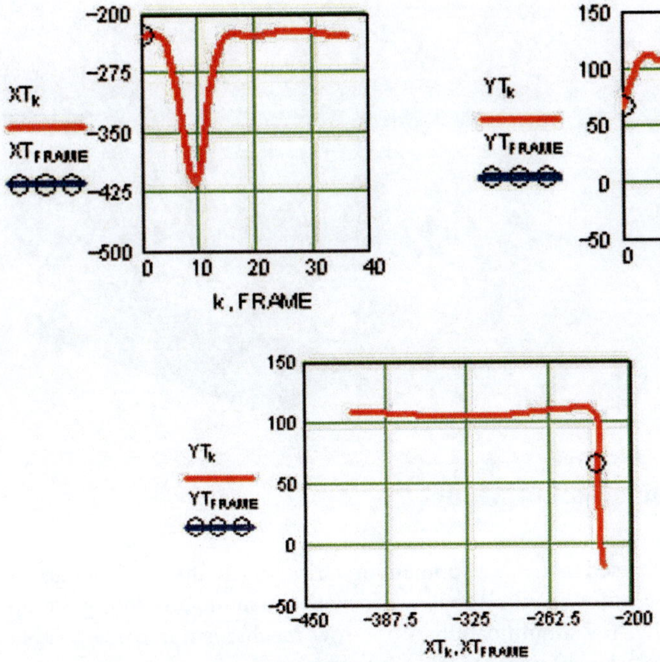

Fig. 7.5 Kinematic parameters of the execution element extreme point

Table 7.5 Equilibrium moment

Equilibrium moment determination	
$RT = 20$	$CM_k = \frac{-(RT \cdot Y1T_k)}{\omega 1}$

Fig. 7.6 Variation of equilibrium moment for a kinematic cycle

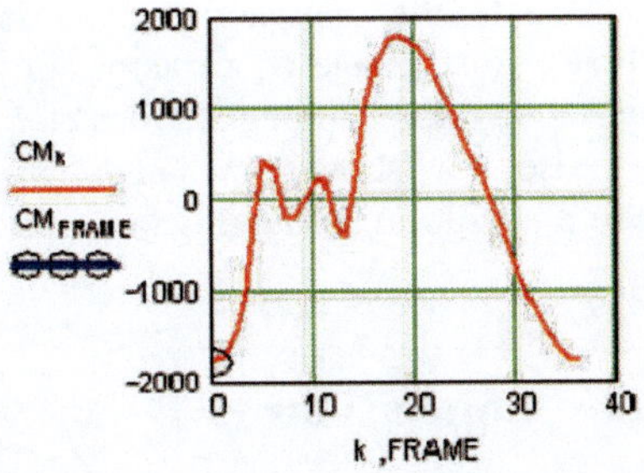

The trajectory of the extreme point T of the effector must satisfy the technological process necessities and a technological process adequate tool will be attached in this point.

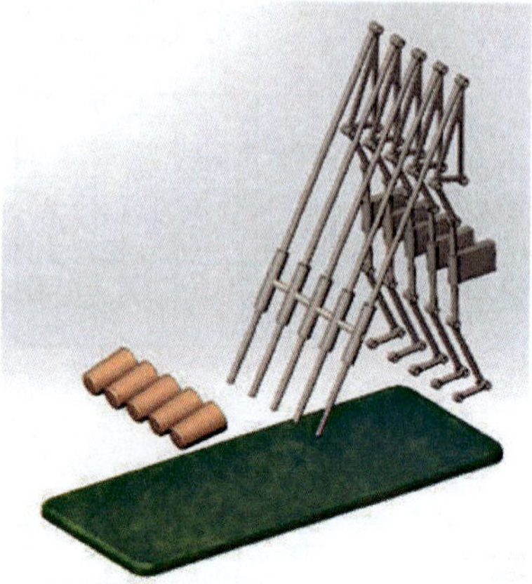

Fig. 7.7 Multi-row seeding/planting mechanism

7.5 Fixed Sequence Robot Model

This mechanism can be used in the development of various agricultural machines. A specific use can be for seeding/planting applications where a multi-row configuration can be used. In Fig. 7.7, for simplification, a five-row mechanism is presented but this can be extended to a 10–20 rows system if required.

An end-effector can be attached to kinematic element 6 (i.e. gripper for planting applications or pneumatic suction pad for seeding) which can be controlled remotely. A hopper/bucket/tray system can be used to allow the mechanism to perform the pick-and-place operation for either seeds or plants. Having in view the kinetic and dynamic modelling of the mechanism (Figs. 7.1 and 7.3), which performances are previously presented (Sects. 7.3 and 7.4) the new constructive model of the fixed sequence robot (Fig. 7.7) is achieved.

Due to the compact construction the entire system can be attached to the three-point linkage of a tractor allowing additional equipment to be towed: water tank/sprayer, fertilizing stations etc.

Also, if a "shovel type" end-effector is attached, the mechanism can be used for soil preparation, in digging holes prior to planting (Fig. 7.8).

7.6 Conclusions

The designed robot with fixed sequence characteristics can be used for various agricultural operations, such as preparation of germination beds for seedlings [7, 8], planting, verification of agricultural crops, harvesting, or it can be integrated in an equipment with the attributes of a fixed sequence mobile robot.

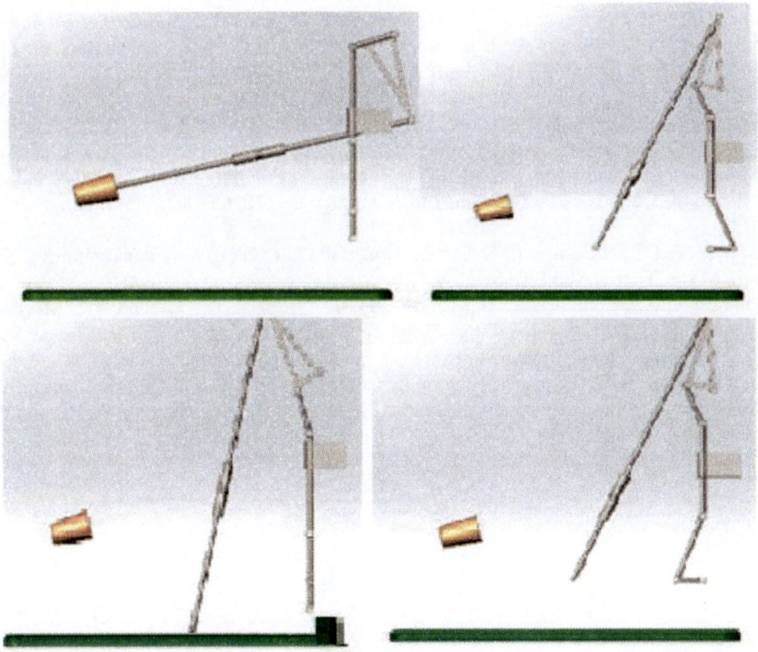

Fig. 7.8 Various working orientations/positions

Having in view the planar linkages [5, 9, 10] with three loops (16 ones) another mechanism may be found and integrated in other robot construction. For such robots it is absolute necessary that the extreme point of the effector to have a trajectory imposed by the technological process.

The main contribution of the authors consists in mechanism modelling with the terminology and methods specific to Mechanisms and Robots Theory and including of the mechanism in a fixed sequence robot.

References

1. https://www.researchgate.net/figure/Kinematic-Model-of-Transplanting-Finger-Mechanism-14-shows-the-mathematical-modelof_fig1_327234353
2. Comanescu, A., Comanescu, D., Dugaesescu, I.: About the Fixed Sequence Walking Micro-robots Modelling, Buletinul Institutului Politehnic din Iaşi publicat de Universitatea Tehnică „Gheorghe Asachi "din Iaşi, Tomul LVI (LX), Fasc. 4, 2010, Secţia Construcţii de maşini (2010)
3. Comanescu, A., Comanescu, D., Dugaesescu, I., Ungureanu, L., Alionte, C., Classical Modules Applied to the Mechanisms Structural Analysis and Synthesis. The 14th IFToMM World Congress, Taipei, Taiwan, October 25–30, 2015 (2015). https://doi.org/10.6567/IFToMM.14TH.WC.OS8.018.

4. Comanescu, A., Comanescu, D., Dugaesescu, I., Ungureanu, L., Alionte, C.: Modelling and Simulation of Mechanisms and Biomorph Systems. POLITEHNICA Press, Bucharest (2019). ISBN: 978-606-515-857-3
5. Crossley, F.R.E.: Contributions to grubler's theory in the number synthesis if plane mechanics. J. Eng. Ind. (1964). ASME
6. Dugaeşescu, I., Comanescu, D., Comanescu, A.: Structural and Kinematic Particularities of the Micro-robot Mechanisms; ICAMaT & Robotics & Polcom, 2014, *Trans Tech Periodicals* published by Trans Tech Publications Ltd., Churerstrasse 20, CH-8808 Pfaffikon, Switzerland (2014)
7. Sonawale, K.H., McCarthy, J.M.: Synthesis of eight-bar linkages by constraining a 6R loop. Mech. Mach. Theory **105** (2016)
8. Comanescu, A., Comanescu, D., Ungureanu, L., Boblea, D., Boblea, L.: Mechanisms-Analysis and Synthesis, Modelling, Simulation and Optimization of Mechanical Systems. POLITEHNICA Press, Bucharest (2021)
9. Kerle, H., Pittschellis, R.: Einführung in die Getriebelehre. B.G.Teubner, Stuttgart (1998)
10. Pelecudi, C.: Bazele Analizei Mecanismelor. Editura Academiei Române, Bucureşti (1967)

Chapter 8
Bi-mobile Mechanisms Modelling for Variable Sequence Robots

A. Comanescu, I. Dugaesescu, D. Boblea, and A. Gheorghita

Abstract The paper puts into evidence the structural and constructive modelling of bi-mobile mechanical systems applied in various equipment such as variable sequence robots, which may be used in different areas. The theoretical concepts, modeling and programming of mechanisms allow to illustrate their characteristics. A new theoretical structural design method for bi-mobile mechanisms which may be frequently integrated in robot construction was developed. All of these being main contributions to the paper.

Keywords Variable sequence robot · Bi-mobile mechanism · Structural model · Kineto-dynamic algorithm · Design structural theoretical method

8.1 Introduction

The three-mobile robot (Fig. 8.1) selected from the specialized literature [1] brought into attention the main structural characteristics of such systems.

The robot components achieved by using Fusion 360 and converted to STL were 3D printed with PLA bio-degradable material and assembled [2].

The robot has three mobilities, two of them belonging to the arm mechanism and the third one is given by the mechanism basis rotation along the vertical axis.

The kinematic scheme of the planar bi-mobile arm mechanism is given in Fig. 8.2, which is used to achieve its direct structural model (Fig. 8.3). One may identify the links according to those mentioned in Fig. 8.1 and its active pairs [(1,2), 2] and [(1,8), 8] marked in the structural model. The mechanism has $n = 9$ links and $c = 11$ kinematic rotation pairs, so that the degree of freedom is $L = 5$ ($L = 3n - 2c$) and the degree of mobility becomes $M = 2$ ($M = L - 3$, $M = 3m - 2c$, $m = n - 2$) the link 1 being selected as basis of the mechanism.

According to the direct structural model (Fig. 8.3) the modular group connection (Fig. 8.4) puts into evidence two active initial modular groups, respectively AIMP

A. Comanescu (✉) · I. Dugaesescu · D. Boblea · A. Gheorghita
University Politehnica of Bucharest, Bucharest, Romania
e-mail: adrianacomanescu@yahoo.com

I. Doroftei et al. (eds.), *Proceedings of SYROM 2022 & ROBOTICS 2022*, Mechanisms and Machine Science 127, https://doi.org/10.1007/978-3-031-25655-4_8

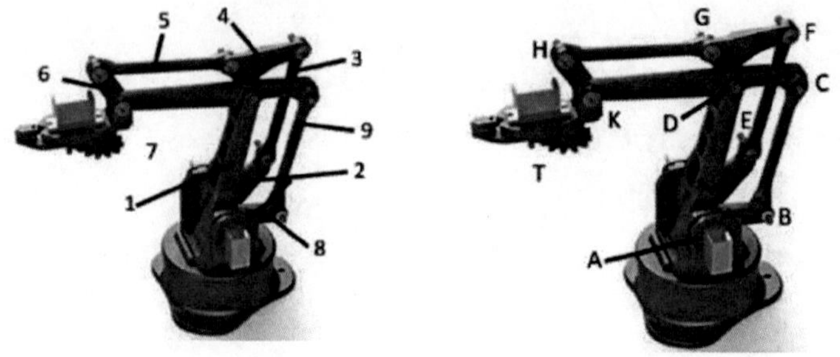

Fig. 8.1 Three-mobile robot

Fig. 8.2 The kinematic scheme of bi-mobile arm mechanism

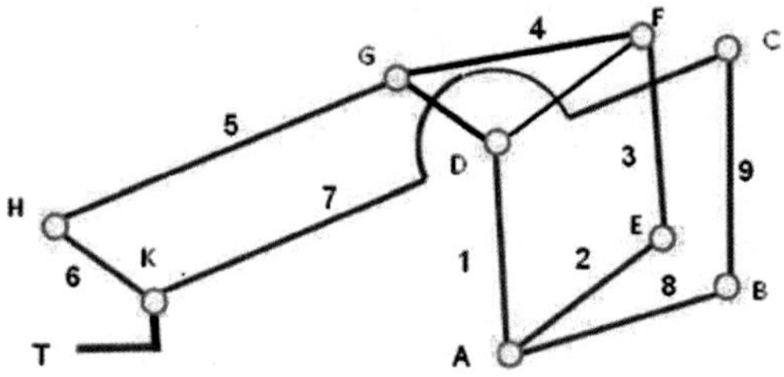

Fig. 8.3 The direct structural model

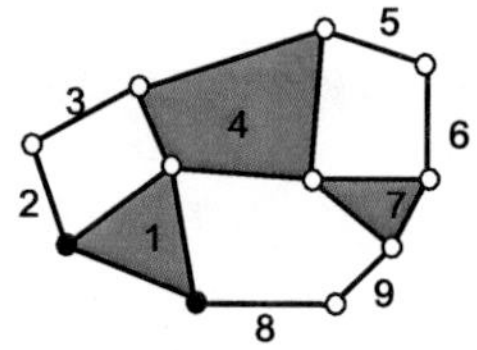

[(1,2), 2] and AIMP [(1,8), 8] and three passive modular groups PMG, the effector being chosen the link 7.

The active pairs (1,2) and (1,8) may work simultaneously or successively. So, when the pair (1,2) is blocked the structural model of mechanism (Fig. 8.5) has the

Fig. 8.4 The modular group connection for the direct structural model

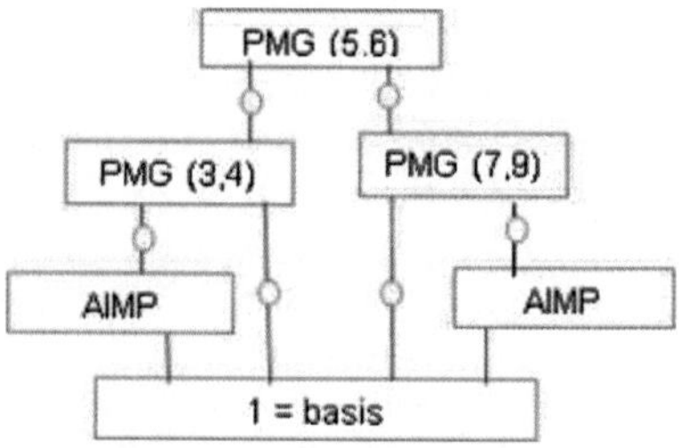

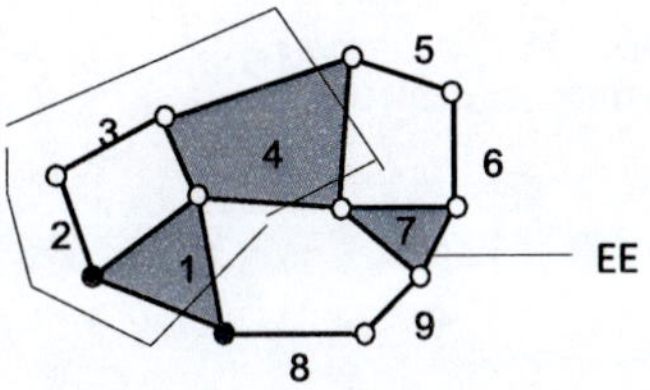

Fig. 8.5 The structural model with the active pair (1,2) blocked

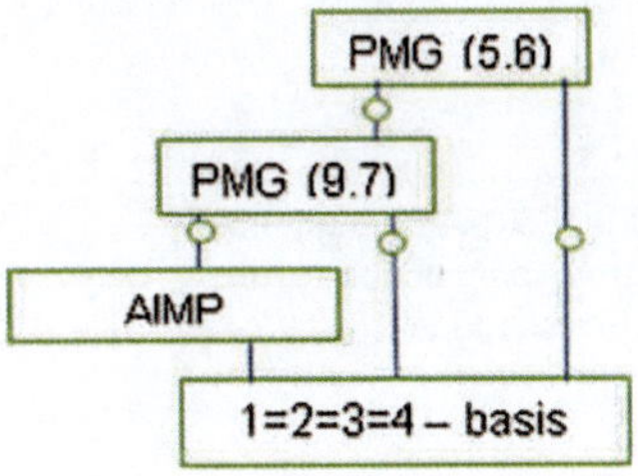

Fig. 8.6 The modular group connection when the active pair (1,2) is blocked

modular group connection shown in Fig. 8.6. Consequently, when the pair (1,8) is eliminated, the structural model becomes that presented in Fig. 8.7, for which the modular group connection is given in Fig. 8.8.

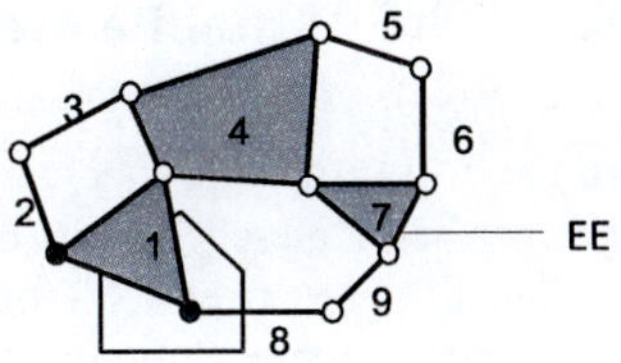

Fig. 8.7 The structural model when the active pair (1,8) is blocked

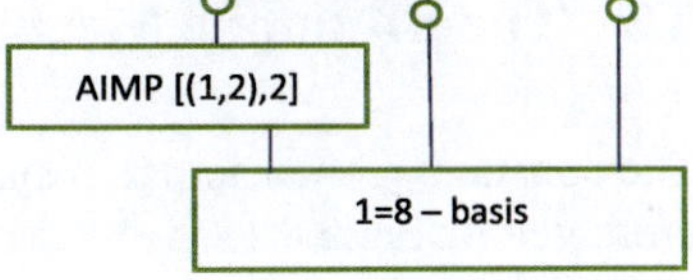

Fig. 8.8 The modular group connection when the active pair (1,8) is blocked

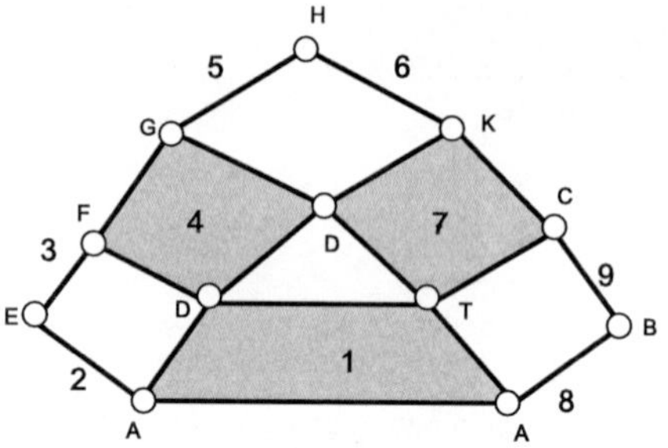

Fig. 8.9 The inverse structural model

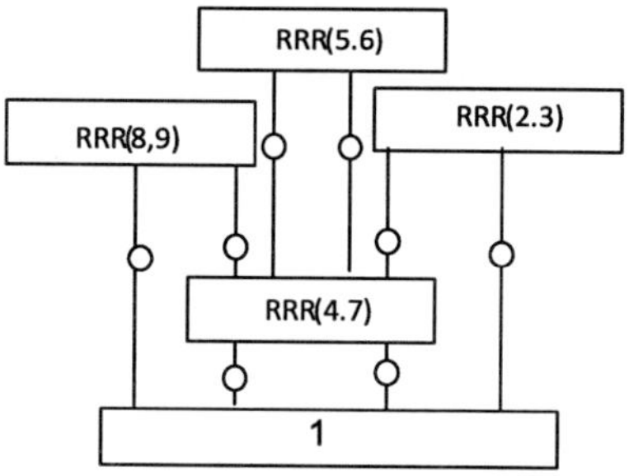

Fig. 8.10 The modular group connection for the inverse model

8.2 The Bi-mobile Mechanism Inverse Structural Model

In the arm bi-mobile mechanism (Fig. 8.1) the link 7 is nominated as effector, the gripper being attached to its extremity.

The gripper extremity T may described any planar trajectory necessary to realize the inverse structural model (Fig. 8.9) by introducing a virtual pair between the basis and the previously mentioned point. This structural model has three structural passive modular groups PMG, the first one including the effector 7 connected to the basis 1 through its virtual pair T (Fig. 8.10).

By using the modular group connection (Fig. 8.10) the angular characteristics of the active pairs (1,2) and (1,8) may be determined for any trajectory of the effector extremity T.

8.3 The Bi-mobile Mechanism Kinematic Modelling

The constant geometric parameters of the bi-mobile arm mechanism (Fig. 8.2) are presented in Table 8.1.

Table 8.1 Geometric parameters

Constant geometric parameters
$a = 20\ AD = 3 \cdot a\ AB = a\ BC = 3 \cdot a\ DC = a$
$XT0 = -4 \cdot a$
$AE = \frac{a}{2}\ DF = \frac{a}{2}\ DG = 0.5 \cdot a\ GF = 2 \cdot a\ YT0 = AD$
$DK = 4 \cdot a\ GH = 4 \cdot a\ KH = 0.5 \cdot a\ EF = AD\ p = 0.5$
$LK = \frac{a}{4} TL = \frac{a}{4}$

The point T placed at the element 7 extremity may have any open or close trajectory with a movement in both directions, such as presented in Fig. 8.11 and with characteristics given in Table 8.2. Through k ascendent and descendent positions, point T has a closed trajectory (Fig. 8.11a. initial position, b. next the extreme ascendent position, c. next the final descendent position).

The dependent parameters of each mentioned passive modular group in their inverse structural model (Fig. 8.9) and their modular group connection (Fig. 8.10) shown in Table 8.3 were generated using the fundamental theoretical notions from the theory of mechanisms and robots [3–5].

The algorithm and its successive steps are presented in Table 8.3 and the results calculated using Mathcad.

Consequently, the point T trajectory (Fig. 8.11) is ensured by the angular characteristics $\phi 20$ and $\phi 80$ (Fig. 8.12) of the active pairs A ≡ (1,2) and A ≡ (1,8).

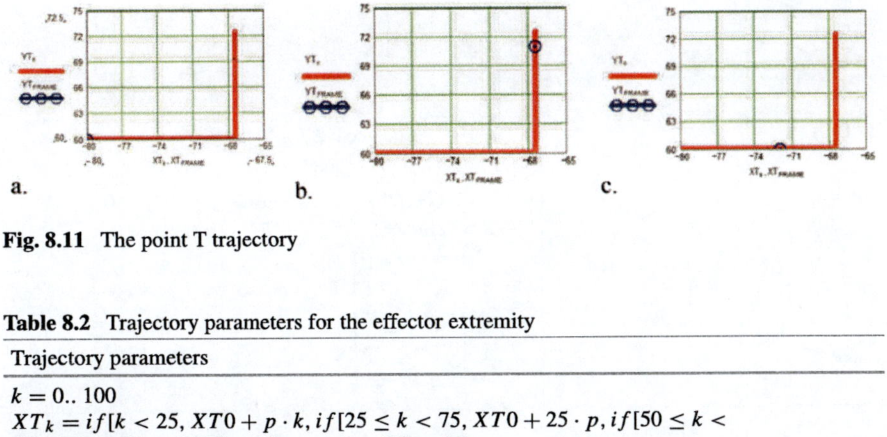

Fig. 8.11 The point T trajectory

Table 8.2 Trajectory parameters for the effector extremity

Trajectory parameters
$k = 0..\ 100$
$XT_k = if[k < 25, XT0 + p \cdot k, if[25 \leq k < 75, XT0 + 25 \cdot p, if[50 \leq k <$ $75, XT0 + 25 \cdot p, XT0 + 25 \cdot p - (k - 75) \cdot p]]]$
$YT_k = if[k < 25, YT0, if[25 \leq k < 50, YT0 + p \cdot (k - 25), if[50 \leq k <$ $75, YT0 + p \cdot 25 - p \cdot (k - 50), YT0]]]$
$XK_k = XT_k\ YK_k = YT_k$

Table 8.3 Dependent parameters

Modulus	Dependent parameters
MPG(7,4) 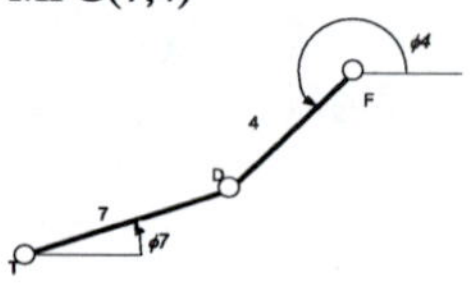 	$XK_k + DK \cdot \cos(\varnothing 7) - 0 - DF \cdot \cos(\varnothing 4) = 0$ $YK_k + DK \cdot \sin(\varnothing 7) - AD - DF \cdot \sin(\varnothing 4) = 0$ 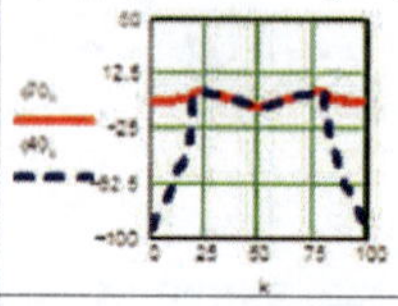
BPT(F)	$XF_k = 0 + DF \cdot \cos(\varnothing 4_k)$ $YF_k = AD + DF \cdot \sin(\varnothing 4_k)$
MPG(2,3) 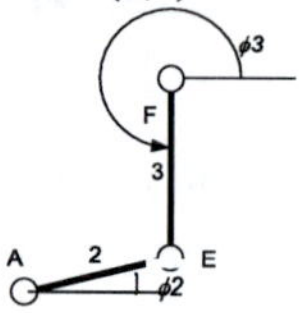 	$0 + AE \cdot \cos(\varnothing 2) - XF_k - EF \cdot \cos(\varnothing 3) = 0$ $0 + AE \cdot \sin(\varnothing 2) - YF_k - EF \cdot \sin(\varnothing 3) = 0$ 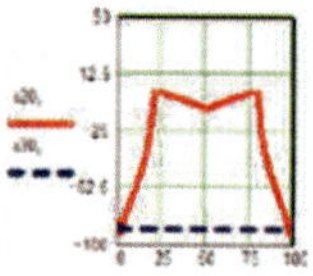
BPT(G)	$XG_k = 0 + DG \cdot cos\left(\varnothing 4_k + 2 \cdot \frac{\pi}{3}\right)$ $YG_k = AD + DG \cdot sin\left(\varnothing 4_k + 2 \cdot \frac{\pi}{3}\right)$
MPG(6,5) 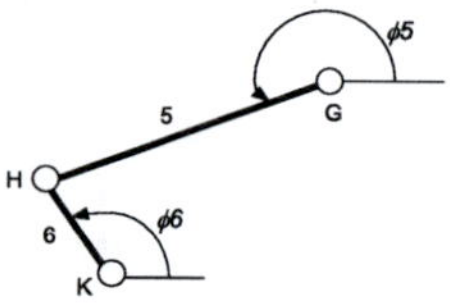 	$XK_k + KH \cdot \cos(\varnothing 6) - XG_k - GH \cdot \cos(\varnothing 5) = 0$ $YK_k + KH \cdot \sin(\varnothing 6) - YG_k - GH \cdot \sin(\varnothing 5) = 0$ 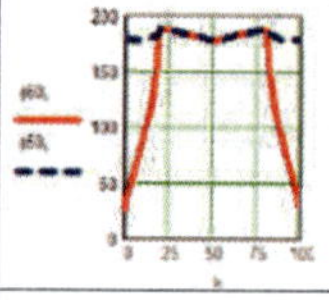
BPT(C)	$XC_k = 0 + DC \cdot \cos(\varnothing 7_k)$ $YC_k = AD + DC \cdot \sin(\varnothing 7_k)$
MPG(8,9) 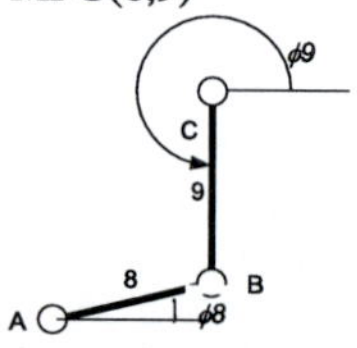 	$0 + AB \cdot \cos(\varnothing 8) - XC_k - BC \cdot \cos(\varnothing 9) = 0$ $0 + AB \cdot \sin(\varnothing 8) - YC_k - BC \cdot \sin(\varnothing 9) = 0$ 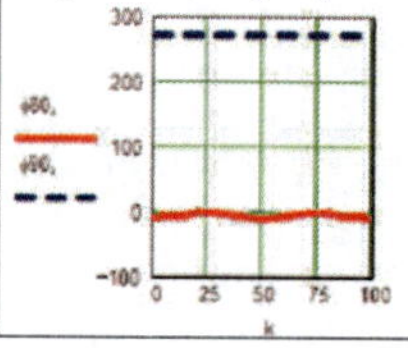

8.4 Bi-mobile Mechanism Basis and Effector Selection

The bi-mobile mechanism (Fig. 8.2) previously analyzed and included in the robot construction (Fig. 8.1) is based on the planar linkage (Fig. 8.3) in which there are selected the basis 1, the effector 7 and the active pairs. This linkage is selected from 40 ones with the same characteristics found in the literature [3–5]. In the chain with

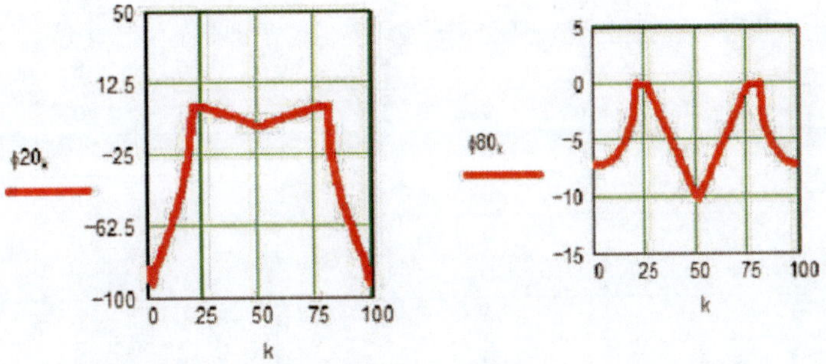

Fig. 8.12 The active pairs angular characteristics

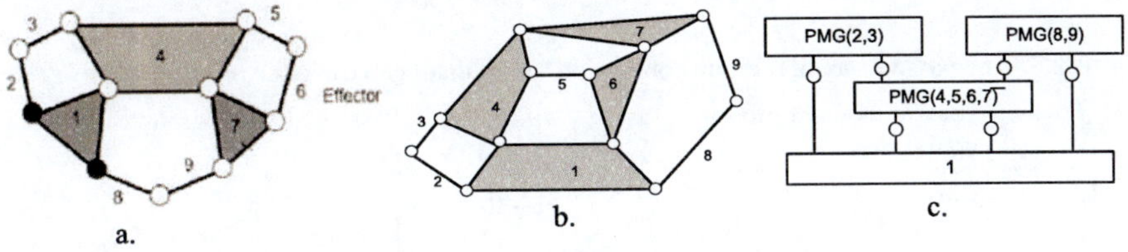

Fig. 8.13 New structure characteristics

three independent contours (the first contour—1,2,3,4, the second—4,5,6,7 and the third one—1,4,7,9,8) the base and the effector do not belong to a structure with four elements or Watt and Stephenson ones, detached from the chosen linkage [3–5]. In the same linkage preserving the active pairs (Fig. 8.13a) one may adopt the base 1 and the effector 6. Its associated inverse model (Fig. 8.13b) and inverse passive modular connection (Fig. 8.13c) includes the PMG with four links [3, 4].

As to the active pairs are maintained, the bi-mobile mechanism has the direct structural model and modular connection given in Figs. 8.3 and 8.4.

The possible bases and their associated effectors were nominated and marked in Table 8.4 by using the Boolean values.

In each 21 distinct cases the passive modular group connection [3, 4] for its corresponding inverse structural model is given in Table 8.5.

8.5 New Bi-mobile Mechanism Structural Design

New bi-mobile mechanisms may be obtained based on any bi-mobile linkage [3–5]. The structural algorithm is shortly presented in this section (Table 8.6).

So, for relevance the previous linkage (Fig. 8.3) is adopted. The designer can rename the elements number, select the basis 0 and the effector 4 (Table 8.5) and

Table 8.4 The distinct possible bases and effectors from the linkage

Link	1	2	3	4	5	6	7	8	9
1	0	0	0	0	1	1	1	0	1
2	0	0	0	0	1	1	1	1	1
3	0	0	0	0	1	1	1	1	1
4	0	0	0	0	0	0	0	1	1
5	1	1	1	0	0	0	0	1	1
6	1	1	1	0	0	0	0	1	1
7	1	1	1	0	0	0	0	1	0
8	0	1	1	1	1	1	1	0	0
9	1	1	1	1	1	1	0	0	0

Table 8.5 The passive modular group connection for all distinct basis-effector solutions

A[*i*,*j*]	Passive modular groups connection	A[*i*,*j*]	Passive modular groups connection
A[1,5]	PMG1(4,5) + PMG1(6,7) + PMG1(2,3) + PMG1(8,9)	A[1,5]	PMG1(1,4) + PMG1(8,3) + PMG1(6,7) + PMG1(8,9)
A[1,6]	PMG3(4,5,6,7) + PMG1(2,3) + PMG1(8,9)	A[6,1]	PMG2(1,4,5,7) + PMG1(2,3) + PMG1(8,9)
A[1,7]	PMG1(4,7) + PMG1(2,3) + PMG1(5,6) + PMG1(8,9)		
A[1,9]	PMG1(8,9) + PMG1(4,7) + PMG1(2,3) + PMG1(5,6)	A[9,1]	PMG1(1,8) + PMG1(4,7) + PMG1(2,3) + PMG1(5.6,)
A[2,5]	PMG2(1,4,3,5) + PMG1(6,7) + PMG1(8,9)	A[2,5]	PMG2(1,2,3,4) + PMG1(6,7) + PMG1(8,9)
A[2,6]	PMG7(1,3,4,5,6,7) + PMG1(8,9)		
A[2,8]	PMG1(1,8) + PMG1(3,4) + PMG1(7,9) + PMG1(5,6)	A[8,2]	PMG1(1,2) + PMG1(3,4) + PMG1(7,9) + PMG1(5,6)
A[2,9]	PMG8(1,3,4,7,8,9) + PMG1(5,6)	A[9,2]	PMG8(1,2,3,4,7,8) + PMG1(5,6)
A[3,5]	PMG1(4,5) + PMG1(1,2) + PMG1(6,7) + PMG1(8,9)		
A[3,8]	PMG2(1,2,4,8) + PMG1(7,9) + PMG1(5,6)	A[8,3]	PMG3(1,2,3,4) + PMG1(7,9) + PMG1(5,6)
A[3,9]	PMG8(1,2,4,7,8,9) + PMG1(5,6)	A[9,3]	PMG8(1,2,3,4,7,8) + PMG1(5,6)
A[4,8]	PMG1(1,8) + PMG1(2,3) + PMG1(3,9) + PMG1(5,6)	A[8,4]	PMG1(1,4) + PMG1(2,3) + PMG1(3,9) + PMG1(5,6)

place the active prismatic pairs (1,2) and (6,5). The new kinematic schematic can be obtained with a new inverse structural model and its modular structural connection.

Point T placed at the extremity of the effector 4 (Table 8.6) has a similar alternative trajectory (Fig. 8.11) as the previous mechanism (§ 3). The full trajectory of point T is ensured by the active prismatic pairs (1,2) and (6,7) with the following characteristics FE_k and FG_k (Fig. 8.14).

Table 8.6 Structural characteristics for a new mechanism

Direct model and its modular connection	New bi-mobile mechanism kinematic scheme	Inverse model and its modular connection
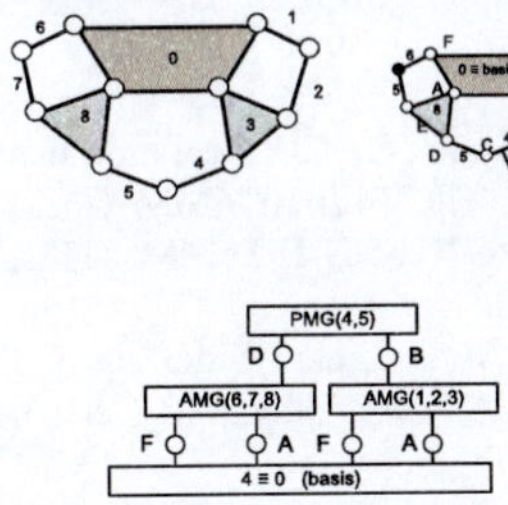	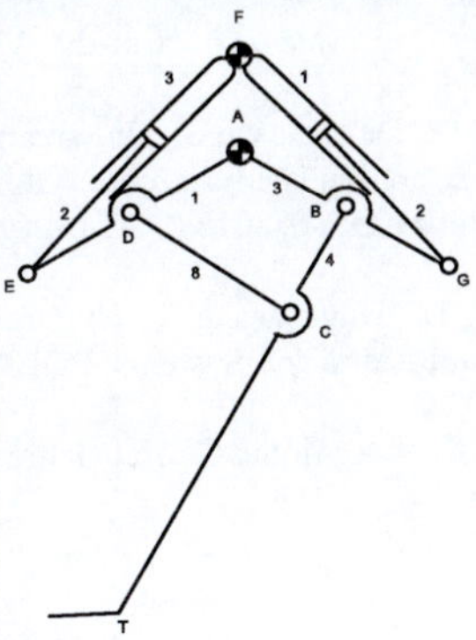	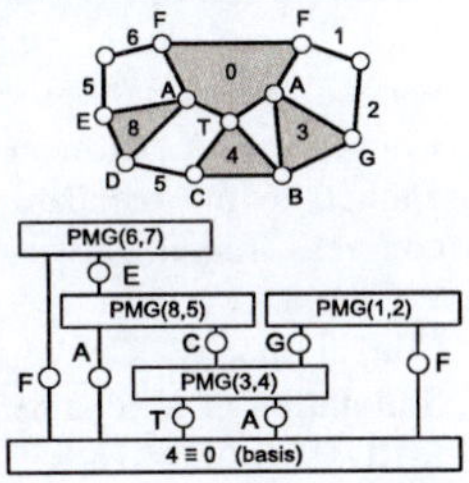

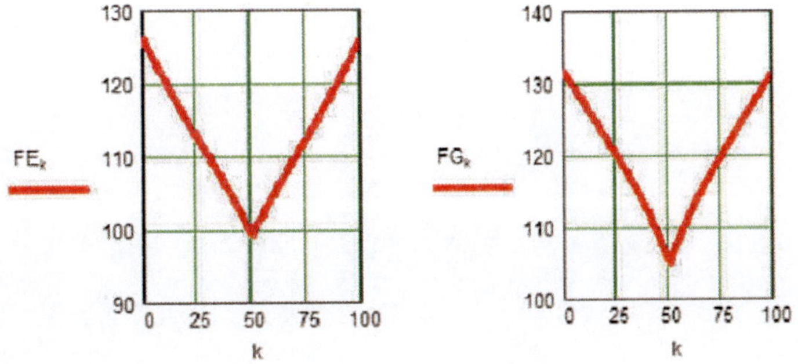

Fig. 8.14 The prismatic active pairs (1,2) and (6,7) characteristics

8.6 Conclusions

The bi-mobile mechanisms present undoubted advantages, given the possibility of modification and flexibility of the effector extremity trajectory depending on the technological process or executed operation, such mechanisms improving and giving new capabilities to robot's design.

To create new mechanisms which may be used in robots' construction is the main aim of any designer and researcher. The knowledge of mechanisms and robots' theory presented in this paper give the possibility to obtain 21 new distinct structures (Table 8.5) only by nominating the possible bases and effectors in the linkage (Fig. 8.13).

This article reveals the contribution made regarding the application of some fundamental theorems in the structural design of the bi-mobile mechanisms used in the construction of robots. These theorems have been analyzed and highlighted in some previous papers.

References

1. https://articulo.mercadolibre.com.mx/MLM-573640818-brazo-robot-diy-incluye-4-servosarduino-uno-100-envio-grat-_JM
2. https://www.3dprintersonlinestore.com/eezybotarm-the-3d-printed-robotic-arm-you-always-wanted-to-make
3. Comanescu, A., Comanescu, D., Dugaeşescu, I., Ungureanu, L., Alionte, C.: Classical modules applied to the mechanisms structural analysis and synthesis. In: The 14th IFToMM world congress, Taipei, Taiwan, October 25–30, 2015 (2015). https://doi.org/10.6567/IFToMM.14TH.WC.OS8.018
4. Comanescu, A., Comanescu, D., Dugaeaescu, I., Ungureanu, L., Alionte, C.: Modelling and Simulation of Mechanisms and Biomorph Systems. POLITEHNICA Press, Bucharest (2019). ISBN: 978-606-515-857-3
5. Kerle, H., Pittschellis, R.: Einführung in die Getriebelehre. B.G.Teubner, Stuttgart (1998)

Chapter 9
Elevating Platforms with Hydraulic/Pneumatic Actuation for Automotive Maintenance

O. Antonescu, D. Valeanu, and D. Antonescu

Abstract The paper presents the main constructive and kinematic schemes of the planar mechanisms used as elevating platforms for automotive maintenance, which are actuated by means of hydraulic or pneumatic linear actuators. These linkages are designed in a single or double X-shape. Usually, two of these X-shape planar mechanisms are assembled in parallel and in vertical plane. In the second part of the paper a structural analysis of the lifting mechanisms in single and double X-shape has been accomplished. For two kinematic schemes of X-shape mechanisms the mobility has been analytically calculated. Also, a static computing methodology of the kinematic joint forces including the driving force in the hydraulic/pneumatic linear actuator has been presented.

Keywords X-shape elevating platform · Mechanism · Structural analysis · Mobility

9.1 Introduction

The elevating platforms in X-shape (scissor lifts) are limited regarding the height and the lifting force. The main advantage of these mechanisms is the reduced space required in the automotive maintenance workshop. Due to the folded position when they are not operating, these mechanisms require a minimum space.

The hydraulic and pneumatic actuation systems form a vast and important domain. In the context of tremendous expansion of the transmissions with pressurized fluids into the most varied industrial domains, the term "synthesis" (system conception) has a wide content [1, 2, 5].

The rational synthesis requires a profound knowledge on the construction and operating of the pressurized fluid system elements. An ideal synthesis methodology

O. Antonescu (✉) · D. Valeanu
University Politehnica of Bucharest, Bucharest, Romania
e-mail: oval33@hotmail.com

D. Antonescu
"Iuliu Maniu" High School, Bucharest, Bucharest, Romania

I. Doroftei et al. (eds.), *Proceedings of SYROM 2022 & ROBOTICS 2022*, Mechanisms and Machine Science 127, https://doi.org/10.1007/978-3-031-25655-4_9

has to be analytical [4, 6] and to lead to a direct or iterative output of some equation or inequation systems having the unknowns as the optimum system parameters (geometric, hydraulic, mechanical, electric, etc.).

The establishing of the constructive parameters of a hydraulic or pneumatic system implies the solving of the differential equation system which describes its dynamic behavior. The nonlinear character of the usual equations imposes the using of numerical computers.

The constructive parameters are adjusted until the computed performances (stability reserve, reaction velocity, accuracy, etc.) are satisfactory.

The elevating platforms in single X-shape (Figs. 9.1, 9.2 and 9.3) are used for both light and heavy loads. Figure 9.4a shows a twin double X-shape mechanism and Fig. 9.4a mobile double X-shape elevating work platform. The mechanisms shown in Figs. 9.1, 9.3 and 9.4 are driven by means of one or two hydraulic or pneumatic linear actuators.

The elevating platforms shown in Figs. 9.3 and 9.4a are working in parallel which means that their linear actuators are driven in the same time. We called them "twin X-shape elevating platforms" and they are designed for lifting automobiles. Depending on which load they were designed for, they could be equipped with two (Fig. 9.5a, b) or four hydraulic linear actuators, two-by-two linked.

Fig. 9.1 Single X-shape elevating platform

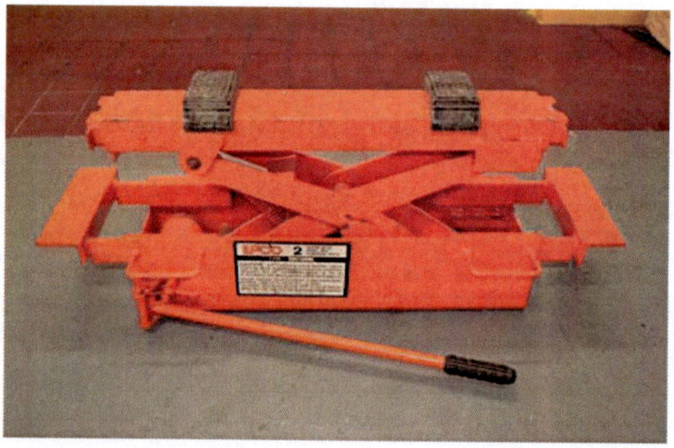

Fig. 9.2 Manual single X-shape elevating platform

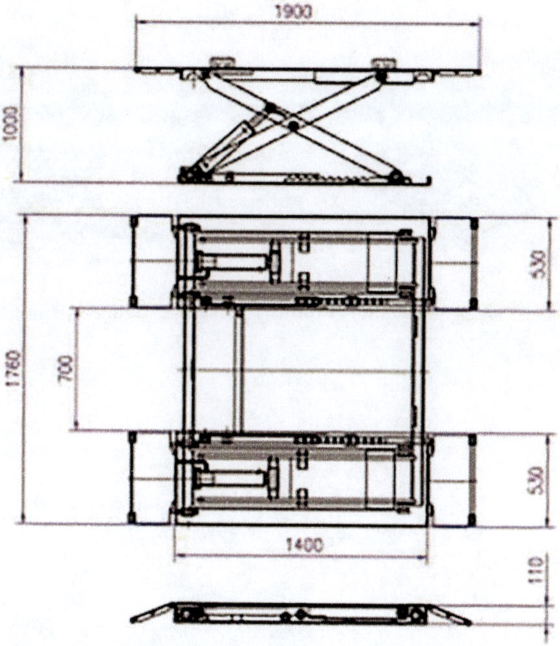

Fig. 9.3 Twin single X-shape elevating platform (top and side-views)

(a) (b)

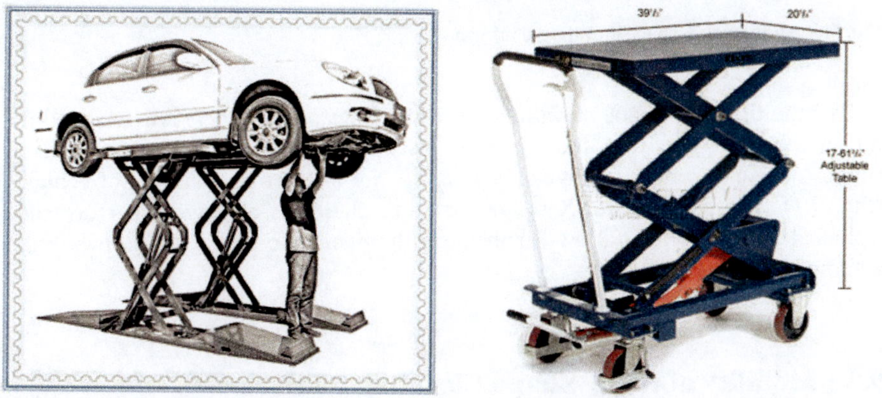

Fig. 9.4 **a** Twin double X-shape elevating platform. **b** Mobile double X-shape elevating work platform (scissor lift)

(a) (b)

Fig. 9.5 Twin single X-shape elevating platforms with two (**a**) or one platform (**b**)

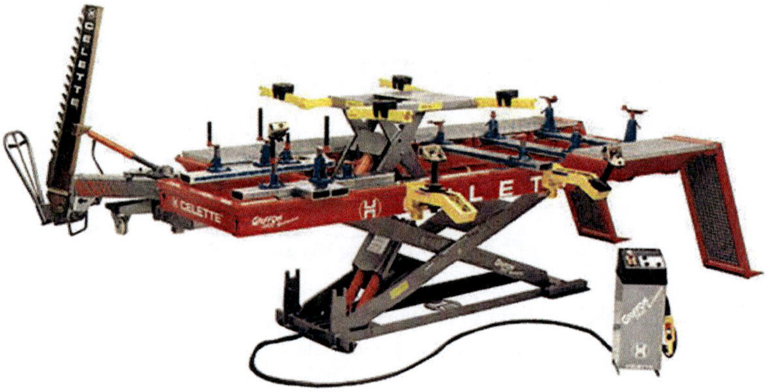

Fig. 9.6 Multilevel X-shape elevating platform

The multilevel X-shape elevating platforms (Figs. 9.6 and 9.7) are designed for higher lifting heights than the previous. Usually these mechanisms have a fixed support which in some cases it can be "buried" for a better accessibility of vehicles (Fig. 9.7). These multilevel X-shape elevating platforms are highly performance technical systems which allow for instance the dismantle of the engine while the automobile is lifted [5–8].

9.2 Mobility of the X-Shape Elevating Platforms

9.2.1 The Single X-Shape Mechanism

The kinematic scheme of the X-shape elevating platform (Fig. 9.8) shows the bars 1 and 2 as kinematic elements articulated in joint B, and the platform 3 as the element articulated in joint D to bar 2 and guided in point E of the bar 1.

Fig. 9.7 Multilevel twin X-shape elevating platform

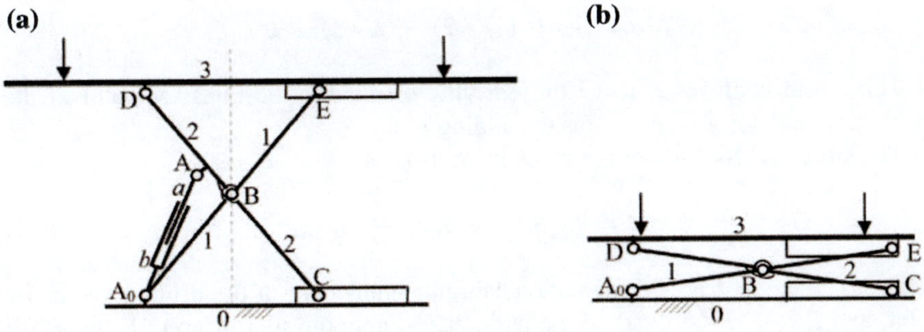

Fig. 9.8 Kinematic scheme of the single X-shape elevating platform in unfolded (a) and folded (b) positions

The elevating manipulator in X-shape [8] allows lifting and lowering of the platform 3 which is maintained in horizontal position due to the geometrical property of the X-shape mechanism ($A_0C = DE$). This mechanism (Fig. 9.8a) operates in vertical plane and it is driven by means of a linear actuator which can consists of a piston (a) and an oscillating cylinder (b).

The technological resisting force acts as the vehicle weight and the driving force acts in cylinder b on the piston a along direction A_0A (Fig. 9.8a). The topological structure of the planar mechanism [3, 4] outlines five rotate kinematic joints, one sliding kinematic joint (piston – cylinder) and two superior kinematic joints (rotation + sliding between elements 2 and 0, and between 1 and 3).

The mobility of the mechanism is calculated by means of the formula [1, 2]:

$$M_b = \sum_{1}^{5} m * C_m - \sum_{2}^{6} r * N_r \quad (9.1)$$

In the formula (9.1) the following denotations have been used:

m—the mobility of the kinematic joint (number of the allowed independent; movements);

C_m—the number of the kinematic joints of class m;

r—the order of the closed independent kinematic contour;

N_r—the number of the independent kinematic contours.

Observing the analyzed kinematic scheme (Fig. 9.8a), the following parameters have been identified:

$$m = 1,\ C_1 = 6;\ m = 2,\ C_2 = 2;\ r = 3,\ N_3 = 3.$$

Substituting these numerical values in formula (9.1) we obtained:

$$M_b = (1 * 6 + 2 * 2) - (3 * 3) = 1 \quad (9.2)$$

This result confirms that the only driving element is the linear actuator of the lifting manipulator where there is the sliding joint.

The structural formula of the mechanism is [2, 4]:

$$M = LCm(1, 2, a, b) + LD(3) \quad (9.3)$$

where *LCm* is the driving kinematic chain, the equivalent of the driving kinematic joint, and *LD*(3) is the dyadic kinematic chain, noticing that in point E there is a superior joint of class 4 with two mobilities ($m = 2$).

The static calculation of X-shape mechanism implies the following phases:

- the isolation of the platform 3 together with the external weight forces, and the reaction forces in joints D and E (Figs. 9.8a and 9.9);
- the isolation of the open kinematic chain (1, 2, *a*, *b*) (Fig. 9.8a) with the known forces in points D and E from previous phase, and also the weight forces of the bars 1 and 2;
- the introduction of the reaction forces (two components) in points A and B (as simple articulations), in A_0 (as double articulation), in the sliding joint between the piston *a* and the cylinder *b*, but also in the joint C;

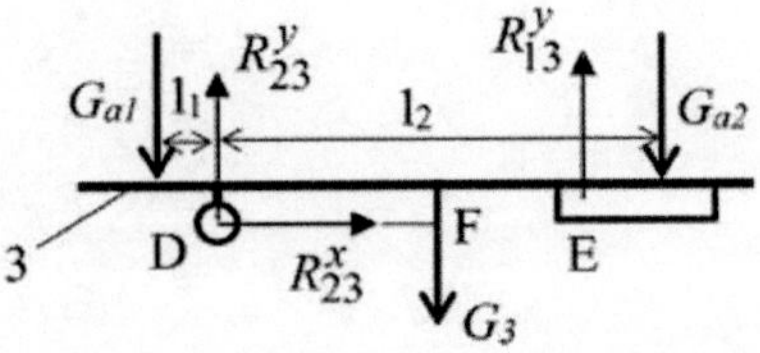

Fig. 9.9 Platform 3 loaded with external and reaction forces

- the writing of the three planar static balancing equations for the platform 3 (Fig. 9.9) and the solving of the three scalar equation system;
- the writing of the specific equations for each kinematic element for the static balancing of the kinematic chain (1, 2, *a*, *b*), resulting $4 \times 3 = 12$ scalar equations with 12 unknowns of which one of them is the driving force of the linear actuator.

The static calculation is under hypothesis of non-friction operation. In the first phase of this calculation we have isolated the platform 3 on which G_1 and G_2 represent the weight forces of the vehicle on the front and rear axles (Fig. 9.9).

The rotate joint in point D implies two unknown components R^x_{23}, R^y_{23} of the reaction force of the bar 2 on the platform 3, and the unknown reaction force R^y_{13} is in superior joint (point E) acting from bar 1 to the horizontal platform 3.

The static equilibrium equations of the platform 3 are:

$$\begin{aligned}
&\Sigma F^x_3 = 0;\; R^x_{23} = 0\\
&\Sigma F^y_3 = 0;\; R^y_{23} + R^y_{13} = G_{a1} + G_{a2} + G_3\\
&\Sigma M^3_D = 0;\; R^y_{13} * DE = G_{a2} * l_2 - G_{a1} * l_1 + G_3 * DF
\end{aligned} \tag{9.4}$$

The three reactions from the system (9.4) are obtained:

$$\begin{aligned}
&R^x_{23} = 0\\
&R^y_{13} = \frac{1}{DE}[G_{a2} * l_2 - G_{a1} * l_1 + G_3 * DF]\\
&R^y_{23} = G_{a1}\left[1 + \frac{l_1}{DE}\right] + G_{a2}\left[1 - \frac{l_2}{DE}\right] + G_3\left[1 - \frac{DF}{DE}\right]
\end{aligned}$$

To be noticed that the distance DE between the two supports is variable.

The other components of the reactions from the articulations A_0, A and B, the sliding joint between the piston *a* and the cylinder *b*, the rotation + sliding joint C, and also the driving force from the linear actuator (Fig. 9.8a), are deduced through numerical calculation of the system with 12 equations and 12 unknowns:

$$\begin{aligned}
&\Sigma F^x_1 = 0;\; R^x_{01} + R^x_{21} = 0\\
&\Sigma F^y_1 = 0;\; R^y_{01} + R^y_{21} = 0 - R^y_{13} = G_1\\
&\Sigma F^x_2 = 0;\; R^x_{a2} + R^x_{23} = 0 - R^x_{21} = 0\\
&\Sigma F^y_2 = 0;\; R^y_{a2} + R^y_{02} - R^y_{23} - R^y_{21} = G_2
\end{aligned}$$

$$\begin{aligned}
\Sigma F_a^x &= 0;\ R_{ba}^x - R_{a2}^x + F_{ba}^{mx} = 0 \\
\Sigma F_a^y &= 0;\ R_{ba}^y - R_{a2}^y + F_{ba}^{my} = G_a \\
\Sigma F_b^x &= 0;\ R_{0b}^x - R_{ba}^x + F_{ba}^{mx} = 0 \\
\Sigma F_b^y &= 0;\ R_{0b}^y - R_{ba}^y + F_{ba}^{my} = G_b
\end{aligned} \tag{9.5}$$

To these 8 projection Eqs. (9.5) we add 4 torque equations, one for each of the 4 kinematic elements $(1, 2, a, b)$. Therefore, by resolving the 12 equation system with 12 unknows, we can deduce the joint reaction forces including the driving force F_{ba}^m required to actuate the elevating platform.

9.2.2 The Double X-Shape Mechanism

The kinematic scheme of the double X-shape elevating platform (Fig. 9.10) shows that the two "X" are one on top of the other (the bars 1 and 2, and the bars 3 and 4 respectively) being articulated in D and E. The bars 3 and 4 are joined in F while the platform 5 is articulated in G to the bar 4 and guided in H to the bar 3.

In this case the lifting height is double compering with the single X-shape mechanism meaning that for the same actuator stroke the manipulator stroke is almost double. We are analyzing two types of kinematic actuator chain mounting, the first as in the single X-shape mechanism when the actuator piston a is linked to bar 2 (Fig. 9.10a) and the second when a is linked to bar 4 (Fig. 9.10b). In the first case there is a recommendation [10] that the actuator piston a should be linked to bar 2 near to the joint B. In both cases of the double X-shape lifting manipulator the mechanism structure is obtained by means of the amplification of the single X-shape mechanism

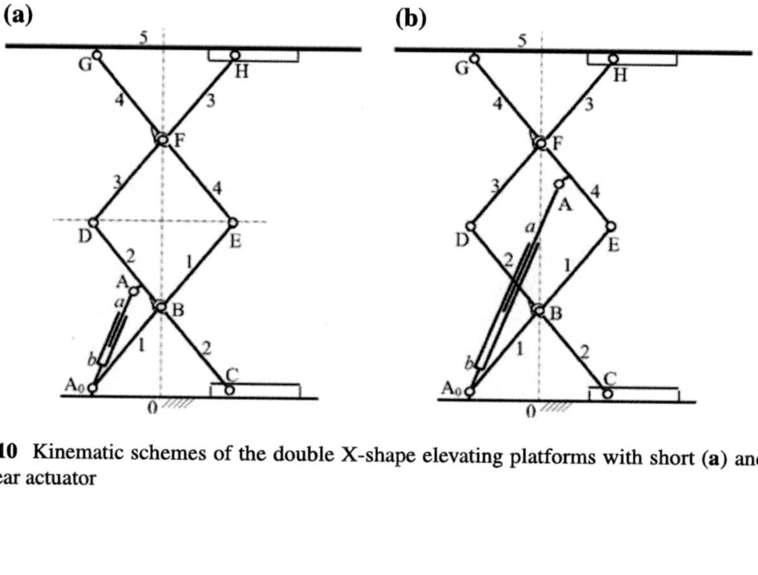

Fig. 9.10 Kinematic schemes of the double X-shape elevating platforms with short (**a**) and long (**b**) linear actuator

(Fig. 9.8) by two neutral kinematic chains [2]: DFE (3,4) of type dyad RRR and GH (5) of type dyad RRT. The mobility of both planar mechanisms (Fig. 9.10) is calculated by means of the same formula (9.1):

$$M_b = (1 * 9 + 2 * 2) - (3 * 4) = 1 \tag{9.6}$$

The driving kinematic chain (consisting of the piston *a* and the cylinder *b*) is a dyad chain type RTR with the mobility equal by zero [3, 4]. This kinematic chain can be mounted to any two distinct kinematic elements, for instance to the bar 2 (Fig. 9.10a) or bar 4 (Fig. 9.10b). We mention that in the case of static calculation [2, 9] of the lifting manipulator (Fig. 9.10b), this implies the isolation, in the last phase, of a kinematic chain with one closed contour (1, 2, 3, 4, *a*, *b*).

For each of the six kinematic elements three planar static balancing equations have to be written, obtaining 18 scalar equations, not taking in consideration the loading of the platform 5 which is the same as in the case of previous mechanism (Fig. 9.9) analyzed at Sect. 2.1.

The 18 unknowns are: 14 components (horizontal and vertical) of the reaction forces in the rotate joints A_0 (double), A, B, D, E, F; 2 components (force and torque) of the reaction force in the sliding joint between the piston *a* and the cylinder *b*; 1 reaction force perpendicular on the fixed guide (point C); 1 driving force between the cylinder *b* and the piston *a*. The resolving of the 18 linear equations with 18 unknowns can be accomplished by a programming sequence.

9.3 Conclusions

The advantages of using such an equipment in workshops are multiple. First of all, we have a better accessibility to any automotive part as in the case of using a ditch. Secondly, we do not need to dig a ditch in the workshop floor. More than that, we do not need to use too much light to see under the vehicle.

But the disadvantage of using a lifting manipulator is that we have to make sure that it is stable and secure to the payload that we designed it for. The actuators are the most vulnerable parts of these mechanisms. They have to be reliable and secure. In the case of actuating system collapse we could design a backup device that will stand and insure the mechanism.

References

1. Antonescu, P.: Mechanisms – course book (in Romanian). Printech Press, Bucharest (2003)
2. Antonescu, O., Antonescu, P.: Mechanism and Machine Science – course book. Politehnica Press, Bucharest (2016)
3. Antonescu, P., Antonescu, O.: Standardization of terminology of MMS and graphical symbols. In: The Third National Seminar on Mechanisms, Craiova (2008)

4. Antonescu, O., Antonescu, P.: Mechanisms and Manipulators. Printech Press, Bucharest (2006)
5. *** Lift, Instruction manual - HL-32 F, First Printing, Korea (2002)
6. *** Elevating work platform, 21 Apr. 2001
7. *** Elevating work platform (2011). www.wikipedia.com
8. *** Understanding Scissor Lift Deflection (2011). www.autoquip.com
9. Adagio, E.M.: A project on scissor design for use in the automotive industry. (2012)
10. *** Pont double ciseaux a poser, Art. nr. TW S3-19. www.twinbusch.fr

Part II
Theoretical and Experimental Mechanics

Chapter 10
Hypercomplex Dual Lie Nilpotent Algebras and Higher-Order Kinematics of Rigid Body

D. Condurache, M. Cojocari, and I. Popa

Abstract This paper proposes a novel representation for a higher-order acceleration vector field in the case of rigid body motion, using the properties of dual and multi-dual nilpotent algebras. It is proved that all information regarding the properties of the distribution of higher-order accelerations is encapsulated in the specified multi-dual (MD) homogeneous matrix or the hyper- multidual (HMD) orthogonal tensor. Moreover, the results are in closed-form and coordinate-free.

Keywords Higher-order kinematics · Multidual nilpotent algebra · Lie group · Lie algebra

10.1 Introduction

The recent development of high-precision robotic systems, artificial vision systems, and molecular dynamics, require new procedures to calculate higher-order accelerations. The key to the proposed approach starts with the property of rigid body displacements group of forming a Lie group, accompanied by its Lie algebra. A previous result offers an isomorphic representation of the Lie group SE (3) with the group of the orthogonal dual tensors and Lie algebra se (3) of the Lie algebra of dual vectors [1]. The results obtained using dual algebras completely solve the problem of finding the field of higher-order accelerations using a set of results obtained by the previous papers [2, 3]. This paper describes a novel general method for studying the vector field of arbitrary higher-order accelerations in the case of rigid body motion. It is proved that all information regarding the properties of the distribution of higher-order accelerations is contained in the specified homogenous multidual (MD) matrix [4] or the hyper-multidual (HMD) orthogonal tensor.

D. Condurache (✉) · M. Cojocari · I. Popa
"Gheorghe Asachi" Technical University of Iasi, Iasi, Romania
e-mail: daniel.condurache@tuiasi.ro; daniel.condurache@gmail.com

I. Doroftei et al. (eds.), *Proceedings of SYROM 2022 & ROBOTICS 2022*, Mechanisms and Machine Science 127, https://doi.org/10.1007/978-3-031-25655-4_10

10.2 Multidual and Dual Lie Algebra Representations of Higher-Order Kinematics

Let be $\mathbb{R}$ the set of real numbers and $n \in \mathbb{N}$ a natural number. Then, we will introduce the set of multidual number [3–7] by $\widehat{\mathbb{R}} = \mathbb{R} + \varepsilon\mathbb{R} + \cdots + \varepsilon^n\mathbb{R}$; $\varepsilon \neq 0, \varepsilon^{n+1} = 0$. For $n = 1$, on obtain the dual algebra [2, 3].

Two generic elements $\widehat{x}, \widehat{y} \in \widehat{\mathbb{R}}$ will be written as follows:

$$\widehat{x} = x + x_1\varepsilon + \cdots + x_n\varepsilon^n;\, x, x_k \in \mathbb{R}, k = \overline{1, n}, \tag{10.1}$$

$$\widehat{y} = y + y_1\varepsilon + \cdots + y_n\varepsilon^n;\, y, y_k \in \mathbb{R}, k = \overline{1, n}. \tag{10.2}$$

We will denote by $x = Re\widehat{x}$ and $Mu\widehat{x} = \sum_{k=1}^{n} x_k\varepsilon^k$ the real parts and respectively, multidual parts of the multidual number $\widehat{x}$. Also, we will denote by $x_k = \frac{d\widehat{x}}{d\varepsilon^k}$; $k = \overline{1, n}$, the k order multidual component of the multidual number $\widehat{x}$ (10.1).

The operations of additions and, respectively, multiplication of two multidual number we will define by

$$\widehat{x} + \widehat{y} = \sum_{k=0}^{n}(x_k + y_k)\varepsilon^k, \tag{10.3}$$

$$\widehat{x}\widehat{y} = \sum_{k=0}^{n}\left(\sum_{p=0}^{k} x_p y_{k-p}\right)\varepsilon^k. \tag{10.4}$$

In Eqs. (10.3) and (10.4), we will denote by $\varepsilon^0 = 1$, $x_0 = x$, $y_0 = y$.

It can be proved that the set $\widehat{\mathbb{R}}$ with the addition operation (10.3) and multiplication operation (10.4) is a commutative ring with unit. An element $\widehat{x} \in \widehat{\mathbb{R}}$ is invertible if and only if $Re\widehat{x} \neq 0$. The zero divisor in the ring $\widehat{\mathbb{R}}$ is characterized by $Re\widehat{x} = 0$. An element from $\widehat{\mathbb{R}}$ is either invertible or zero divisor.

Let be $f : \mathbb{I} \subseteq \mathbb{R} \to \mathbb{R}$, $f = f(x)$ a n-times differentiable real function. We will define the multidual function $\widehat{f}$ of multidual variable $\widehat{x}$ by equation:

$$f(\widehat{x}) = f(x) + \sum_{k=1}^{n} \frac{\Delta^k}{k!} f^{(k)}(x) \tag{10.5}$$

where we denoted by $\Delta : \Delta = \widehat{x} - x = \sum_{k=1}^{n} x_k\varepsilon^k$.

Using the Eq. (10.5), we will define the following multidual functions of multidual variable:

$$\sin\widehat{x} = \sin x + \sum_{k=1}^{n} \frac{\Delta^k}{k!} \sin\left(x + k\frac{\pi}{2}\right), \tag{10.6}$$

$$\cos\hat{x} = \cos x + \sum_{k=1}^{n} \frac{\Delta^k}{k!} \cos\left(x + k\frac{\pi}{2}\right), \tag{10.7}$$

10.2.1 Multidual Vectors

Let be $\mathbf{V}_3$ the set of free vectors from three-dimensional Euclidean space. We will introduce the set of multidual vectors by: $\widehat{\mathbf{V}}_3 = \mathbf{V}_3 + \varepsilon \mathbf{V}_3 + \cdots + \varepsilon^n \mathbf{V}_3$; $\varepsilon \neq 0$, $\varepsilon^{n+1} = 0$.

Two generic vectors from $\widehat{\mathbf{V}}_3$ will be written as below:

$$\hat{\mathbf{a}} = \mathbf{a} + \mathbf{a}_1\varepsilon + \cdots + \mathbf{a}_n\varepsilon^n;\ \mathbf{a},\ \mathbf{a}_k \in \mathbf{V}_3,$$

$$\hat{\mathbf{b}} = \mathbf{b} + \mathbf{b}_1\varepsilon + \cdots + \mathbf{b}_n\varepsilon^n;\ \mathbf{b},\ \mathbf{b}_k \in \mathbf{V}_3. \tag{10.8}$$

The addition of two multidual vectors and, respectively, the multiplication of a multidual vector by a scalar $\hat{x} \in \widehat{\mathbb{R}}$ is defined by:

$$\hat{\mathbf{a}} + \hat{\mathbf{b}} = \mathbf{a} + \mathbf{b} + (\mathbf{a}_1 + \mathbf{b}_1)\varepsilon + \cdots + (\mathbf{a}_n + \mathbf{b}_n)\varepsilon^n. \tag{10.9}$$

$$\hat{x}\hat{\mathbf{a}} = \sum_{k=0}^{n}\left(\sum_{p=0}^{k} x_p \mathbf{a}_{k-p}\right)\varepsilon^k. \tag{10.10}$$

It can be easily proven that the set of vectors $\widehat{\mathbf{V}}_3$ is a free module of rank 3 over the multidual number ring $\widehat{\mathbb{R}}$.

We will define the scalar product respectively the cross product of two vectors from $\widehat{\mathbf{V}}_3$ by:

$$\hat{\mathbf{a}} \cdot \hat{\mathbf{b}} = \sum_{k=0}^{n}\left(\sum_{p=0}^{k} \mathbf{a}_p \cdot \mathbf{b}_{k-p}\right)\varepsilon^k, \tag{10.11}$$

$$\hat{\mathbf{a}} \times \hat{\mathbf{b}} = \sum_{k=0}^{n}\left(\sum_{p=0}^{k} \mathbf{a}_p \times \mathbf{b}_{k-p}\right)\varepsilon^k. \tag{10.12}$$

The triple scalar vector product of three multidual vectors $\hat{\mathbf{a}}$, $\hat{\mathbf{b}}$, $\hat{\mathbf{c}}$ is defined by:

$$\langle \hat{\mathbf{a}}, \hat{\mathbf{b}}, \hat{\mathbf{c}} \rangle = \hat{\mathbf{a}} \cdot \left(\hat{\mathbf{b}} \times \hat{\mathbf{c}}\right). \tag{10.13}$$

All three multidual vectors $\hat{\mathbf{e}}_1, \hat{\mathbf{e}}_2, \hat{\mathbf{e}}_3$ represent a basis in the free module $\widehat{\mathbf{V}}_3$ if and only if $Re\left(\hat{\mathbf{e}}_1, \hat{\mathbf{e}}_2, \hat{\mathbf{e}}_3\right) \neq 0$.

10.2.2 Multidual Tensors

An $\widehat{\mathbb{R}}$—linear application of $\widehat{\mathbf{V}}_3$ into $\widehat{\mathbf{V}}_3$ is called a Euclidean multidual tensor:

$$\mathbf{T}\left(\widehat{\lambda}_1\hat{\mathbf{v}}_1 + \widehat{\lambda}_2\hat{\mathbf{v}}_2\right) = \widehat{\lambda}_1\mathbf{T}(\hat{\mathbf{v}}_1) + \widehat{\lambda}_2\mathbf{T}(\hat{\mathbf{v}}_2);\ \widehat{\lambda}_1, \widehat{\lambda}_2 \in \widehat{\mathbf{R}};\ \forall\hat{\mathbf{v}}_1, \hat{\mathbf{v}}_2 \in \widehat{\mathbf{V}}_3. \tag{10.14}$$

Let $\mathbf{L}\left(\widehat{\mathbf{V}}_3, \widehat{\mathbf{V}}_3\right)$ be the set of tensors, then any $\widehat{\mathbf{T}} \in \mathbf{L}\left(\widehat{\mathbf{V}}_3, \widehat{\mathbf{V}}_3\right)$, the multidual transposed tensor denoted by $\widehat{\mathbf{T}}^T$ is defined by $\hat{\mathbf{v}}_1 \cdot \left(\widehat{\mathbf{T}}\hat{\mathbf{v}}_2\right) = \hat{\mathbf{v}}_2 \cdot \left(\widehat{\mathbf{T}}^T\hat{\mathbf{v}}_1\right)$; $\forall\hat{\mathbf{v}}_1, \hat{\mathbf{v}}_2 \in \widehat{\mathbf{V}}_3$.

For any dual vector $\hat{\mathbf{a}} \in \widehat{\mathbf{V}}_3$, the associated skew-symmetric dual tensor will be denoted by $\hat{\mathbf{a}}\times$ and will be defined $(\hat{\mathbf{a}}\times)\widehat{\mathbf{b}} = \hat{\mathbf{a}} \times \widehat{\mathbf{b}}$, $\forall\widehat{\mathbf{b}} \in \widehat{\mathbf{V}}_3$.

10.2.3 Multidual Differential Transform

So, being $f : I \subseteq \mathbb{R} \to \mathbb{R}$, $f = f(t)$, a real function of real time variable, n-th differentiable, $n \in \mathbb{N}$. To this function, it will be associated the multidual function of real variable $\breve{f}$ given by the following equation:

$$\breve{f} = f + \varepsilon\dot{f} + \cdots + \frac{\varepsilon^n}{n!}f^{(n)} = e^{\varepsilon \mathrm{D}}f, \tag{10.15}$$

where $e^{\varepsilon\mathrm{D}} = 1 + \varepsilon\mathrm{D} + \cdots + \frac{\varepsilon^n}{n!}\mathrm{D}^n$ with $\mathrm{D} = \frac{d}{dt}$ the derivative operator with respect to time.

Theorem 10.1 [3] *Being f and g two real function of class $C^n(I)$. The following properties take place:*

$$\overbrace{f+g} = \breve{f} + \breve{g}, \tag{10.16}$$

$$\overbrace{fg} = \breve{f}\breve{g}, \tag{10.17}$$

$$\overbrace{\lambda f} = \lambda\breve{f}, \forall\lambda \in \mathbb{R}, \tag{10.18}$$

$$\overbrace{f(\alpha)} = f(\breve{\alpha}), \alpha \in C^n(I), \tag{10.19}$$

$$\mathring{f} = \breve{f}. \tag{10.20}$$

10.3 Higher-Order Kinematics with Multidual Transform

Let be a rigid motion given by a curve in Lie group of the rigid displacements SE (3), given by the homogeneous matrix $\mathbf{g} = \begin{bmatrix} \mathbf{R} & \boldsymbol{\rho} \\ \mathbf{0} & 1 \end{bmatrix}$ where $\mathbf{R} \in \mathbb{SO}_3$ is a proper orthogonal tensor [2–4], $\mathbf{R} = \mathbf{R}(t)$, and $\boldsymbol{\rho} = \boldsymbol{\rho}(t)$ a vector functions of a time variable. As shown in [2, 3], the vector field of the higher-order accelerations is given by the below matrix:

$$\boldsymbol{\Psi}_n = \begin{bmatrix} \boldsymbol{\Phi}_n & \mathbf{a}_n \\ \mathbf{0} & 1 \end{bmatrix}, \tag{10.21}$$

where the tensor $\boldsymbol{\Phi}_n$ and the vector $\mathbf{a}_n$ are given by the below equations:

$$\boldsymbol{\Phi}_n = \mathbf{R}^{(n)}\mathbf{R}^T, \tag{10.22}$$

$$\mathbf{a}_n = \boldsymbol{\rho}^{(n)} - \boldsymbol{\Phi}_n\boldsymbol{\rho}. \tag{10.23}$$

The n-th order acceleration of a point of the rigid body given by the position vector $\mathbf{r}$ can be computed with the following relation:

$$\mathbf{a}_{\mathbf{r}}^{[n]} = \mathbf{a}_n + \boldsymbol{\Phi}_n\mathbf{r};\ n \in \mathbb{N}. \tag{10.24}$$

Next, a new non-iterative procedure that permits the determination of the higher-order acceleration field using the multidual differential transformation will be presented. In [2–4], is described an iterative procedure used to determine the instantaneous tensor $\boldsymbol{\Phi}_n$ using the time derivative of the spatial twist of rigid body motion.

So, for $\mathbf{R} = \mathbf{R}(t) \in \mathbb{SO}_3$ and $\boldsymbol{\rho} = \boldsymbol{\rho}(t) \in \mathbf{V}_3$; $\forall t \in I \subseteq \mathbb{R}$, can be defined:

$$\breve{\mathbf{R}} = e^{\varepsilon \mathrm{D}}\mathbf{R} \tag{10.25}$$

$$\breve{\boldsymbol{\rho}} = e^{\varepsilon \mathrm{D}}\boldsymbol{\rho}. \tag{10.26}$$

Theorem 10.2 *For any $\mathbf{R} = \mathbf{R}(t) \in \mathbb{SO}_3$, ; $\forall t \in I \subseteq \mathbb{R}$, n-th order differentiable function, the below relation takes place:*

$$\breve{\mathbf{R}}\breve{\mathbf{R}}^{\mathbf{T}} = \breve{\mathbf{R}}^{\mathbf{T}}\breve{\mathbf{R}} = \breve{\mathbf{I}}, \tag{10.27}$$

$$\det \breve{\mathbf{R}} = 1. \tag{10.28}$$

If $\mathbf{R} = \mathbf{R}(\alpha, \mathbf{u})$ *where the angle* α *and the unit vector* $\mathbf{u}$ *are the natural invariants of the orthogonal tensor* $\mathbf{R}$, *takes place a Rodrigues-like formula:*

$$\breve{\mathbf{R}} = \mathbf{I} + \sin\breve{\alpha}(\breve{\mathbf{u}}\times) + (1 - \cos\breve{\alpha})(\breve{\mathbf{u}}\times)^2, \tag{10.29}$$

where

$$\breve{\alpha} = e^{\varepsilon \mathrm{D}}\alpha, \tag{10.30}$$

$$\breve{\mathbf{u}} = e^{\varepsilon \mathrm{D}}\mathbf{u}. \tag{10.31}$$

Let be the following homogenous multidual matrix capture the properties of higher-order accelerations vector field:

$$\breve{\Psi} = \begin{bmatrix} \breve{\mathbf{R}} & \breve{\boldsymbol{\rho}} \\ \mathbf{0} & 1 \end{bmatrix} \begin{bmatrix} \mathbf{R}^T & -\mathbf{R}^T\boldsymbol{\rho} \\ \mathbf{0} & 1 \end{bmatrix} = \begin{bmatrix} \breve{\boldsymbol{\Phi}} & \breve{\mathbf{a}} \\ \mathbf{0} & 1 \end{bmatrix} \tag{10.32}$$

$$\breve{\boldsymbol{\Phi}} = \breve{\mathbf{R}}\mathbf{R}^T = \mathbf{I} + \boldsymbol{\Phi}_1\varepsilon + \cdots + \frac{\boldsymbol{\Phi}_n}{n!}\varepsilon^n, \tag{10.33}$$

$$\breve{\mathbf{a}} = \breve{\boldsymbol{\rho}} - \breve{\mathbf{R}}\mathbf{R}^T\boldsymbol{\rho} = \mathbf{a}_1\varepsilon + \frac{\mathbf{a}_2}{2}\varepsilon^2 \ldots + \frac{\mathbf{a}_n}{n!}\varepsilon^n \tag{10.34}$$

For $n = 4\left(\varepsilon^5 = 0\right)$, the velocity, acceleration, jerk, and jounce vector fields of rigid body motion will be simultaneously described. In the case of helical rigid body motion ($\mathbf{u} = \text{const.}$, $\breve{\mathbf{u}} = \mathbf{u}$), from Eqs. (10.22), (10.23), and Theorem 10.2, after some algebra, results that:

$$\begin{cases} \breve{\boldsymbol{\Phi}} = \mathbf{I} + \sin\Delta\breve{\alpha}(\mathbf{u}\times) + (1 - \cos\Delta\breve{\alpha})(\mathbf{u}\times)^2 \\ \breve{\mathbf{a}} = \Delta\breve{\boldsymbol{\rho}} - \sin\Delta\breve{\alpha}\,\mathbf{u}\times\boldsymbol{\rho} - (1 - \cos\Delta\breve{\alpha})\,\mathbf{u}\times(\mathbf{u}\times\boldsymbol{\rho}) \end{cases} \tag{10.35}$$

where with $\Delta\breve{x}$ was denoted the multidual part of the time function $\breve{x}$.

10.4 Hyper-Multidual Algebra and Higher-Order Kinematics

Let be $\underline{\mathbb{R}} = \mathbb{R} + \varepsilon_0\mathbb{R}$; $\varepsilon_0 \neq 0$, $\varepsilon_0^2 = 0$, the set of dual numbers, and $n \in \mathbb{N}$ a natural number. We will introduce the set of hyper-multidual (HMD) numbers by: $\widehat{\underline{\mathbb{R}}} = \underline{\mathbb{R}} + \varepsilon\underline{\mathbb{R}} + \ldots \varepsilon^n\underline{\mathbb{R}}$; $\varepsilon \neq 0$, $\varepsilon^{n+1} = 0$. For $n = 1$, on obtain hyper-dual numbers [5]. It can be easily proved that the set $\widehat{\underline{\mathbb{R}}}$ with the addition operation and multiplication is a commutative ring with unit. An element from $\widehat{\underline{\mathbb{R}}}$ is either invertible or zero divisor [6]. The linear $\mathbb{R}$-algebra $\widehat{\underline{\mathbb{R}}}$ is the direct product of dual algebra $\underline{\mathbb{R}}$ and multidual algebra $\widehat{\mathbb{R}}$. $\widehat{\underline{\mathbb{R}}}$ has a structure of $2(n+1)$—dimensional associative, commutative and unitary generalized Clifford Algebra, and $\widehat{\underline{\mathbb{R}}} = \widehat{\mathbb{R}} + \varepsilon_0\widehat{\mathbb{R}}$; $\varepsilon_0 \neq 0$, $\varepsilon_0^2 = 0$.

Let be $f : \underline{I} \subseteq \underline{\mathbb{R}} \to \underline{\mathbb{R}}$, $f = f(\underline{x})$ a n-times differentiable dual function of dual variable. We will define the HMD function $\hat{\underline{f}}$ of HMD variable $\hat{\underline{x}}$ by equation:

$$\hat{\underline{f}}(\hat{\underline{x}}) = f(\underline{x}) + \sum_{k=1}^{n} \frac{\underline{\Delta}^k}{k!} f^{(k)}(\underline{x}) \tag{10.36}$$

here we noted by $\underline{\Delta} : \underline{\Delta} = \hat{\underline{x}} - \underline{x} = \sum_{k=1}^{n} \underline{x}_k \varepsilon^k$, and $\Delta^p = 0; p \geq n+1$.

The HMD vectors, tensors and hyper-multidual transform are introduced as in Sect. 10.2. Due to the ring character of the set of dual numbers $\underline{\mathbb{R}}$, all properties presented in the Sect. 10.2 are retained. A rigid body motion is given by a curve in Lie group of orthogonal dual tensor $\underline{SO}_3$, [1]. So, for $\underline{\mathbf{R}} = \underline{\mathbf{R}}(t) \in \underline{SO}_3$, $\forall t \in I \subseteq \mathbb{R}$, the orthogonal dual tensor, can be defined by HMD tensor:

$$\breve{\underline{\mathbf{R}}} = e^{\varepsilon D}\underline{\mathbf{R}} \tag{10.37}$$

$$\breve{\underline{\mathbf{R}}} = \underline{\mathbf{I}} + (\sin\breve{\underline{\alpha}})\breve{\underline{\mathbf{u}}} \times +(1 - \cos\breve{\underline{\alpha}})(\breve{\underline{\mathbf{u}}}\times)^2 = \exp(\breve{\underline{\alpha}}\breve{\underline{\mathbf{u}}}\times) \tag{10.38}$$

where:

$$\breve{\underline{\alpha}} = e^{\varepsilon D}\underline{\alpha} \tag{10.39}$$

$$\breve{\underline{\mathbf{u}}} = e^{\varepsilon D}\underline{\mathbf{u}} \tag{10.40}$$

Orthogonal HDM tensor $\breve{\underline{\Psi}} = \breve{\underline{\mathbf{R}}} \cdot \underline{\mathbf{R}}^T = (\mathbf{I} + \varepsilon_0\breve{\mathbf{a}}\times)\breve{\Phi}$ corresponding by isomorphism of the multidual homogenous matrix $\breve{\underline{\Psi}} = \breve{\mathbf{g}}\mathbf{g}^{-1} = \begin{bmatrix} \breve{\Phi} & \breve{\mathbf{a}} \\ \mathbf{0} & 1 \end{bmatrix}$, that captures all the information about the acceleration field of order $k = \overline{1,n}$.

If $\underline{\mathbf{u}} = \text{const}$ ($\breve{\underline{\mathbf{u}}} = \underline{\mathbf{u}}$) on obtain:

$$\underline{\breve{\Psi}} = \underline{\breve{\mathbf{R}}} \cdot \underline{\mathbf{R}}^T = \exp(\underline{\breve{\alpha}}\, \underline{\mathbf{u}} \times) \exp(-\underline{\alpha}\, \underline{\mathbf{u}} \times) = \exp[(\Delta \underline{\breve{\alpha}}) \underline{\mathbf{u}} \times] \tag{10.41}$$

With Rodrigues's like formula from Eq. (10.41) results:

$$\underline{\breve{\Psi}} = \underline{\mathbf{I}} + (\sin \Delta \underline{\breve{\alpha}}) \underline{\mathbf{u}} \times + (1 - \cos \Delta \underline{\breve{\alpha}})(\underline{\mathbf{u}} \times)^2. \tag{10.42}$$

For $n = 4\left(\varepsilon^5 = 0\right)$, the velocity, acceleration, jerk, and jounce vector fields of rigid body motion will be simultaneously described.

In the previous equation the mapping's $\sin \Delta \underline{\breve{\alpha}}$ and $\cos \Delta \underline{\breve{\alpha}}$ are polynomial and not transcendent, considering that $\Delta \underline{\breve{\alpha}}^p = 0;\ p \geq n + 1$.

10.5 Conclusions

A general method for the study of the vector field of arbitrary higher-order accelerations field of rigid body motion is described. It is proved that all information regarding the properties of the distribution of higher-order accelerations field is contained in the specified multidual homogenous matrix or hyper-multidual orthogonal tensor. The properties for velocity, acceleration, jerk, and hyper-jerk (jounce) fields are given in specific cases. The results interest the theoretical kinematics, higher-order kinematics analysis in the case of a serial manipulator [4, 8], control theory, and multibody kinematics.

References

1. Condurache, D.: Higher-order kinematics of rigid bodies. A tensors algebra approach. In: Interdisciplinary Applications of Kinematics. Mechanisms and Machine Science, vol. 71, pp. 215–225 (2019)
2. Condurache, D.: Higher-order relative kinematics of rigid body motions: a dual lie algebra approach. In: Advances in Robot Kinematics 2018, vol. 8, pp. 83–91 (2018)
3. Condurache, D.: Multidual algebra and higher-order kinematics. In: New Trends in Mechanism and Machine. Mechanisms and Machine Science, vol. 89, pp. 48–55. Springer, Cham (2020)
4. Condurache, D.: Hyper-multidual algebra and higher-order kinematics. In: Advances in Robot Kinematics, Springer Proceedings in Advanced Robotics, vol. 24, pp. 57–64. Springer, Cham (2022)
5. Cohen, A., Shoham, M.: Application of hyper-dual numbers to multibody kinematics. ASME. J. Mech. Robot. **8**(1), 011015 (2016)
6. Messelmi, F.: Differential calculus of multidual functions. Annu. Rev. Chaos Theory Bifurc. Dyn Syst. **10**, 1–15 (2021)
7. Müller, A.: An overview of formulae for the higher-order kinematics of lower-pair chains with applications. Mech. Mach. Theory **142** (2019)
8. Condurache, D.: Higher-order relative kinematics of rigid body, and multibody systems. A novel approach with real and dual lie algebras. Mech. Mach. Theory **176** (2022)

Chapter 11
Study of Vibrations of a Cantilever Beam with Uniformly Distributed Mass

S. Alaci, F.-C. Ciornei, I.-C. Romanu, A.-M. Velniciuc, and C. Bujoreanu

Abstract The paper presents the study by finite element analysis and by two experimental methods of the vibrations of a cantilever beam. The beam is made of homogenous material and has constant cross-section. The fundamental period of flexural vibrations is first calculated from analytical relation. The beam is modelled in CATIA and by Finite element analysis, the first ten frequencies of vibrations are obtained. The numerical and theoretical results are in good agreement. The experimental tests were performed via two methods: first, the oscillatory motion is video-recorded and the image is processed to obtain the period of vibration; second, a small acceleration sensor is attached to the free end of the beam and it provides a signal with the same period as the vibration of the system.

Keywords Vibrations · Cantilever beam · Distributed mass · Experimental period

11.1 Introduction

The mechanical vibration motions occur any time when elements characterised by inertia properties (mass and/or mass moments) are associated with elastic elements [2, 4, 9]. There is a large range of vibration motions and characteristic to all of them is the property that each point of the vibratory body performs motions in a limited

S. Alaci · F.-C. Ciornei · I.-C. Romanu
Stefan Cel Mare, University of Suceava, Suceava, Romania
e-mail: stelian.alaci@usm.ro

F.-C. Ciornei
e-mail: florina.ciornei@usm.ro

I.-C. Romanu
e-mail: ionutromanucristian@usm.ro

A.-M. Velniciuc · C. Bujoreanu (✉)
"Gheorghe Asachi" Technical University of Iasi, Iasi, Romania
e-mail: c.bujorea@tuiasi.ro

A.-M. Velniciuc
e-mail: adi-mihaita.velniciuc@student.tuiasi.ro

I. Doroftei et al. (eds.), *Proceedings of SYROM 2022 & ROBOTICS 2022*, Mechanisms and Machine Science 127, https://doi.org/10.1007/978-3-031-25655-4_11

space and additionally, it passes successively through a position of equilibrium. To classify the vibratory motions considering all criteria of classification is clearly an impossible task. One of the most important classification principles is based on the type of the differential equations which describe the motion, and here there are linear and non-linear vibrations [6]. The study of non-linear vibrations is more complicated [3]. For the case of linear systems with one degree of freedom, the motion is described by linear differential equations [5]. If additionally, the coefficients of the differential equation are constants, the motion is a periodical one, characterised by the period T corresponding to the shortest interval of time after which the motion repeats. The period is and intrinsic characteristic of a vibratory motion and depends on the inertial and elastic properties of the system [7]. This remark is the basis of numerous applications aiming to find the inertia and elastic characteristics of the system [2]. The present work proposes as vibratory system a homogenous cantilever beam of constant cross-section which performs flexural oscillations in the fundamental mode of vibration. The fundamental period of oscillation is required and this goal is reached by direct calculus using the formula from technical literature, by a programme using finite element analysis and by two experimental methods.

11.2 Analytical and Numerical Results

A cantilever beam with constant rectangular cross-section is considered, loaded by the distributed force $q(x)$, as in Fig. 11.1. There are known the following geometrical and elastic parameters: the dimensions of the beam L, b, l, the density ρ and the Young modulus E.

The period of oscillation of the beam corresponding to the "k" mode of vibration is found using the equation [7]:

$$T_k = \frac{2\pi}{\left(\frac{\alpha l_k}{L}\right)^2}\sqrt{\frac{\rho A}{E I_z}} \tag{11.1}$$

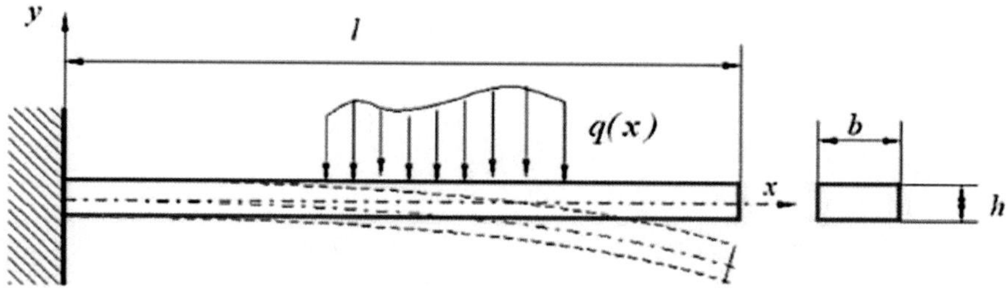

Fig. 11.1 Prismatic beam with transversal vibrations

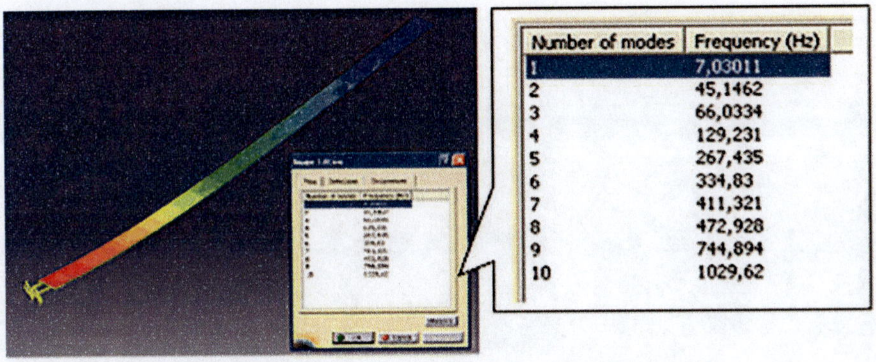

Fig. 11.2 Numerical solution of the frequency of the first 10 modes of vibrations

where $I_z = \frac{bh^3}{12}$ is the principal moment of inertia of the cross section with respect to the Oz horizontal axis and αl_k is the "k" root of the transcendental equation:

$$\cos(\alpha l)_k \cos h(\alpha l)_l = -1 \tag{11.2}$$

For the actual case, there were considered: $L = 0.5\,\text{m}$; $b = 20\,\text{mm}$; $h = 2.11\,\text{mm}$; $E = 2.11 \cdot 10^{11}\,\text{Pa}$; $\rho = 2710\,\text{kg/m}^3$. The Eq. (11.2) is solved numerically and the first two values of the solutions are $(\alpha l)_0 = 1.875091$, $(\alpha l)_1 = 4.694091$. To these two values, the values of the first two periods of oscillation correspond: $T_0 = 0.144\,\text{s}$; $T_1 = 0.023\,\text{s}$, and the first two frequencies of oscillation, respectively: $\nu_0 = 6.929\,\text{Hz}$; $\nu_1 = 43.425\,\text{Hz}$.

The beam was modelled using the CATIA/Analysis&Simulation module [10] with the parameters mentioned above and the frequencies of the first 10 modes of vibrations were found, as shown in Fig. 11.2.

One can observe that the value of the fundamental frequency $\nu_0 = 7.030\,\text{Hz}$ obtained numerically is close to the exact one that is $\nu_0 = 6.929\,\text{Hz}$.

11.3 Experimental Tests for Finding the Fundamental Period of Oscillation of the Beam

11.3.1 Finding the Frequency of Oscillation by Video-Capturing the Motion

For obtaining the experimental fundamental frequency of vibration, the beam 1 was clamped in a vise 2, fixed on the base plate 3 from the work-bench. The length of the cantilever beam was adjusted to the value $L = 500\,\text{mm}$. The beam is disturbed from the position of static equilibrium by bending carefully as the deformation should not

Fig. 11.3 The laboratory test-rig

exceed the limit value from the elastic domain. The motion of the beam is filmed using a high-speed camera 4 with the acquisition velocity of 210 frames/sec, Fig. 11.3.

The experiment follows the next steps:

- The beam is set to vibration motion and the motion is filmed with the high-speed camera
- The movie is split into frames and the period of vibration is found considering several values for the number of oscillations. In Fig. 11.4 there are presented the extreme positions (frames from the software of film analysis) of the beam for 5 complete oscillations.

The following formula is applied for finding the experimental period of oscillation:

$$T^{(n)} = \frac{1}{n}\frac{F^{(n)} - F^{(0)}}{f} \tag{11.3}$$

Fig. 11.4 Finding the period of oscillation from n = 5 oscillations

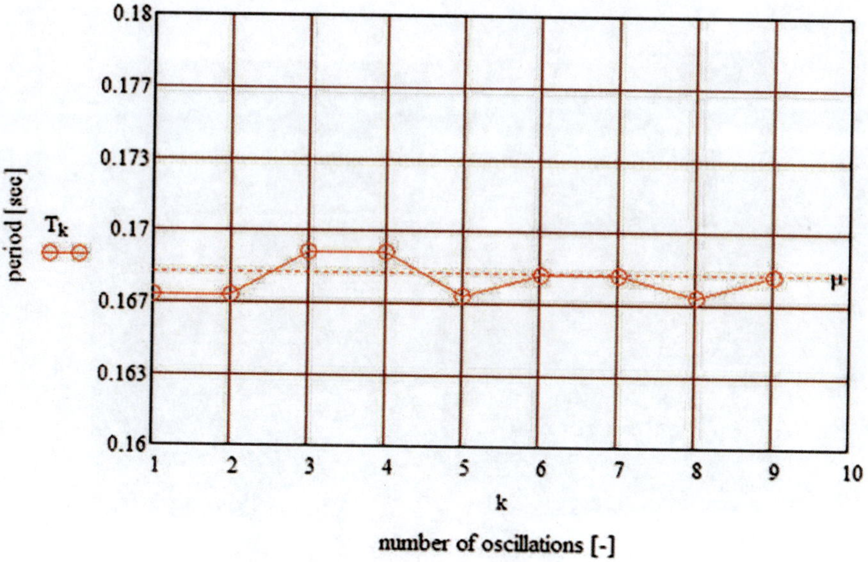

Fig. 11.5 The period of oscillation found by experimental method after k oscillations, $k = 1 \div 10$

where $T^{(n)}$ is the period found from n complete oscillations of the beam, $F^{(n)}$ is the number of the frame corresponding to the last image, $F^{(0)}$ is the number of the frame corresponding to the initial position, f is the frequency of acquisition of the images, that is 210 frames/sec. The fundamental period of oscillation for $n = 1.10$ complete oscillations of the beam was found and the values are presented in Fig. 11.5.

A first conclusion is that based on the data from Fig. 11.5, the fundamental period of oscillation is $T_{exp} = 0.168 \pm 1.28 \cdot 10^{-3}$(s) and differs from the values obtained by analytical and numerical methods.

11.3.2 The Experimental Oscillation Frequency Obtained with Acceleration Sensors

An alternative method for experimental finding the period of oscillation is proposed subsequent. The principle of the technique consists in attaching on the beam an acceleration sensor with negligible mass and dimensions. Under the assumption of small oscillations, the acceleration is proportional to the displacement [8] and thus, the period of the acceleration will coincide to the period of deformation of the beam. The test rig is presented in Fig. 11.6 and contains the aluminium beam 1, the vise 2, the base plate 3, the acceleration sensor 4, the d.c. power source 5, the electronic memory oscilloscope 6 connected to the computer 7. The beam is set into oscillation and using the oscilloscope the signal generated by the acceleration sensor is captured.

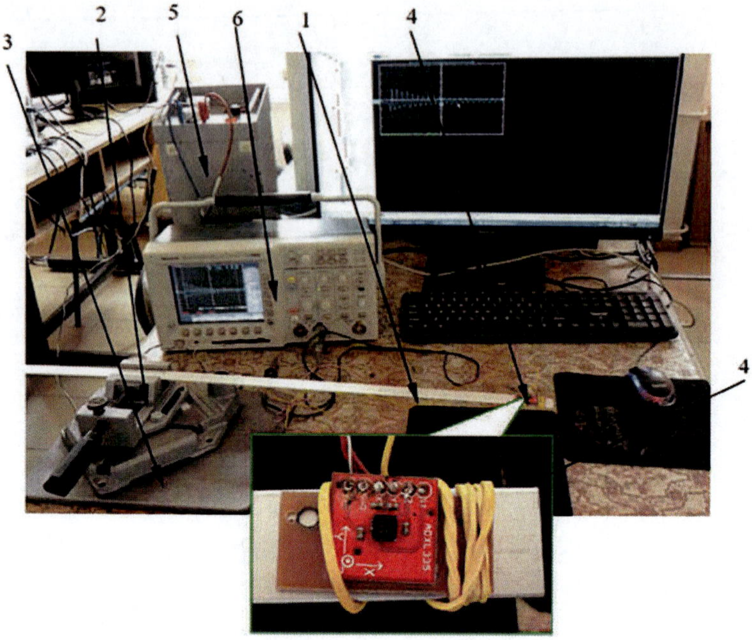

Fig. 11.6 Finding the period of oscillation with the acceleration sensors

The experimental data were saved in matrix form and afterwards they were exported to Mathcad software.

For several experiments, the files with experimental data were exported and processed in Mathcad, where the time variation of the acceleration of the sensor was represented. Then, two instants, corresponding to two maxima of acceleration were identified for each experiment. In Fig. 11.7 there is presented one of the plots from Mathcad from which the values of the experimental periods were found: $T_{exp} = (890 + 810) \cdot 10^{-4}(\text{s})$; $T_{exp} = 0.17\,\text{s}$.

One can notice that the values obtained for the period of oscillation by the two experimental methods are very close, but differ from the values found by theoretical and numerical ways. A possible cause of these difference could be that the values accepted for the mechano-elastic properties of the material of the beam in theoretical and numerical calculus are not the actual ones of the beam used in experiments. Thus, considering that the density of the material is $\rho = 2710\,\text{kg/m}^3$, (as in the library of CATIA software) and applying the relation (11.1) with the value of the period from experimental tests $T_0 \cong 0.17\,\text{s}$ it is obtained a value of Young modulus $E_{exp} = 0.5045 \cdot 10^{11}\,\text{Pa}$. This value is 28% smaller than the one considered initially, given by CATIA library. This significant difference should be confirmed by tests based on alternative methods and for future work, an identic beam will be supposed to torsion and bending tests for finding the Young modulus. It is also known the

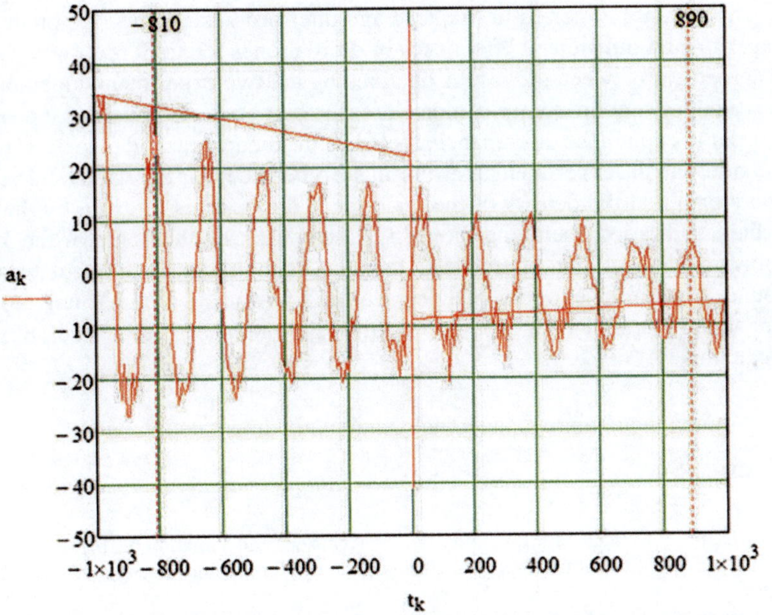

Fig. 11.7 Finding the experimental period of oscillation for a test (a_k is the signal from oscilloscope, proportional to acceleration and t_k is the time, in milli-seconds)

indentation test developed by Oliver and Pharr. The tests from laboratory weighting and measuring the rod, showed a similar value for density.

11.4 Conclusions

The work presents a proposed device for the study of vibrations of systems with distributed mass. The designed system consists in a straight beam of constant cross-section, made of aluminium and homogenous. The supporting scheme is the horizontal cantilever beam with fixed support. For a stipulated length of the free end of the beam, the theoretic period of flexural oscillation is found. Then, the beam is modelled in a CAD software and imposing the boundary conditions, the frequencies for the first ten modes of vibrations are found by finite element analysis method. An excellent concordance between the values of fundamental frequencies obtained via analytical and numerical methods is remarked.

The experimental study is performed using two methods: in the first case, the beam is fixed at one end, then is set into oscillation, the motion is filmed and afterwards the movie is processed for getting the fundamental frequency; in the second case, a small acceleration sensor is fixed on the free end of the beam and the signal generated is captured using a memory oscilloscope, the result being saved in matrix form. The data are imported in Mathcad and after processing, they are plotted and the experimental fundamental frequency of oscillation is found. It is observed a very good agreement between the results obtained by the two experimental methods but there is a significant difference compared to the theoretical and numerical results.

In order to explain the discrepancies between the theoretical and numerical results on one side, and the experimental results on the other side, it is accepted (and verified in laboratory) that the density of the material of the beam is as given by the most of technical libraries, including the CATIA software, but the value of the Young modulus may explain the differences. Then, the Young modulus is calculated using the found value of the experimental period. It results a value of the Young modulus for the actual material with 28% smaller than the one accepted in theoretical and FEA calculus.

References

1. Blevins, R.D.: Formulas for Dynamics. Wiley, Acoustics and Vibration (2016)
2. Buzdugan, G.: Strength of Materials, (in Romanian: Rezistenta materialelor) Ed. Tehnica (1980)
3. Cveticanin, L.: Strongly Nonlinear Oscillators-Analytical Solutions. Springer (2014)
4. Jazar, R.N.: Advanced Vibrations-A Modern Approach. Springer (2013)
5. Lalanne, C.: Mechanical Vibration and Shock Analysis, Vol. I, Sinusoidal Vibration, 2nd ed. Wiley (2009)
6. Nagle, R., Saff, E., Snider, A.: Fundamentals of Differential Equations, 9th ed. Pearson (2017)
7. Rades, M.: Mechanical Vibrations (in Romanian: Vibratii mecanice) Printech (2008)
8. Sinclair, I.R.: Sensors and Transducers, 3rd ed. Elsevier (2001)
9. Vance, J., Zeidan, F., Murphy, B.: Machinery Vibration and Rotor Dynamics. Wiley (2010)
10. Zamani, N.: CATIA FEA Tutorials, Release 17. SDC Publications (2018)

Chapter 12
Considerations on the Proper Selection of Sensors for Vibroacoustic Study of the Vehicles

N. A. Niţu, J. Machado, and C. Bujoreanu

Abstract Automotive industry is a very fast growing one, with a lot of smart solution for customers operation needs, like remote control of the vehicle by the mobile phone, route planning for the electric vehicles according to availability of the charging stations. Maintenance and repairs requirements involve methods of diagnosing and predicting failures that ensure a short downtime from customer point of view and efficiency from the workshop side. In addition to the classic limitative solutions for diagnosis, the vibroacoustic method, which presume the spectral analysis of main components of the vehicle, offers the possibility to raise the level of success in diagnosing of all type of faults: mechanical, electrical, pneumatic and hydraulic. For these it is necessary to establish on the basis of scientific and technical criteria, a matrix of sensors able to record sound/noise and vibrations of the vehicle as a whole and of the main components. The aim of our work is to identify and describe the main features of these sensors in order to be selected for a correct vibroacoustic diagnose of the vehicles.

Keywords Automotive · Vibroacoustic diagnosis · Sensors selection · Main features

12.1 Introduction

Modern automotive customers have as a main need the high degree of dynamism, which is reflected on *Research & Development* processes for new designed products. An example would be a convenience feature built into a mobile phone application

N. A. Niţu · C. Bujoreanu (✉)
"Gheorghe Asachi" Technical University of Iasi, Iasi, Romania
e-mail: cbujorea@tuiasi.ro

N. A. Niţu
e-mail: nicolae-adrian.nitu@student.tuiasi.ro

J. Machado
University of Minho, Braga, Portugal
e-mail: jmachado@dem.uminh.pt

I. Doroftei et al. (eds.), *Proceedings of SYROM 2022 & ROBOTICS 2022*, Mechanisms and Machine Science 127, https://doi.org/10.1007/978-3-031-25655-4_12

that allows the owner, after certain security checks, to remotely control the basic functions of the vehicle, as follows [1]:

- navigation services: route planning according to availability of the fuel and electrical charging stations on the map;
- assistance and support: maintenance management, telediagnosis, accident recovery and remote vehicle diagnosis;
- vehicle info: locking/unlocking by the mobile phone app, speed alert and trip data combined with geofencing.

Another already full filed requirement, that already exists, comes from the area of commercial vehicles, which must manage fleets intelligently [2]:

- mapping and trip records of vehicles from the fleet (tractor and trailer either);
- performance analysis and reports for individual vehicle of the fleet;
- driver card and mass memory data download;
- logistics for order planning and optimization of transport potential.

The challenge comes from the diagnostic / telediagnostic area, as most of the detected faults are of the electrical type and in some cases the use of interpolation could detect a small number of mechanical failures. The gearbox displacement sensor could detect wear over the limit of the shift mechanism by electrical measurement and electronic interpolation. But there are major sources of defects in the gearbox, such as damage to the roller bearings and fatigue of the gear teeth, which could not be detected, let say, by the classical method of electrical measurement.

From these observations we see the opportunity to broaden the methods of diagnosis by studying the vibroacoustic fingerprint for the entire vehicle and also for the main components, which is different in the mode of malfunction from normal operation. This vibroacoustic method offers the possibility to observe abnormal function of the vehicle from the mechanical, hydraulic and pneumatic among to the electrical nature of the faults. To observe the vibroacoustic behavior of the vehicle it is necessary to set up a matrix of microphones and accelerometers selected according to certain operating conditions and position on the vehicle under study (such as laboratory or on-road measurements, large variations in temperature, humidity and exposure to shocks).

12.2 Goal of Study

The method of vibroacoustic diagnostic of the vehicle involves comparing the actual noise and vibration behavior with the reference one for the healthy operation of the main components. This requires the creation of a database as a reference for the proper functioning of the vehicle components.

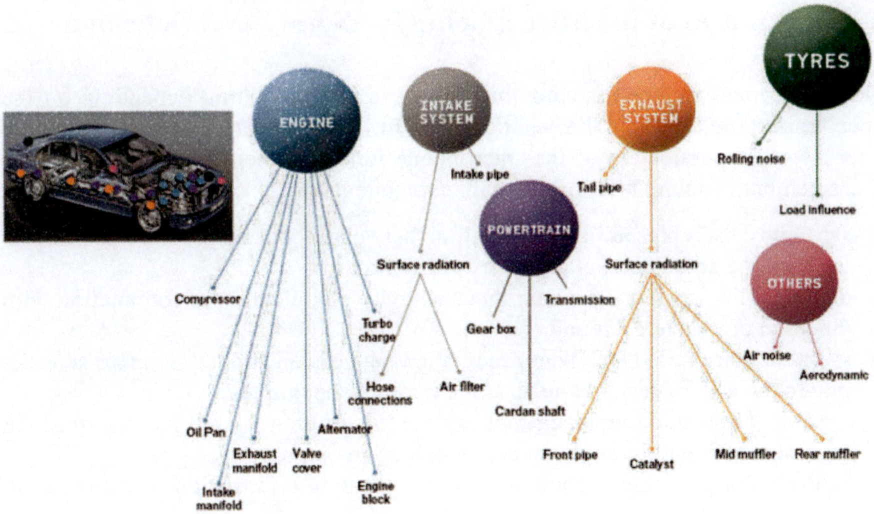

Fig. 12.1 Noise sources distribution on passenger car [3]

In the following figures are presented the main sources of noise and vibration, explained on passenger car (Fig. 12.1). The main sources of noise and corroborated vibrations are: engine, powertrain, tires, intake system, exhaust system and others. The differences come from their contribution to the overall vibroacoustic fingerprint according to the vehicle class.

The vibroacoustic database should be organized by specific criteria based on frequency analysis of the vibroacoustic fingerprint of the main components of the vehicle. The aim of this research is to identify the frequency range specific to each main source of vehicle noise and vibration, correlated with the amplitude of this frequency, specific to the normal mode of operation of these components. A complement to this research would be the creation of relevant faults, for the daily operation, on each main component of the vehicle and, accordingly, the recording of the specific amplitude correlated with this new condition. In this way, we could observe the incipient status of fault of the component with level of noise and vibration outside of regular interval.

All of the above must define a set/matrix of sensors suitable for each relevant frequency range of noise and vibration of each main component, adapted to their operating conditions (temperature range, interference, mounted position, power supply, wind and humidity). On the other hand, the correct distribution of sensors allows the exact source of abnormal vibroacoustic behavior to be determined to avoid misdiagnosis of components. This could save time and resources.

12.3 Features of the Microphones for Noise Level Detection

Dedicated microphones for automotive offer different solution depending on the specific purpose, such as the testing phase in development or in the production site. Acoustic parameters of the microphone influence their selection for proper measurement of sound level of the main components of vehicle [4]:

- sensitivity is representing the relation between acoustic pressure applied to microphone and its output as electrical information.
- directionality is used to explain the sensitivity modification in conjunction with direction of registered sound.
- signal to noise ratio (SNR) defines relation between an applied standard acoustic info to the sum of residual noise at the microphone output.
- dynamic range and overload point explain the difference from the lowest to the highest signal that microphone could emit as an output.
- equivalent input noise is showing that every output information is equivalent to specific acoustic input.
- frequency response, related to magnitude, presents the sensitivity fluctuation over the acoustic band.
- total harmonic distortion (THD) counts the distortion of the output electrical signal of the microphone to an applied undistorted acoustic reference.

Table 12.1 summarizes different solutions for MEMS microphones according to the specific application, characterized by their range of operation for main parameters.

The types of power supplies for dedicated microphones are an important point in choosing the right solution and we could review two main types:

- externally polarized which requires a voltage applied directly to the backplate;
- prepolarized using an equivalent permanent charge voltage fixed into the electret from the backplate.

Figure 12.2 depicts the schemes for externally polarized and prepolarized microphones that we intend to use in future experimental research. The prepolarized microphone, schematically shown in Fig. 12.2b, offers the features of an integrated preamplifier, low cost and possibly water and dust resistant because it could be encapsulated with the included battery.

Those used in the microphone array should have excellent phase characteristics (some examples are included in Table 12.2) and can be combined with appropriate software to evaluate the spatial transformation of compound sound fields to figure out the hotspot, in our case, abnormal sound behavior of fault or incipient fault component of the vehicle.

Externally polarized microphones, schematically presented in Fig. 12.2a, have the features of higher operating interval (frequency range, dynamic range and temperature range) and without additional signal conditioning in data acquisition system. They are summarized in Table 12.3.

Table 12.1 Technical characteristics of MEMS microphones [5]

Parameter	MP45DT02-M	MP34DT04-C1	MP23AB01DM	MP23AB01DH
Sensitivity	−26 dBFS	−26 dBFS	−38 dBV	−38 dBV
Directivity	Omnidirectional	Omnidirect	Omnidirect	Omnidirect
SNR	61 dB	64 dB	64 dB	65 dB
AOP	120 dBSPL	120 dBSPL	130 dBSPL	135 dBSPL
EIN	33 dB	30 dB	30 dB	29 dB
THD+N	<5% @ 115 dBSPL	<5% @ 115 dBSPL	<10% @ 130 dBSPL	<5% @ 130 dBSPL
PSR	−70 dB	−70 dB	−85 dB	−100 dB
Max. current consumption	650 µA	700 µA	250 µA	250 µA
Package dimensions	3.7 × 4.7 × 1.2 mm	3 × 4 × 1.095 mm	2.5 × 3.35 × 0.98 mm	2.5 × 3.3 × 0.9 mm
Port location	Top port	Top port	Bottom port	Bottom port
Operating temperature	−30 °C < T <+ 85 °C	−40 °C < T <+ 85 °C	−40 °C < T <+ 85 °C	−40 °C < T <+ 85 °C

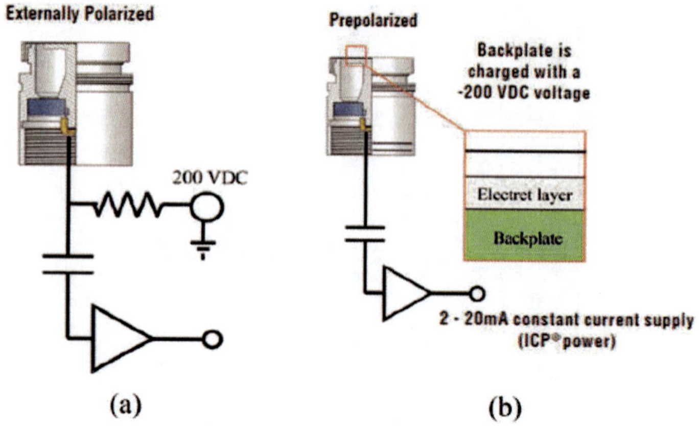

Fig. 12.2 Schematic of externally polarized and prepolarized microphones [6]

Table 12.2 Characteristics of prepolarized matrix microphones with integral preamplifier [7]

Model number	(New) 130A24	130A23	130F21	130B40
Microphone diameter	1/2″	1/4″	1/4″	1/4″
Response	Free-field	Free-field	Free-field	Pressure
Sensitivity (±3 dB at 250 Hz)	10 mV/Pa	14 mV/Pa	45 mV/Pa	8.5 mV/Pa
Frequency response (±2 dB)	20 Hz to 16 kHz [2]	20 Hz to 20 kHz	10 Hz to 20 kHz [1]	20 Hz to 10 kHz [2]
Dynamic range	<30 dBA to >143 dB [3]	<30 dBA to >143 dB [3]	24 dBA to >122 dB	<32 dBA to >142 dB [3]
Polarized voltage	0 V	0 V	0 V	0 V
Temperature range (°F)	−14 to +122 °F	−14 to +122 °F	+14 to +122 °F	−40 to +176 °F
Temperature range (°C)	−10 to +50 °C	−10 to +50 °C	−10 to +50 °C	−40 to +80 °C
Connector	BNC Jack	SMB Socket	10–32 Jack	10–32 Jack
Features	Rugged water and dust resistant	High frequency and high amplitudes	General purpose	Low profile and surface mount to minimize wind

Notes [1] ±4 dB. [2] 20 Hz to 10 kHz ± 3 dB, 20 to 20 kHz ± 6 dB. [3] 150 dB max without clipping. [4] 1/2″ preamplifier diameter

Table 12.3 Characteristics of externally polarized microphone [8]

Model number	Pressure		Random incidence	
	2530	2575	2559	2560
Diameter	1/4″	1″	1/2″	5 mV/μ 1/2″
Open Circuit sensitivity	1.4 mV/Pa	45 mV/Pa	12.9 mV/Pa	45.2 mV/Pa
Frequency range (±2 dB)	4 Hz to 70 kHz	2.6 Hz to 80 kHz	4 Hz to 25 kHz	2.6 Hz to 10 kHz
Dynamic range upper limit-3% distortion limit[a]	172 dB	146 dB	160 dB	146 dB
Dynamic range lower limit-cartridge thermal noise[a]	31 dB (A)	10 dB (A)	18 dB (A)	15 dB (A)
Temperature range	−40 to +302 °F −40 to +150 °C	−40 to +302 °F −40 to +150 °C	−40 to +302 °F −40 to +150 °C	−40 to +302 °F −40 to +150 °C

[a] re 20 μPa

From type of output, the actual microelectromechanical (MEMS) microphone could be analog or digital [9]. MEMS microphones with analogic output present the feature of lower energy consumption and a condensed ASIC structure (Application Specific Integrated Circuit). Digital MEMS microphone incorporates an analogic to digital converter (ADC) for converting the analogic information into a pulse density modulation signal (PDM) for gaining a vigorous data transmission and convenient rejection of electromagnetic interference (EMI).

12.4 Features of the Accelerometers for Vibration Detection

Solution for vibration measurement, on the automotive area, are basically based on MEMS accelerometers. The detection principle of the MEMS accelerometer is based on seismic mass hold by a spring, and with a dampened movement, overseen by specific techniques. In the case of an acceleration or deceleration movement applied to MEMS accelerometer, the seismic mass changes its position comparative to the housing and generates an electrical signal correspondingly. The mathematical model used is a second order differential equation type [10]. The relevant information for sensing element can be detected by [10]:

- piezoelectric effect: when this piezo element is under strain it generates an electrical voltage proportional with acceleration.
- piezoresistive effect: is manifested in materials such as silicon, which produce a change in electrical resistance due to the modification of its physical geometry as an effect of applied acceleration.

- capacitive effect: presumes the detection of capacitance variation of sensing elements, generated by movement of a seismic mass under external effects of acceleration.

From the response criterion, two categories of accelerometers are identified:

- AC—response: means that the output is alternating current type. This alternating current device isn't usually use to measure gravity and centrifugal acceleration, there are static parameters. It is proper for study of dynamic events only.
- DC—response: means that the output is direct current type, and could react until zero hertz frequency. As consequence, it is used for measuring the static or dynamic acceleration. The measurement of static acceleration must be corroborated to other features in order to choose a DC response accelerometer in the vibration diagnosis.

AC accelerometers could be used as a *charge output* or *voltage output* type. The *charge mode* type of piezoelectric accelerometers is characterized by wide range of temperature (−200 °C to +640 °C), broad frequency dynamic range (>120 dB) and wide frequency bandwidth (>10 kHz) and they are well suited for use in the study of the vibration at extreme temperatures conditions (e.g. for turbine engine monitoring). Concerning the *voltage mode* type of piezoelectric accelerometers, the main characteristic expresses their range predetermined by the internal amplifier imposed by the producer which cannot be changed, therefore for dynamic measurements they are proper in lightweight structures.

The most used DC accelerometers are capacitive or piezorezistive. First one, the capacitive type, is used in industry as an air-bag sensor, and has the features of lower range of acceleration measurement (up to 1500 Hz), high output stability and good linearity. Because of its large bandwidth capability (>7000 Hz), piezoresistive type accelerometers are mostly used for measurements of impulse or impact where are dealing with typically high frequency range and g level (>10,000 g), as in automotive safety testing [11]. When selecting the appropriate accelerometers, we should consider the mono or multiaxial measuring capabilities associated with the size, mass, detection technology, and others that are described in Fig. 12.3.

Usually, in automotive vibroacoustic measurement the single-axle or triaxle accelerometers are used. At the same time, the properties of the sensors (AC/DC power supply, prepolarization/external polarization, analogic/digital output, etc.) should be corelated with data acquisition hardware which could have integrated software with synchronization of the signals in case of advanced version, or for the basic version is necessary to be developed specific software for data acquisition with saving into the files which will be subjected to a subsequent filtration.

Data acquisition hardware could have the following features: 2 ÷ 1.000 input channels; scalable configuration: single module, multi modules, distributed systems or multiple separate distributed systems; LAN interface; GPS synchronization; conditioning of all sound and vibration transducers; automatic detection of hardware and transducers [13].

		Type nr.	Sensing technology	Range [g]	Frequency response [Hz] (±5%)	Operating tempertaure [°C]	Mass [g]	TEDS option	Ground isolation option	Application: Powertrain	Application: Structural testing	Application: General purpose	Description
Triaxial accelerometer		8766A050	PiezoStar	±50	1…6,000	-55…165	7	yes	with pad	x			Miniature PiezoStar, high temperature, high thermal stability
		8766A500	PiezoStar	±500	0.5…10,000	-55…165	4.5	yes	with pad	x			
		8763B50	Ceramic	±50	0.5…7,000	-55…100	4.5	yes	with pad		x	x	Miniature ceramic shear
		8763B500	Ceramic	±500	1…10,000	-55…120	3.6	yes	with pad		x	x	
		8688A5	Ceramic	±5	0.5…3,000	-40…55	6.7	yes	with pad		x		Miniature PiezoBeam, high output
		8688A50	Ceramic	±50	0.5…5,000	-40…65	6.5	yes	with pad		x		
		8762A50	Ceramic	±50	0.5…6,000	-55…80	23	yes	yes		x	x	Annular ceramic shear, rugged
		8764A250	Ceramic	±250	1…10,000	-55…100	6	yes	yes		x		PiezoStar, throughhole, high temperature, thermal stability
		8765A250M5	PiezoStar	±250	1…9,000	-55…165	6.4	no	yes	x			
Single-axis accelerometer		8703A50	PiezoStar	±50	0.5…10,000	-196…165	10	no	with pad	x	x	x	PiezoStar, high temperature, high thermal stability
		8705A50	PiezoStar	±50	0.5…10,000	-55…165	7.6	no	with pad	x	x	x	
		8702B500	Quarz	±500	1…10,000	-196…165	8.2	yes	with pad		x	x	Quartz shear, high temperature
		8704B500	Quarz	±500	1…10,000	-55…120	7.1	yes	with pad		x	x	Quartz shear, high temperature
		8714B500	Ceramic	±500	1…10,000	-55…165	4.2	yes	yes	x			Annular ceramic shear, through hole, high temperature
		8715A250	PiezoStar	±250	1…10,000	-55…165	2.4	yes	yes	x			Miniature PiezoStar, through hole, high temperature, high thermal stability
		8715A5000	PiezoStar	±5,000	2…10,000	-55…165	2.1	yes	yes	x			
	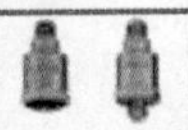	8774B50	Ceramic	±50	0.5…8,000	-55…100	2.9	no	yes		x	x	Ceramic shear, general vibration
		8776B50	Ceramic	±50	0.5…8,000	-55…100	3.3	no	yes		x	x	

Fig. 12.3 Characteristics for selection of suitable accelerometer [12]

12.5 Conclusions

Selection process of the microphones and accelerometers used for vibroacoustic study of the vehicle must take in consideration specific required features, adapted to the studied component of the vehicle (high temperature exposure, low/high frequency range, bandwidth, or exposure at shocks/dust/ humidity). The sensors characteristics and the data acquisition hard and soft must be compatible. On top it must take into account the type of use for testing noise and vibration if the measurements are made on laboratory, field or mix measurements. All these requirements lead to correct signals acquiring and processing to create a vibroacoustic database for automotive. It is a difficult task and our future research intends to apply this study on automotive classes to determine their vibroacoustic fingerprints.

References

1. Mercedes-Benz USA: Mercedes Me Connect Services (2022)

2. Daimler Truck, A.G.: Fleetboard, Daimler Truck AG (2022)
3. ACEA: Vehicle Noise–Setting the Right Sound Levels, European Automobile Manufacturers Association, Brussels, pp 10–12 (2021)
4. Application Note AN4426: Tutorial for MEMS Microphones, STMicroelectronics, pp 11–16 (2017)
5. Application Note AN4426: Tutorial for MEMS Microphones, STMicroelectronics, p 18 (2017)
6. PCB Piezoelectronics: Understanding Sound Power and Microphone Response a Complete Guide to Acoustic Fundamentals. Amphenol Company, 11, p. 10 (2021)
7. PCB Piezoelectronics: Acoustic Measurement Sensors & Instrumentation. Amphenol Company, 08, p. 13 (2021)
8. PCB Piezoelectronics: Acoustic Measurement Sensors & Instrumentation. Amphenol Company, 09, p. 14 (2021)
9. MEMS Microphones: STMicroelectronics (2022). https://www.st.com/en/mems-and-sensors/mems-microphones.html
10. Swingler, J.: MEMS for Passenger Safety in Automotive Vehicles. Heriot-Watt University, UK, Woodhead Publishing Limited, pp. 9–10 (2013)
11. TE Connectivity Sensors: Choosing the Right Type of Accelerometer. TE Connectivity Ltd., pp 07, 3–4 (2017)
12. Kistler 961-324e-2.0: Powertrain NVH–from Sensor to Answer. Kistler Group Eulachstrasse 22 8408 Winterthur Switzerland, pp 4–5 (2019)
13. Bruel & Kjaer BP 2215–37: Data Aquisition Hardware, Hottinger Brüel & Kjær A/S, Denmark, 08, 1 (2021)

Part III
Serial and Parallel Manipulators

Chapter 13
Experimental Identification of Singularities in Parallel Manipulators

J.-L. Pulloquinga, M. Ceccarelli, V. Mata, and A. Valera

Abstract In Robotics, a singular configuration could be dangerous due to instantaneously changing the number of degrees of freedom. This paper presents a general experimental procedure to identify singularities in Parallel Kinematic Mechanisms based on end-effector movements and the actuator's forces. Next, this work presents three experimental options for dealing with singularities. Finally, a sensor-based procedure to escape from a singularity is outlined. As an example, the experimental identification of singularities is applied to L-CaPaMan parallel robot to verify the feasibility of the proposed procedure. In this case, the end-effector movements are measured using an Inertial Measurement Unit, and a set of current sensors measures the actuator's forces.

Keywords Robotics · Experimental analysis · Singularities · Kinematics · Parallel manipulators

13.1 Introduction

A Parallel Kinematic Mechanism (PKM) uses at least two closed kinematic chains to drive an end-effector or mobile platform. The close chain architecture provides a PKM with high accuracy at high working velocity with lower power consumption

---

J.-L. Pulloquinga (✉) · A. Valera
Department of Systems Engineering and Automation, Universitat Politècnica de València, València, Spain
e-mail: jopulza@doctor.upv.es

A. Valera
e-mail: giuprog@sia.upv.es

M. Ceccarelli
Department of Industrial Engineering, University of Rome Tor Vergata, Roma, Italy
e-mail: marco.ceccarelli@uniroma2.it

V. Mata
Department of Mechanical Engineering and Materials, Universitat Politècnica de València, València, Spain
e-mail: vmata@mcm.upv.es

I. Doroftei et al. (eds.), *Proceedings of SYROM 2022 & ROBOTICS 2022*, Mechanisms and Machine Science 127, https://doi.org/10.1007/978-3-031-25655-4_13

[1–3]. In addition, a PKM is stiffer and with more load capacity than its serial counterparts. However, the PKM's architecture generates possible singularities within the workspace, reducing its kinematic performance and the size of the effective workspace.

Initially, Gosselin and Angeles [4] investigated the singularities of a PKM based on Jacobian matrices obtained from constraint equations and classified them into two types. In an inverse kinematic singularity or Type I, a PKM is not able to reach the desired location with any movement of the actuators, i.e., at least one degree of freedom (DOF) is lost [2]. In a forward kinematic singularity or Type II, a PKM could lose control of the end-effector despite having all the actuators locked, i.e., at least one DOF is gained [5]. Type II singularities could be particularly critical because the mobile platform at the singularity cannot bear the external forces, losing control of the PKM motion [6].

In general, the singularities of a PKM, due to the instantaneous change in its DOF, are a problem to avoid. A singular configuration is identified when the determinant of the Jacobian matrix is zero. However, it requires an accurate theoretical kinematic model of the PKM. Ottaviano and Ceccareli [7] proposed an experimental characterization of singularities on a six-wire parallel robot based on measuring the location and orientation of the end-effector. The experimental procedures from [7] require particular tracking systems, making it difficult to generalize.

In this paper, attention is addressed to identifying singularities in PKMs through an experimental procedure using singularity effect sensing. An example with L-CaPaMan [8] is discussed to prove the feasibility of the procedure and highlight the peculiarities of its implementation. Moreover, a sensor-based experimental procedure for escaping from a singularity is proposed.

13.2 Parallel Manipulators and Singularities

A PKM drives the location and orientation $\boldsymbol{x}$ of the end-effector using at least two serial chains connected by a set of joints. The motion of the PKM is controlled by a subset of active or actuated joints $\boldsymbol{q}_{ind}$. Figure 13.1 shows a general scheme of a PKM where q_{ij} stand for active and passive joints and $\{O - x_0 y_0 z_0\}$, $\{P - x_p y_p z_p\}$ stand for the fixed and mobile reference frames, respectively.

Based on serial kinematic chains, a set of constraining equations $\boldsymbol{\Phi}$ are defined as

$$\boldsymbol{\Phi}(\boldsymbol{x}, \boldsymbol{q}_{ind}) = 0 \tag{13.1}$$

Taking time derivatives from (1), the PKM motion can be defined as

$$\boldsymbol{J}_D \dot{\boldsymbol{x}} + \boldsymbol{J}_I \dot{\boldsymbol{q}}_{ind} = 0 \tag{13.2}$$

where $\boldsymbol{J}_D$ and $\boldsymbol{J}_I$ are the Forward and Inverse Jacobian matrices, respectively.

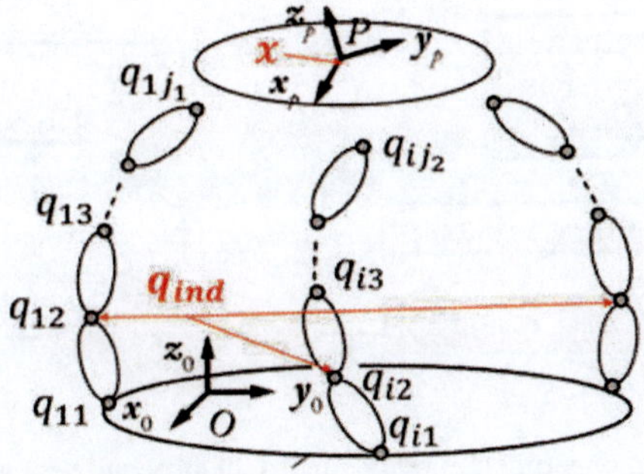

Fig. 13.1 A general scheme for a PKM with main parameters

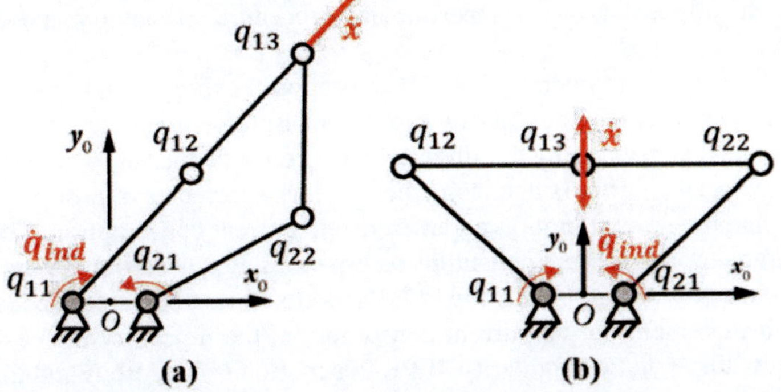

Fig. 13.2 Example of a five-bar PKM in a singularity: **a** type I; **b** type II

Based on (2), there is two possible degeneration of PKM motion. The first case appears calculating $\dot{\boldsymbol{q}}_{ind}$ with a rank deficient $\boldsymbol{J}_I$ matrix, i.e., $det(\boldsymbol{J}_I) = 0$. In this case, $\dot{\boldsymbol{x}}$ is zero despite having a set of non-zero $\dot{\boldsymbol{q}}_{ind}$. Figure 13.2a shows a five-bar PKM that cannot move in the direction of $\dot{\boldsymbol{x}}$ despite its actuators moving, in other words, at least one DOF is lost. The second motion degeneration takes place by solving $\dot{\boldsymbol{x}}$ with $det(\boldsymbol{J}_D) = 0$, generatig a non-zero $\dot{\boldsymbol{x}}$ despite $\dot{\boldsymbol{q}}_{ind}$ is zero. Figure 13.2b shows an example of this singularity where the PKM can move in the direction despite the actuators being locked, i.e., at least one DOF is gained.

13.3 Experimental Identification of Singularities

Type I and Type II singularities show an instantaneous change in PKM's DOF that a motion sensor can detect. In addition, the degeneracy of the PKM motion requires an increment in the force applied to the actuators. Since force is proportional to the

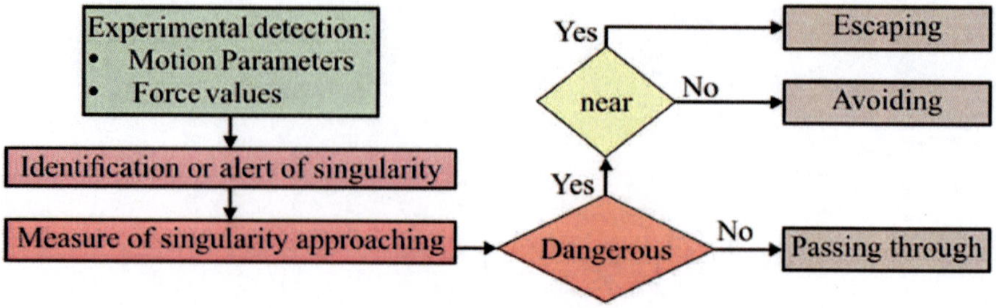

Fig. 13.3 A scheme for an experimental procedure to identify and deal with singularities

current consumed by the actuators, a current sensor could measure the increment of the force. Based on suddenly changes detected by motion or current sensors, this research proposes a novel experimental procedure to identify and deal with singularities.

Figure 13.3 shows an overall view of the proposed experimental procedure. The first step is to identify a singularity by motion measurement and increase in current consumption. Next, measuring the singularity approach defines the dangerousness of a singularity. A singularity is detected when the measurements of motion or current are lower than the normal motion parameters and current consumption. The normal motion parameter and current consumption are calculated according to the PKM.

In the procedure of Fig. 13.3, if the PKM's motion does not increase considerably with a high increment on the current consumption, the trajectory is not modified, letting it pass through the singularity. If PKM's motion is dangerous and the current consumed by the actuators continues to increase, PKM is near a singularity, and an avoidance method is required. The singularity avoidance can be performed by changing the dynamic model of the PKM [9] or modifying the motion trajectory [10]. On the other hand, if PKM's motion is dangerous and the current consumed by the actuators remains constant or reaches the feasible limits, the PKM is in a singularity, and an escaping method is required.

Figure 13.4 present a practical strategy to escape from a singularity using motion and current measurements. The first step is to modify the direction of the actuator's forces to decrease the dangerous motion of the PKM. If the change in the actuator's forces does not work, a controllable external force can be applied to the end-effector to modify its trajectory. If these two actions do not work, the PKM is trapped in a singularity, and an assemble mode change must be applied.

13.4 An Experimental Layout

Figure 13.5 shows a layout to implement the experimental identification to deal with singularities in the case of study with L-CaPaMan [8]. The experimental identification and the strategy for dealing with singularities are performed in Matlab, while the

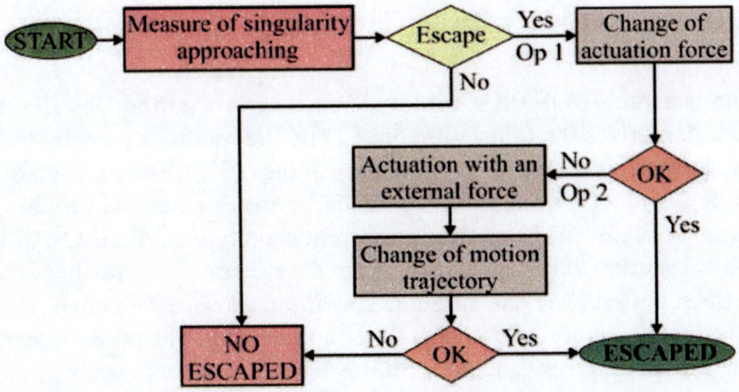

Fig. 13.4 An experimental strategy to escape from a singularity

Fig. 13.5 A scheme of experimental layout for an example with L-CaPaMan

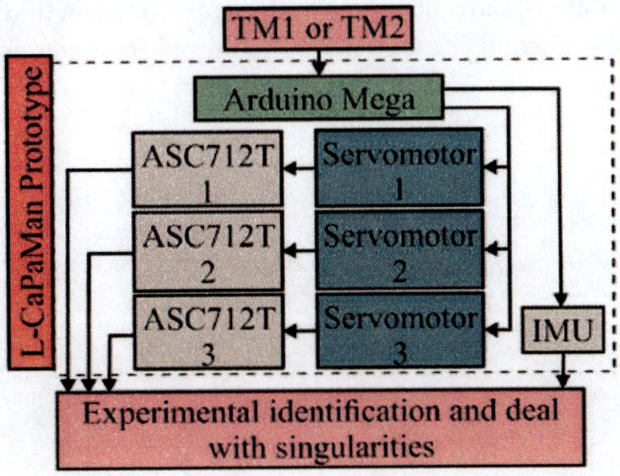

controller works in Arduino. The experimental scheme in Fig. 13.5 is applied during the execution of two trajectories:

- TM1: Starts from a non-singular configuration and goes throw a Type I singularity.
- TM2: Starts from a Type I singularity and go to a non-singular configuration.

The planning of trajectory TM1 or TM2 is sent from Matlab to Arduino Mega using serial communications with a baud rate of 115,200. Arduino board sets the desired actuator's location by means of three servomotors. Contemporary, an IMU measures the linear acceleration and the angular velocity of the centre of the mobile platform, and three current sensors (ASC712T) measure the energy consumed by the actuators. Finally, the measurements are sent from Arduino Mega to Matlab for analysis. The experimental procedure for identifying singularities requires an isolated environment for the PKM to avoid increasing current measured due to external forces.

13.5 A Case of Study with L-CaPaMan

L-CaPaMan is a three-DOF PKM prototype designed for earthquake simulation [8, 11]. L-CaPaMan drives the translation on Z_p and the rotations φ, ψ by means a set of three actuators represented as α_1, α_2 and α_3 in Fig. 13.6a. The location of the end-effector on X_p, Y_p and the length of the thre strokes s_i are passive variables defined by Z_p, φ and ψ. Figure 13.6b shows the current prototype of the L-CaPaMan [8].

Table 13.1 describes the parameters used for the experiments with the L-CaPaMan.

During the reported TM1 execution, L-CaPaMan shows a well controlled motion of the end-effector. Figure 13.7a shows the module of the acquired linear acceleration of the mobile platform's centre. Figure 13.7b shows two peaks in the current consumed by actuator 1 that correspond to a Type I singularity with passing through motion indicated in Fig. 13.7a.

Figure 13.8a shows the location of the end-effector reached in Z_p where the two peaks of current appers at the same location that verifies the correspondence between the current increments and singularity detection. Figure 13.8b shows the location

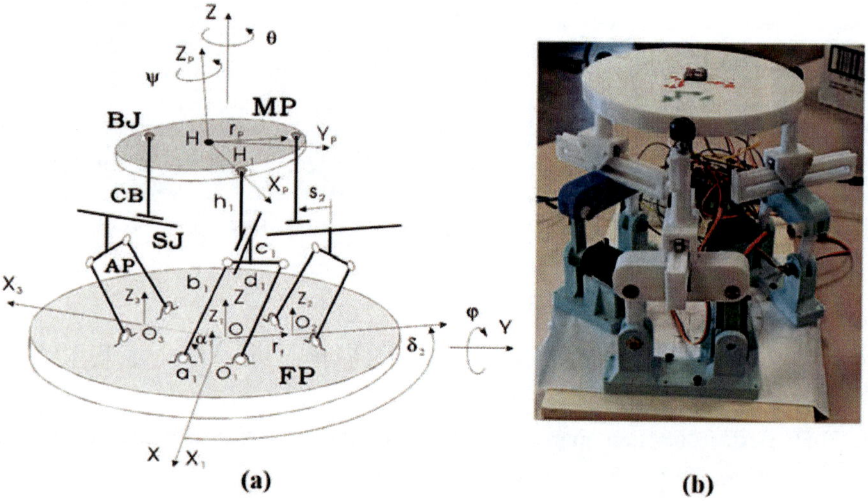

Fig. 13.6 L-CaPaMan design: **a** kinematic [11]; **b** current prototype [8]

Table 13.1 Parameters for the experimental tests with L-CaPaMan in Fig. 13.6b

	Initial point (deg)			Final point (deg)			Waveform
	α_1	α_2	α_3	α_1	α_2	α_3	
TM1	135	135	135	45	45	45	Sinusoidal
TM2	90	90	90	45	45	45	Sinusoidal

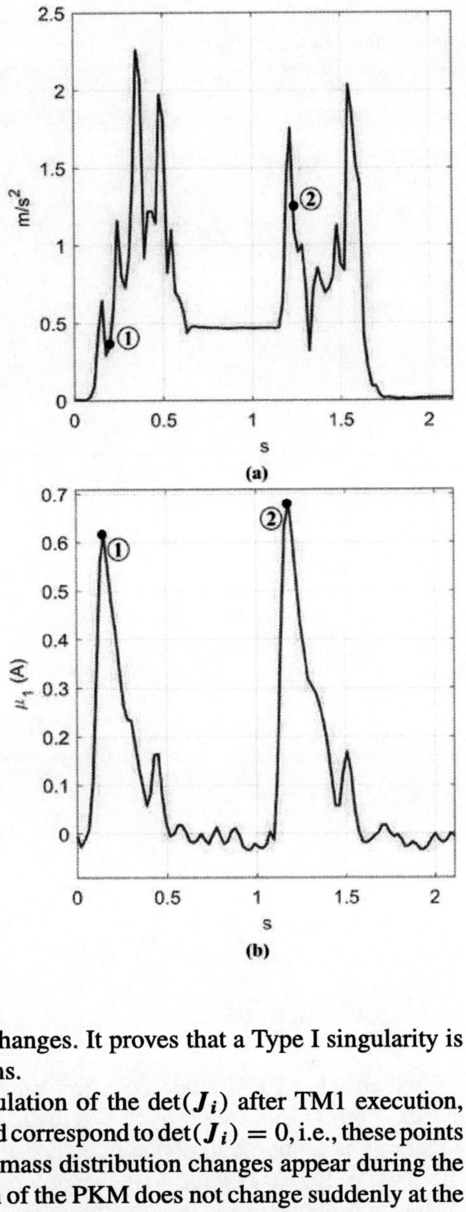

Fig. 13.7 Acquired results of a TM1 test: **a** linear acceleration module; **b** current on actuator 1

reached by actuator 1 with no sudden changes. It proves that a Type I singularity is passed through with no motion problems.

Moreover, Fig. 13.9 shows the calculation of the $\det(\boldsymbol{J}_i)$ after TM1 execution, where the two peaks of current consumed correspond to $\det(\boldsymbol{J}_i) = 0$, i.e., these points are Type I singularities. In addition, no mass distribution changes appear during the Type I crossing because the acceleration of the PKM does not change suddenly at the two singular points (see Fig. 13.7a). No significant acceleration changes are noted during the TM2 test, with a peak current consumption at the beginning. The TM1 and TM2 tests verified that a type I singularity in a PKM motion could be detected by sensing the motion and the current consumed by the actuators.

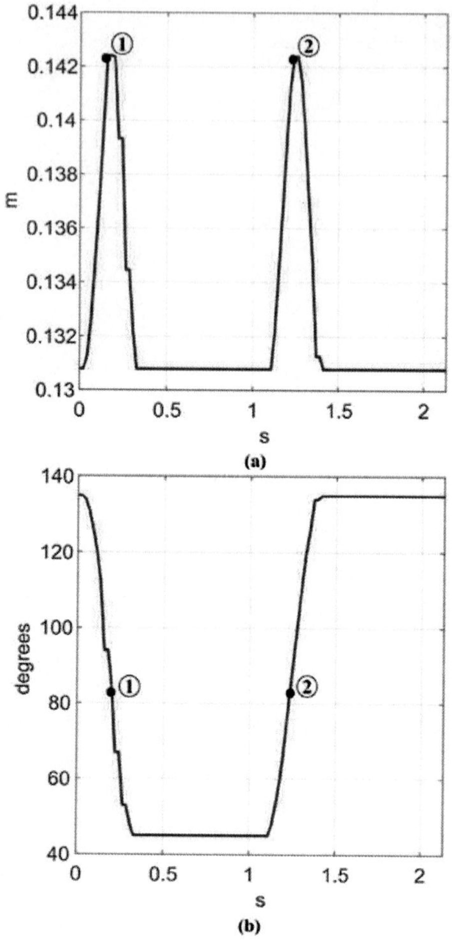

Fig. 13.8 Acquired end-effector location of TM1 test: **a** Z_p **b** actuator 1

13.6 Conclusions

A singularity in a PKM motion can be identified successfully by sensing the motion and the current consumed by the actuators. An experimental identification of a singularity is proposed to define three possible strategies to deal with a singularity. The strategy for passing through a singularity was successfully implemented on L-CaPaMan prototype as a case of study. Based on the results, this paper proposed a practical method to escape from a singularity using only sensor information. The increment of current consumed appears at the same location in working and joint space, which verifies the possibility of detecting Type I singularity. The uncontrolled effects of external force have been avoided by isolating the PKM, while the mass distribution changes are discarded by the motion measured by the IMU sensor.

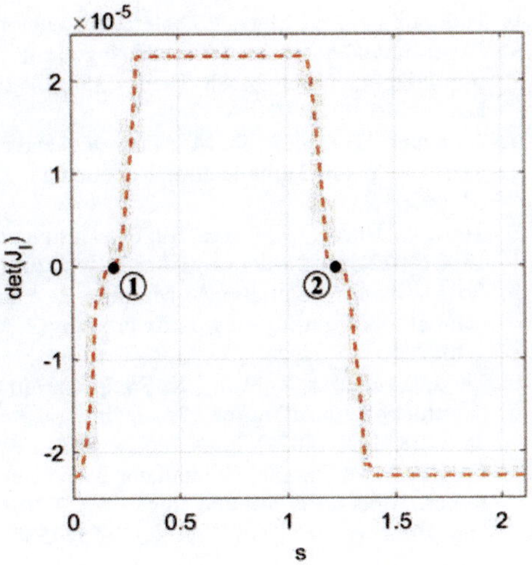

Fig. 13.9 Computed $\det(\boldsymbol{J}_i)$ after post-processing results of a TM1 test in Fig. 13.7

However, the experimental procedure must be implemented in several PKMs with different architectures to verify the advantages of identifying singularities based on sensors.

In future work, the sensor-based singularity identification will be applied in a PKM with two translational and two rotational DOFs. Moreover, the strategy proposed for sensor-based escaping from a singularity will be developed for practical implementation.

Acknowledgements This work was partially funded by Fondo Europeo de Desarrollo Regional (PID2021-125694OB-I00), by Programa de Ayudas de Investigación y Desarrollo de la Universitat Politècnica de València, Spain (PAID-01-19), and cofounded by Programa de Ayudas para movilidad de estudiantes de doctorado de la Universitat Politècnica de València en 2022 from Spain and by Erasmus+ Student Mobility for Traineeship 2020 permitting a research stay of the first author at University of Rome Tor Vergata.

References

1. Mata Amela, V., Valero Chuliá, F., Cuadrado Iglesias, J.I.: 1995 Mechanics of robots. ISBN: 84-7721-303-8. Univ Politècnica València
2. Ceccarelli, M.: Fundamentals of mechanics of robotic manipulation. Springer Int. Publ. **112**, 1–381 (2022). https://doi.org/10.1007/978-3-030-90848-5
3. Briot, S., Khalil, W.: Dynamics of parallel robots–from rigid links to flexible elements. ISBN: 978–3–319–19787–6 (2015)
4. Gosselin, C., Angeles, J.: Singularity analysis of closed-loop kinematic chains. IEEE Trans. Robot. Autom. **6**, 281–290 (1990). https://doi.org/10.1109/70.56660
5. Ottaviano, E., Gosselin, C.M., Ceccarelli, M.: Singularity analysis of CaPaMan: a three-degree of freedom spatial parallel manipulator. Proc. IEEE Int. Conf. Robot. Autom. 2, 1295–1300 (2001). https://doi.org/10.1109/ROBOT.2001.932789

6. Pulloquinga, J.L., Mata, V., Valera, Á., Zamora-Ortiz, P., Díaz-Rodríguez, M., Zambrano, I.: Experimental analysis of Type II singularities and assembly change points in a 3UPS+RPU parallel robot. Mech. Mach. Theory. **158**, 104242 (2021). https://doi.org/10.1016/j.mechmachtheory.2020.104242
7. Ottaviano, E., Ceccarelli, M.: Numerical and experimental characterization of singularities of a six-wire parallel architecture. Robotica **25**, 315–324 (2007). https://doi.org/10.1017/S0263574706003110
8. Titov, A., Russo, M., Ceccarelli, M.: Design and performance of L-CaPaMan2. Appl. Sci. **12**, 1380 (2022). https://doi.org/10.3390/app12031380
9. Six, D., Briot, S., Chriette, A., Martinet, P.: A controller avoiding dynamic model degeneracy of parallel robots during singularity crossing. J. Mech. Robot. 9 (2017). https://doi.org/10.1115/1.4037256
10. Zhou, J., Li, Z., Li, X., Wang, X., Song, R.: Human-robot cooperation control based on trajectory deformation algorithm for a lower limb rehabilitation robot. IEEE/ASME Trans. Mechatron. **26**, 3128–3138 (2021). https://doi.org/10.1109/TMECH.2021.3053562
11. Ceccarelli, M.: Parallel manipulator architectures from CAPAMAN design. In: 19th Int. Work. Robot. Alpe-Adria-Danube Reg. RAAD 2010-Proceedings Budapest. pp. 187–192 (2010). https://doi.org/10.1109/RAAD.2010.5524588

Chapter 14
Requirements and Problems for Space Berthing System

A. Titov and M. Ceccarelli

Abstract In this paper a berthing task is introduced with a system which considers a station and a CubeSat-like satellite. In the presented task a manipulator is used to catch, to move the micro-satellite towards berthing port, and to dock it with interfaces activation. The requirements are discussed for the system in terms of issues for design, control, and sensorization. According to the requirements, possible problems are formulated with characteristic parameters. A conceptual CAD solution for the berthing manipulator is presented with a gripper, and a dynamic simulation for basic movements is elaborated for characteristic results.

Keywords Space robotics · Berthing system · Small satellites · CubeSat

14.1 Introduction

Two types of mating spacecrafts exist, such as docking and berthing. Docking considers connection of two free-flying spacecrafts. Berthing task executes a satellite pairing with a manipulator, that is installed on target spacecraft. The International Space Station (ISS) has docking systems for spacecrafts, such as low-impact docking system (LIDS) [1] and IDSS [2], and allows to provide docking operation, whereas all the manoeuvres are made from the satellite and the station. This method requires high accuracy and a lot of consumption of fuel on the satellite; the fuel is a critical need for other operations, and it is necessary to save it. For that reason, another type of mating is used more often, and it is the berthing. Canadarm2 or Space Station Remote Manipulator System (SSRMS) [3], is the main manipulator on ISS, which is used for maintaining the station, berthing, and moving spacecrafts. With a 14.22 m of length and 7 degrees of freedom, it is possible to catch satellites and place them in the docking adapter without additional fuel consumption. Its load capability up

A. Titov (✉) · M. Ceccarelli
LARM2 Laboratory of Robot Mechatronics, University of Rome Tor Vergata, Roma, Italy
e-mail: aleksandr.titov@students.uniroma2.eu

M. Ceccarelli
e-mail: marco.ceccarelli@uniroma2.eu

I. Doroftei et al. (eds.), *Proceedings of SYROM 2022 & ROBOTICS 2022*, Mechanisms and Machine Science 127, https://doi.org/10.1007/978-3-031-25655-4_14

to 116,000 kg for maintain large objects, such as cargo spacecrafts, like SpaceX Dragon. European Robotic Arm (ERA) [4] is attached to the Russian segment of the ISS, and it is used for similar tasks as SSRMS. Japan Experimental Module Remote Manipulator System (JEMRMS) is a part of Japanese science module of the ISS [5, 6], and it is used for logistics tasks. The berthing system Orbital Express Demonstration Manipulator System (OEDMS) [7] was presented as a solution for an autonomous rendezvous, capture, berthing and transfer of a functional part to the chaser. In comparison with other solutions, this system is developed not for ISS, and the dimensions and masses of satellites are comparable.

14.2 The Berthing Task

A new solution is developed for test of berthing of small satellites. It is a system which includes one medium cylindric form satellite called "station" and six nano- and micro-satellites CubeSat called "chasers". According to a general classification of the satellites by mass [8], medium solutions consider a mass from 500 to 1,000 kg, micro–from 10 to 100 kg, and nano–from 1 to 10 kg. Term small is general for all satellites less than 100 kg. The CubeSat standard [8] has six form factors: 1U, 1.5U, 2U, 3U, 6U, 12U, where "U" is a unit defined as a cube with a 10 cm edge. Thus, 1U CubeSat has dimensions 100.0 × 100.0 × 113.5 mm with a mass up to 2.0 kg, 3U is 100.0 × 100.0 × 340.5 mm with a mass up to 6.0 kg, 6U is 100.0 × 226.3 × 366.0 mm with a mass up to 12.0 kg, and 12U is 226.3 × 226.3 × 366.0 mm with a mass up to 24.0 kg.

From [9], the general scheme of a berthing process is presented in Fig. 14.1. To be captured, chaser moves into the defined area, called "berthing box". At the same time manipulator is placed into the position for capturing. When the chaser is in the berthing box, its thrusters stop the chaser and switch off. After that, the capturing is initialized. The manipulator grapples the chaser and transfer to the berthing port with communication interfaces. The process is finished when structural connection between chaser and station is set. The scheme in form of box diagram is presented in Fig. 14.2.

14.3 Requirements

According to tasks, requirements can be subdivided into the requirements to the whole system, to the station, to the manipulator, to the end effector. A box diagram with the scheme for the requirements is shown in Fig. 14.3.

For the described task Low Earth circular orbit [9] with attitude of 400 km can be chosen. Finally, it will be defined with the launcher and the spaceport. Low Earth orbit also allows to finish the mission by burning in the atmosphere to save the orbit from space debris. As mentioned before, the planned mass of the system should

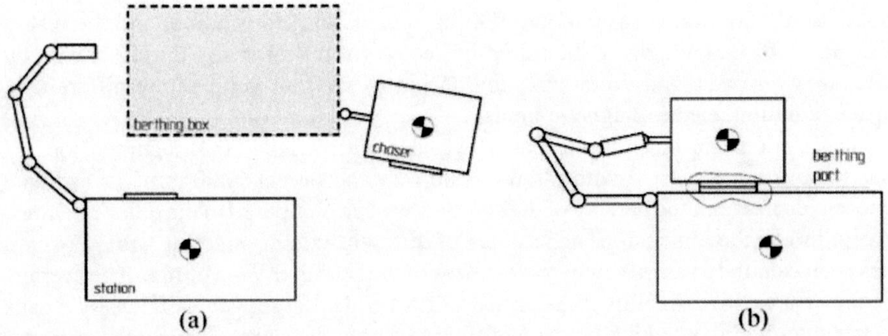

Fig. 14.1 A scheme of a berthing task: **a** the task starting; **b** an insertion of the satellite into the berthing port

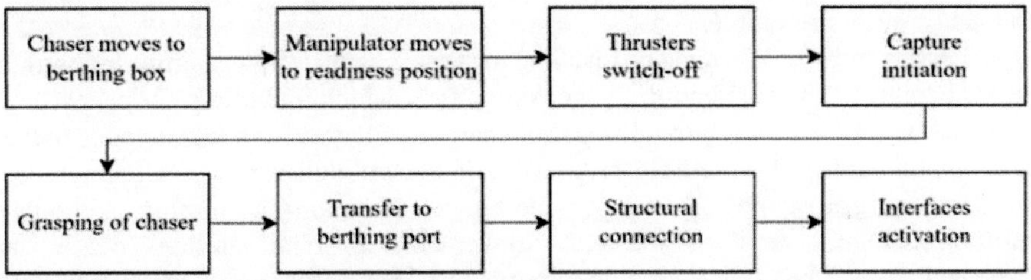

Fig. 14.2 A scheme of a berthing task with main phases

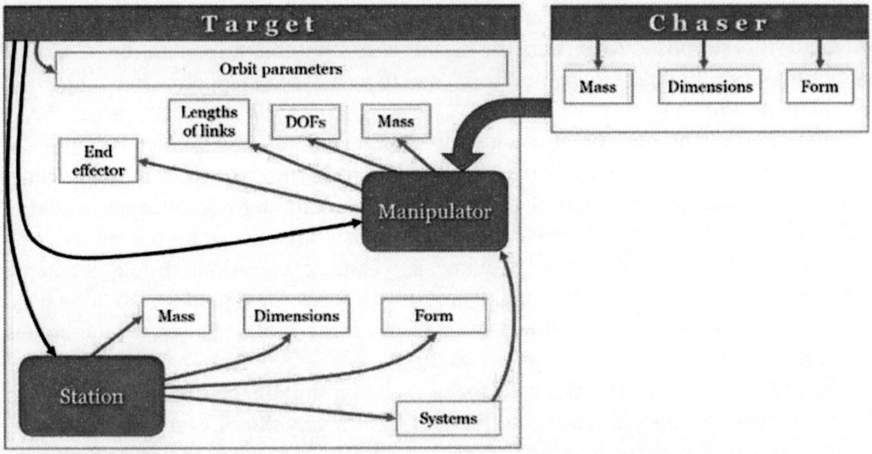

Fig. 14.3 A scheme for the requirements to the berthing manipulator system

not exceed 1000 kg, of which near 900 kg is a mass of the station, and the rest—the mass of manipulator. It should be protected from destroying during launch by heating, overloads, and vibrations, and during work from temperature differences, space radiation, electromagnetic fields.

The station can be presented as a cylinder with mass 900 kg. Dimensions of the stations are assumed with a radius 500 mm and height 2000 mm. To estimate chosen parameters, the density of the station has been compared to the other satellites. According to the dimensions and masses of different existing satellites, the density ρ has been calculated as m/V, where m—mass of the satellite, V—volume. The average density for existing satellites is assumed as 1.5 kg/m^3. Larger value of density means better satellite layout and efficiency of filling its volume. For the developed station $\rho = m/(\pi \cdot r^2 \cdot h) = 573$ kg/m^3, what is almost three times less than average. Therefore, the efficiency is needed to be increased. By assuming radius to 400 mm and height to 1000 mm, the calculated density is equal to 1791 kg/m^3, which is more than average and hence more preferable.

Necessary subsystems must be installed in station, such as power, attitude control, thermal control, communication. Power subsystem, which includes solar panels and chemical batteries, should provide enough energy to all other systems, to control the manipulator during the berthing task. Attitude control subsystem, which includes thrusters and gyroscopes, should provide the reorientation and aiming. Although berthing task in general considers the movements of small satellite—chaser, in presented case CubeSat are not able to move by thrusters. It means the station will move towards chaser. At the end of mission station with berthed chasers should be deorbited to burn in atmosphere. Communication subsystem works both sides, and the possibility to give commands from Earth in emergency situations should be provided. It is supposed to execute the berthing task without teleoperation. Chasers' recognition, optimal path to capture them and to move into respective berthing ports should be provided.

Total mass of the manipulator with end effector should not exceed 100 kg. The beam length must allow to insert every CubeSat to its berthing port. At the same time, during the launch and the end of the mission the manipulator should be in folded position well fixed and not increase the total dimensions of the station in launch state more than 15% in every direction. Moving speed and accelerations should not create undesired forces and torques during the berthing process. Designed for berthing task, the maximum load capacity defined by maximum mass of a chaser, which equals 24.0 kg.

The only function of the end effector is grasping the objects type CubeSat. End effector construction should allow to fix the CubeSats by the ribs and keep it during the process. In comparison to other CubeSats, which have the square section, CubeSat 6U has a rectangular section. Therefore, to grasp it, the end effector should adapt to the rectangular form of the CubeSat 6U. Force feedback is necessary to initialize the grasping. Minimization of mass is desirable as an extra requirement.

14.4 Problems

The box diagram with the problems is presented in Fig. 14.4.

Before starting the mission, the berthing system should go to the orbit. Optimization of the price of the project is one of most important problems. Berthing system can be launched alone or with another payload. Using standard cargo satellites to reach the ISS or its alternatives [10] can be cheapest solution.

Overloads, heating, and vibrations can destroy the whole satellite or its parts, making the system unable to work. After a successful launch space conditions, such as temperature differences, radiation, electromagnetic fields, affect the system.

The most dangerous problem for any satellite on the described orbit—space debris. Large pieces can be avoided in advance, as being tracked. Pieces with dimensions less 10 cm can't be tracked, but their kinetic energy is enough to destroy the berthing system.

To control the station with thrusters, its centre of mass should be predictable and controllable. For this class of satellites, the power consumption is large to receive solar power from solar panels on the surface, therefore, the folding mechanism for the solar panels is necessary.

Sensors of a vision system have a limited angle of observation, usually it is less than 180°. To monitor the chasers, vision system should provide 360° observation angle.

A moving manipulator changes parameters of inertia during any movements. Fast movements create accelerations, which rotate the station.

A position control for the manipulator during the berthing process is a hard computational task. Any movements of the manipulator rotate the station, and, consequently, parts of the manipulator rotate, which complicates positioning related to chaser.

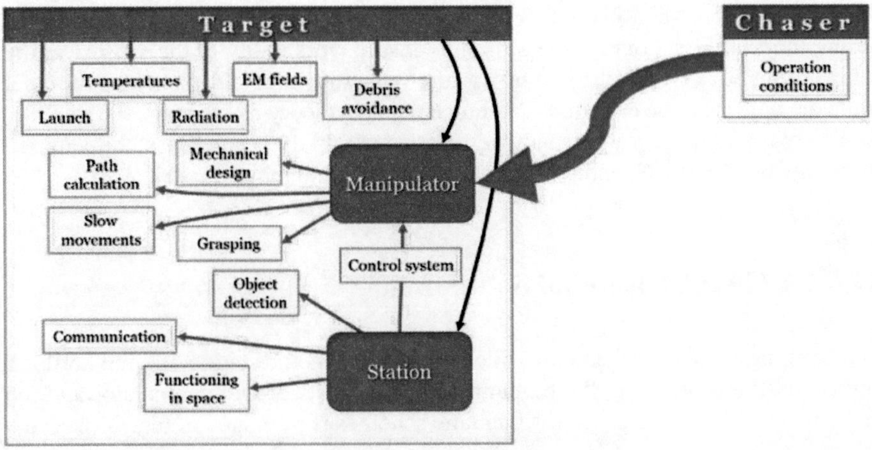

Fig. 14.4 A scheme of the problems for a system for the berthing task

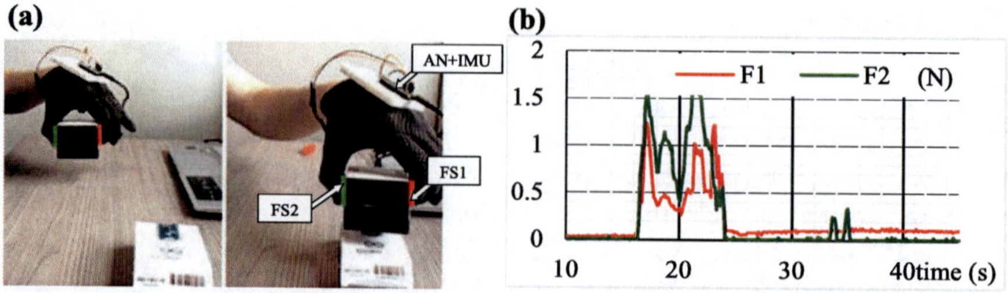

Fig. 14.5 **a** A snapshot of a lab glove test simulating a berthing task: FS1—1st force sensor, FS2–2nd force sensor, AN + IMU—Arduino Nano with built-in IMU sensor, **b** Acquired force data from force sensors in a glove berthing-like test

Grasping of the free-flying object is not a trivial task. The orientation of the chaser is random, even in readiness position misalignments by angles and shifts should be taken into consideration. In CubeSat ribs are more resistant than body, so the contact of fingers of the end effector should be with them.

14.5 A Glove Lab Testing

To characterize the operation and design problem for designing a manipulator for berthing task, a lab experience with a human-like berthing operation was considered as convenient. The snapshot of a test is shown in Fig. 14.5a.

Arduino Nano with built-in accelerometer is used. Two force sensors are installed on the glove and connected to the Arduino to register the contact forces while capturing. The box with mass 0.1 kg imitates a CubeSat body. Berthing task is presented like pick-and-place task. Force data is shown in Fig. 14.5b as acquired during the test. First contact was strong to ensure workability of the sensors in the setup during box capture, the second capture was "soft"—to understand the minimal force the object can be captured. Too much force in theory can damage the satellite, for that reason, the soft capture has been carried out. In the second case, the acquired forces are near to 0.1 N, and it is enough to capture and move the object.

14.6 A CAD Conceptual Solution

The berthing system, presented in this paper, contains the station and the berthing device; on the other hand, the berthing device consists of the manipulator and the end effector. One of the possible kinematic schemes is presented in Fig. 14.6a. This scheme contains a 7 d.o.f. manipulator and a 1 d.o.f. end effector. The extra d.o.f. is used by manipulator to move around the station, and one d.o.f. for end effector allows it to grasp CubeSats. Lengths of links are shown as l_1 and l_2, A–cylindrical

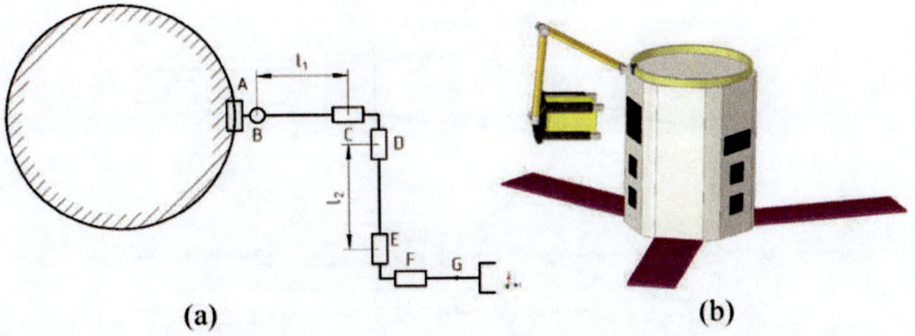

Fig. 14.6 A conceptual design of a berthing manipulator system: **a** a kinematic design, **b** a CAD model

joint, B, C, D, E, F, G—rotational joints. One extra d.o.f. allows manipulator to move around the body of the station, allowing to save fuel and not to re-orient the station for the berthing task. A CAD conceptual solution for berthing system is presented in Fig. 14.6b also to be used for validation and design via simulation.

The station is a cylindric body with six cut planar surfaces; three of them are for keeping the solar panels, two of them contain berthing ports for all types of CubeSats, and the last one is reserved for the manipulator in folding state. The manipulator is presented with a sliding frame guide, six rotation motors, two links and an end effector with four corner-shape fingers. The lengths of links are 600 mm, outer diameter for a sliding body is 750 mm, height of fingers of the end effector—300 mm. A diameter of the station is 800 mm, and its length is 1000 mm. To estimate the mass, all the parts are modelled with PLA material with density 1250 kg/m^3. The overall mass of the system is 623.9 kg.

A dynamic simulation for a berthing task was carried out with a moving the grasped CubeSat type of 12U to the berthing port. The results of displacements, velocities, and accelerations of the end effector are presented in Fig. 14.7a, Fig. 14.7b, c, respectively. The torques of the motors are presented in Fig. 14.7d, where T_A – T_E respect to joints A – E from Fig. 14.7b.

The motors movements are chosen by selection method with the accuracy up to 5°. At the beginning of simulation, the berthing hand is placed in the opposite side of the berthing port in the extremal reachable point. From 0 to 2 s manipulator moves to the opposite side of the station. From 2 to 3 s motor A rotates 30° clockwise, motor C rotates 90° clockwise, motor D rotates 5° counterclockwise, motor E rotates 75° counterclockwise. From 3 to 4 s motor C rotates 45° clockwise, from 4 to 5 s motor B rotates 10° clockwise. The berthing operation was simulated with angular velocities up to 90 deg/s in the motors. With that program, the required torques reach 100 N m. The Dynamixel servomotors, which are widely used in many fields, can reach up to 44.7 N m. To use real components, the berthing system should be optimized in terms of the structure and the working modes.

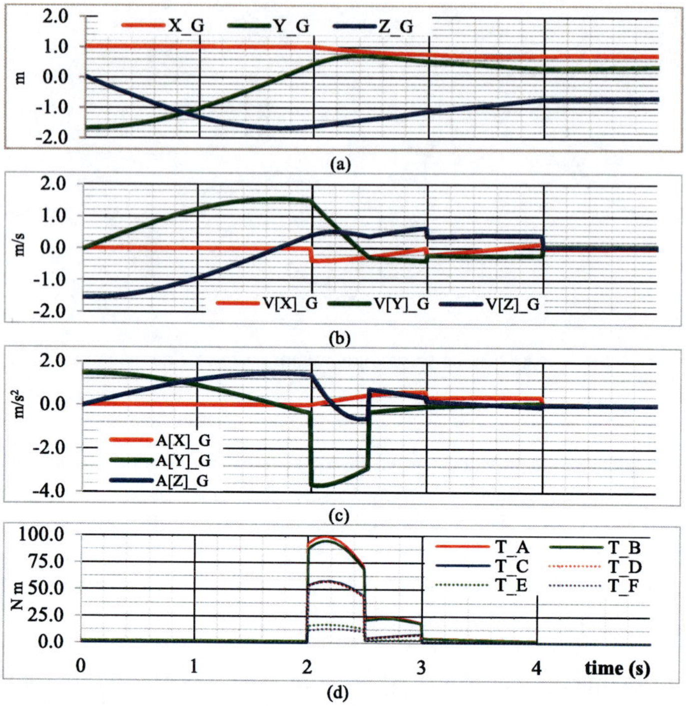

Fig. 14.7 Computed results during simulation of a berthing operation as in terms of: **a** displacements, **b** velocities, **c** accelerations of the end effector, and **d** the torques of the motors

14.7 Conclusions

The berthing task is considered for small satellites of CubeSat type up to 24 kg. For this existing solutions of berthing manipulators, such as SSRMS, ERA, JEMRMS look to be bulky. A general scheme of spacecrafts berthing is briefly outlined, and the objects for berthing are characterized in dimensions and mass. The requirements have been specified for the berthing system, which includes not manipulator only, but also the station where the manipulator should be installed. The problems are outlined in terms of general problems for the satellites and regarding to the berthing task. A glove lab test is carried out as the first attempt of a prototype for berthing. Force sensors installed on the fingers, register the force around 0.1 N enough to grasp the box and overcome the gravity force. According to the task and requirements, the first concept of the berthing system is developed in kinematic scheme and in CAD design. A dynamic simulation is performed in case of moving the CubeSat to the berthing port characterized with a slow motion: velocities do not exceed 1.8 m/s, and accelerations are less than 4 m/s^2.

References

1. Zhang, L., Shao, J., Zou, H.: Control system research and performance prospect of low impact docking system. In: Proceedings of the 3rd International Conference Mechatronics, Robotic Automation, vol. 15, pp. 1026–1030 (2015)
2. NASA: IDSS International Docking System Standard (IDSS) Interface Definition Document (IDD) Revision E open access 1–142. Retrieved from https://www.internationaldockingstandard.com/download/IDSS_IDD_Revision_E_TAGGED.pdf
3. Xu, W., She, Y., Xu, Y.: Analytical and semi-analytical inverse kinematics of SSRMS-type manipulators with single joint locked failure. Acta Astronaut. **105**, 201–217 (2014). https://doi.org/10.1016/j.actaastro.2014.09.003
4. Boumans, R., Heemskerk, C.: The European robotic arm for the international space station. Robot. Auton. Syst. **23**, 17–27 (1998). https://doi.org/10.1016/S0921-8890(97)00054-7
5. Doi, S., Wakabayashi, Y., Hosokavva, T., Sugano, T.: Control aspect of Japanese experiment module remote manipulator system. IFAC Proc. **34**, 141–146 (2001). https://doi.org/10.1016/S1474-6670(17)41695-8
6. Laryssa, P., Lindsay, E., Layi, O., Marius, O., Nara, K., Aris, L., Ed, T.: International space station robotics: a comparative study of ERA, JEMRMS and MSS. In: 7th ESA Work. Adv. Sp. Technol. Robot. Autom. "ASTRA 2002", pp. 1–8 (2022)
7. Ogilvie, A., Allport, J., Hannah, M., Lymer, J.: Autonomous satellite servicing using the orbital express demonstration manipulator system. In: Proceedings of the of the 9th International Symposium on Artificial Intelligence, Robotics and Automation in Space (i-SAIRAS'08), pp. 25–29 (2008)
8. Johnstone, A.: CubeSat design specification Rev. 14.1 the CubeSat program. Cal Poly SLO, pp. 1–34 (2022)
9. Fehse, W.: Automated Rendezvous and Docking of Spacecraft. Cambridge University Press, New-York (2003)
10. Sorokin, I.V., Konoshenko, V.P., Markov, A.V.: Research potential of the ISS Nauka module. Acta Astronaut. (2022). https://doi.org/10.1016/j.actaastro.2022.01.044

Chapter 15
A Kinematic Analysis of the TORVEastro Astronaut Robot Arm

M. Paoloni, G. Cupertino, and M. Ceccarelli

Abstract In this paper are described the forward and inverse kinematics of the arms of the TORVEastro project robot. TORVEastro is a research program dedicated to development of a new service robot that could assist astronauts in Extra-Vehicular-Activities (EVA) aboard orbiting space stations. After a short description of the arm mechanical structure, its forward kinematics is resolved. Then the inverse kinematics is analyzed by an original approach, to study the conditions for the existence and uniqueness of its solution; while performing this analysis, the constraints on the positions and orientations reachable by the arm end-effector emerge.

Keywords Forward kinematics · Inverse kinematics · Service robots · Space robotics · Astronaut robots

15.1 Introduction

The TORVEastro [1, 2] project goal is the development of a new service robot that could assist astronauts in Extra-Vehicular-Activities (EVA) aboard orbiting space stations; its typical tasks could be the surveillance and the maintenance of the external structures. The main challenge is the design of new mechanisms and solutions for the robot actuators that will be compliant with the peculiar requirements of the space environment, but the reliability and accuracy of the robot control system is also important.

M. Paoloni (✉) · G. Cupertino
Robotics and Artificial Intelligence (RIA) Lab, ENEA–Italian National Agency for New Technologies, Energy and Sustainable Economic Development, Roma, Italy
e-mail: marco.paoloni@enea.it

G. Cupertino
e-mail: giacomo.cupertino@enea.it

M. Ceccarelli
Laboratory of Robot Mechatronics (LARM2), Roma Tor Vergata University, Roma, Italy
e-mail: marco.ceccarelli@uniroma2.it

I. Doroftei et al. (eds.), *Proceedings of SYROM 2022 & ROBOTICS 2022*, Mechanisms and Machine Science 127, https://doi.org/10.1007/978-3-031-25655-4_15

This paper describes the initial stages of the analysis of the robot arms kinematics, foreseen by the project for the control system algorithms; after a preliminary description of the mechanical structure exposed in the second section, in the third one the TORVEastro arm is modeled according to the procedures exposed in [3] to describe its forward kinematics. In the fourth section is exposed the inverse kinematics with its solution and the analysis of the constraints that ensure its existence and uniqueness.

15.2 Robot Arm Description

As depicted in Fig. 15.1, the TORVEastro robot consists of a cylindrical central body, on which three identical arms, arranged symmetrically, are mounted; each of the three arms, being constituted by three rigid mechanical links, each articulated by a revolute joint, has three degrees of freedom and is equipped with a gripper as end-effector.

As depicted in Fig. 15.2, each TORVEastro arm constitutes a serial open kinematic chain, where the first link is connected to the central body by the first joint, the second link is connected to the first one by the second joint and the third link is connected to the second one by the third joint. The first joint allows the first link to rotate around an axis perpendicular to the body lateral surface, the second joint allows the second link to revolute around an axis intersecting the first joint one perpendicularly and the third joint allows the third link to revolute around an axis parallel to the second joint one. During the arms movements, the orthogonality of the first and second joint axes and the parallelism of the second and third ones are always conserved. At the end of the chain, the end-effector is rigidly connected to the last link without any other joint.

To describe the kinematics of one of the robot arms (see Fig. 15.3), a reference frame located on the central body and another located on the end-effector were adopted. The position of the end-effector of the arm k will be described by the origin of its reference frame, whose position is expressed in the body frame by the vector $\mathbf{p}_k$. Its orientation will be described by the three unit vectors of its frame axes directed

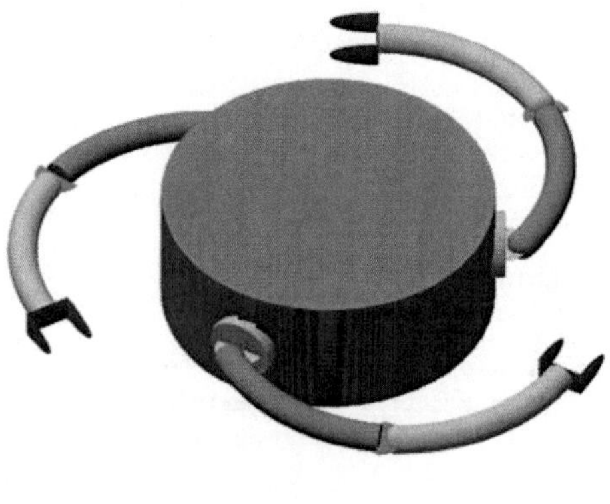

Fig. 15.1 Sketch of the TORVEastro robot

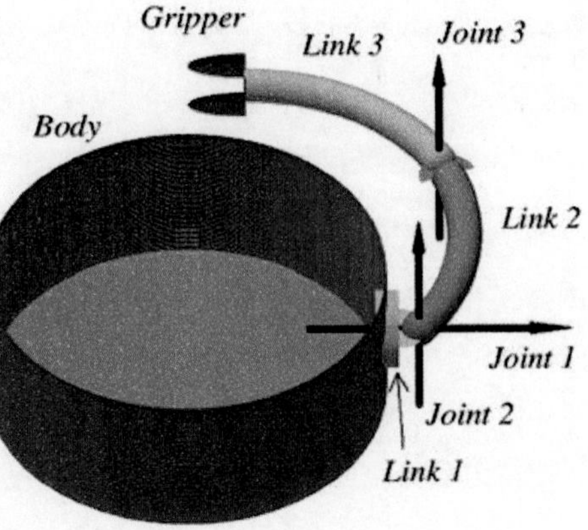

Fig. 15.2 Kinematic structure of the arm

as follows: one axis lies in the direction from which the gripper would approach an object and is expressed in the body frame by the unit vector $\mathbf{a}_k$, another axis is in the direction specifying the orientation of the gripper (from a finger to the other) and is expressed by the unit vector $\mathbf{s}_k$ and the last axis must form a right-handed reference frame and is expressed by the unit vector $\mathbf{n}_k$.

With the introduced four vectors it is possible to define a homogeneous matrix $\mathbf{Q}_k$ that expresses the transformation of the components of a homogeneous vector referred to the end-effector frame into the correspondent ones referred to the central body frame; this matrix will be expressed by

$$\mathbf{Q}_k = \begin{bmatrix} \mathbf{n}_k & \mathbf{s}_k & \mathbf{a}_k & \mathbf{p}_k \\ 0 & 0 & 0 & 1 \end{bmatrix} \tag{15.1}$$

For the body reference frame (see Fig. 15.4), since all three axes of the first joint of each arm lie on the same plane, the origin is placed on this plane at the intersection of these three axes; after having assigned a progressive numbering to the arms, the x_B axis is placed on the same plane, directed as the rotation axis of the first joint of the

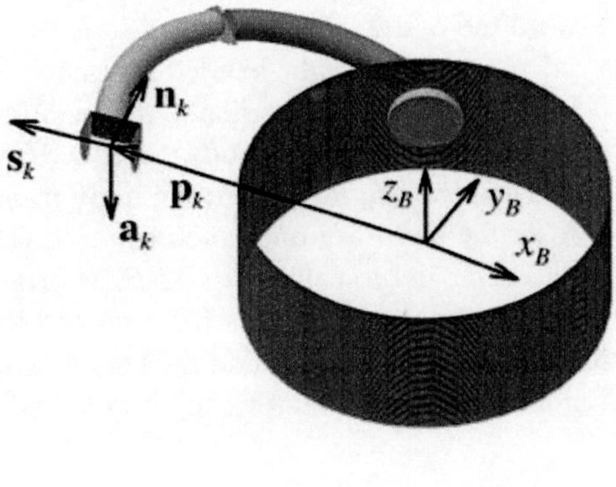

Fig. 15.3 End-effector reference frame

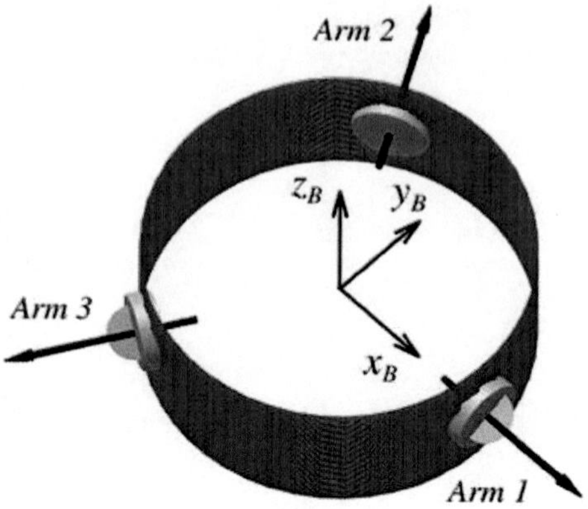

Fig. 15.4 Body reference frame

first arm, the z_B axis is directed as the normal to the common plane with its pointing direction chosen in order to see the three arms numbered counterclockwise and the y_B axis is disposed so that this reference frame is right-handed.

15.3 Forward Kinematics

To solve the arm forward kinematics, the four vectors $\mathbf{n}_k$, $\mathbf{s}_k$, $\mathbf{a}_k$, and $\mathbf{p}_k$ must be expressed as functions of the joint angles. To accomplish this task for a serial open kinematic chain [3], a set of reference frames (one for the central body and one for each link) is defined and for each frame is evaluated the homogeneous matrix that performs the transformation of the components of a homogeneous vector referred to that frame into the correspondent ones referred to the frame embedded in the previous link.

To apply the algorithm to the TORVEastro robot arms, three auxiliary body reference frames are adopted, one for each arm (see Fig. 15.5); for the arm k, the origin of the auxiliary frame is placed at the intersection of the rotation axes of the first and second joints, the z_{0k} axis is aligned with the rotation axis of the first joint pointing outward the body, the y_{0k} axis is aligned with the z_B axis and the x_{0k} axis is chosen in order to obtain a right-handed reference frame.

During the arms movements, the positions and the orientations of the auxiliary frames remain fixed in the body reference frame, the common distance between their origins and the origin of the main body frame is indicated with d_0.

After the auxiliary one, another reference frame is fixed on the first link of the arm k. As it can be seen in Fig. 15.6, its origin is the same of the auxiliary frame, the z_{1k} axis is the rotation axis of the second joint and the x_{1k} axis is aligned with the common normal to the z_{1k} and z_{0k} axes; the positive direction of the x_{1k} axis is chosen in order that its unit vector can be obtained by the cross product of the ones of the

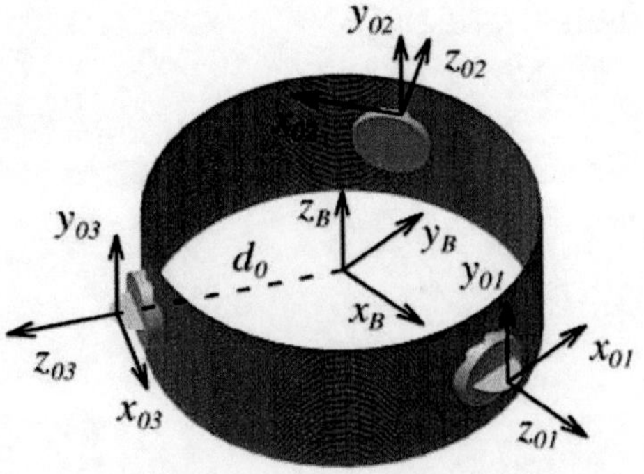

Fig. 15.5 Auxiliary body reference frames

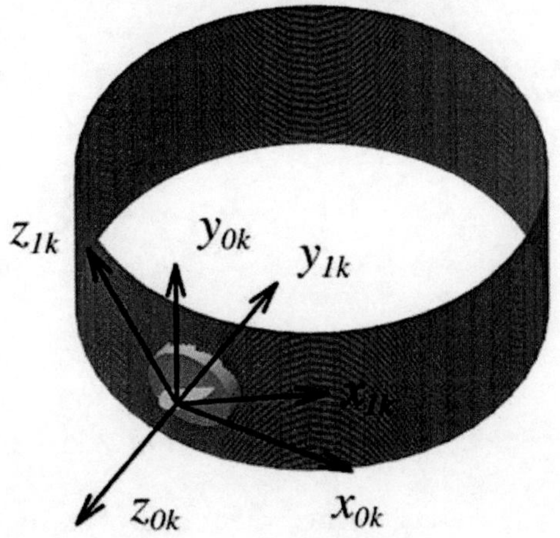

Fig. 15.6 First link reference frame

z_{1k} and z_{0k} axes. Finally, the y_{1k} axis is disposed to obtain a right-handed reference frame.

As it can be seen in Fig. 15.7, for the reference frame embedded in the second link of the arm k, the origin is placed on the intersection of the third joint rotation axis with its normal passing for the origin of the frame embedded on the first link; the z_{2k} axis is the rotation axis of the second joint and the x_{2k} axis is placed along the cited normal with positive direction from z_{1k} to z_{2k} and finally the y_{2k} axis is disposed to obtain a right-handed reference frame. The distance between the origins of the frames embedded in the first and second links is indicated with a_2.

Finally the reference frame embedded in the third link of the arm k is defined (see Fig. 15.8); its z_{3k} axis is aligned with the end-effector unit vector $\mathbf{a}_k$ and its x_{3k} axis is aligned with the normal to the z_{3k} axis passing for the origin of the frame embedded in the second link; the origin of this frame is then placed at the intersection of the x_{3k} and the z_{3k} axes and the positive direction of the x_{3k} axis is from the origin of the second link frame to the one of the third; the y_{3k} axis is disposed to obtain a right-handed reference frame.

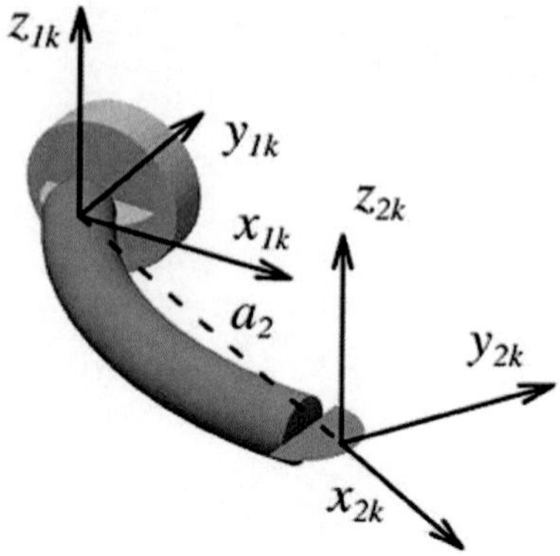

Fig. 15.7 Second link reference frame

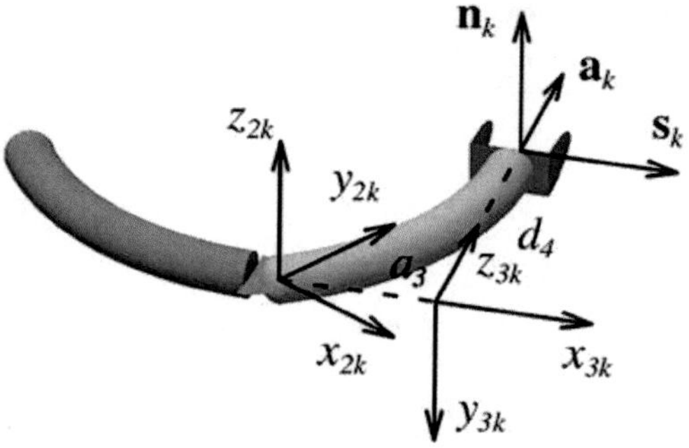

Fig. 15.8 Third link reference frame

The distance between the origin of the frame embedded in third link and the origin of the frame embedded in the second is indicated with a_3, while distance between the same origin and the origin of the frame embedded in the end-effector is indicated with d_4.

After the assignments of the reference frames, it will be possible to evaluate the matrices that expresses the transformation of the components of a homogeneous vector referred to a frame into the ones referred to the previous frame.

The joint angles ϑ_{jk} describing the rotations around the joint j axis of the arm k are adopted as joint coordinates; the links reference frames have been expressly assigned to define a joint angle as the angle from the $x_{(j\text{-}1)k}$ axis to the x_{jk} one, counted as positive for the counterclockwise verse around the $z_{(j\text{-}1)k}$ axis.

The transformation from the auxiliary body reference frame k into the main body reference frame will be performed by the matrix $\mathbf{S}_k$ expressed by

$$\mathbf{S}_k = \begin{bmatrix} -\sin\beta_k & 0 & \cos\beta_k & d_0\cos\beta_k \\ \cos\beta_k & 0 & \sin\beta_k & d_0\sin\beta_k \\ 0 & 1 & 0 & 0 \\ 0 & 0 & 0 & 1 \end{bmatrix} \tag{15.2}$$

where the parameter β_k will have the null value for the first arm, the value $2\pi/3$ for the second and the value $4\pi/3$ for the third.

Posing $c_{jk} = \cos\vartheta_{jk}$ and $s_{jk} = \sin\vartheta_{jk}$, the transformation from the link j frame of the arm k into the previous can be performed by the matrices $\mathbf{A}_j(\vartheta_{jk})$ expressed by

$$\mathbf{A}_1(\vartheta_{1k}) = \begin{bmatrix} c_{1k} & 0 & -s_{1k} & 0 \\ s_{1k} & 0 & c_{1k} & 0 \\ 0 & -1 & 0 & 0 \\ 0 & 0 & 0 & 1 \end{bmatrix} \mathbf{A}_2(\vartheta_{2k}) = \begin{bmatrix} c_{2k} & -s_{2k} & 0 & a_2 c_{2k} \\ s_{2k} & c_{2k} & 0 & a_2 s_{2k} \\ 0 & 0 & 1 & 0 \\ 0 & 0 & 0 & 1 \end{bmatrix} \tag{15.3}$$

$$\mathbf{A}_3(\vartheta_{3k}) = \begin{bmatrix} c_{3k} & 0 & -s_{3k} & a_3 c_{3k} \\ s_{3k} & 0 & c_{3k} & a_3 s_{3k} \\ 0 & -1 & 0 & 0 \\ 0 & 0 & 0 & 1 \end{bmatrix}$$

Finally, the transformation from the end-effector frame into the third link frame can be performed by the matrix $\mathbf{E}$ (the same for all the arms) expressed by

$$\mathbf{E} = \begin{bmatrix} 0 & 1 & 0 & 0 \\ -1 & 0 & 0 & 0 \\ 0 & 0 & 1 & d_4 \\ 0 & 0 & 0 & 1 \end{bmatrix} \tag{15.4}$$

With the use of these five matrices, the homogeneous transformation matrix introduced in Eq. (15.1) can be obtained by successive transformations from the end-effector frame to the body frame passing through the link frames and the auxiliary one, obtaining the expression

$$\mathbf{Q}_k(\vartheta_{1k}, \vartheta_{2k}, \vartheta_{3k}) = \mathbf{S}_k \cdot \mathbf{A}_1(\vartheta_{1k}) \cdot \mathbf{A}_2(\vartheta_{2k}) \cdot \mathbf{A}_3(\vartheta_{3k}) \cdot \mathbf{E} \tag{15.5}$$

that will be the solution for the forward kinematics problem.

Posing $\mathbf{T}_3(\vartheta_{1k}, \vartheta_{2k}, \vartheta_{3k}) = \mathbf{A}_1(\vartheta_{1k}) \cdot \mathbf{A}_2(\vartheta_{2k}) \cdot \mathbf{A}_3(\vartheta_{3k})$, this new matrix will be expressed by

$$\mathbf{T}_3(\vartheta_{1k}, \vartheta_{2k}, \vartheta_{3k}) = \begin{bmatrix} c_{1k}c_k & s_{1k} & -c_{1k}s_k & c_{1k}(a_3 c_k + a_2 c_{2k}) \\ s_{1k}c_k & -c_{1k} & -s_{1k}s_k & s_{1k}(a_3 c_k + a_2 c_{2k}) \\ -s_k & 0 & -c_k & -(a_3 s_k + a_2 s_{2k}) \\ 0 & 0 & 0 & 1 \end{bmatrix} \tag{15.6}$$

where it has been posed $c_k = \cos(\vartheta_{2k} + \vartheta_{3k})$ and $s_k = \sin(\vartheta_{2k} + \vartheta_{3k})$.

15.4 Inverse Kinematics

To solve the arm inverse kinematics, it is necessary to evaluate the joint angles that allow the end-effector position and orientation to assume a desired configuration expressed by the assignment of the four vectors $\mathbf{n}_k$, $\mathbf{s}_k$, $\mathbf{a}_k$, and $\mathbf{p}_k$. To accomplish this task for the one of the TORVEastro arms, Eq. (15.5) is rewritten as

$$\mathbf{T}_3^{-1}(\vartheta_{1k}, \vartheta_{2k}, \vartheta_{3k}) = \mathbf{E} \cdot \mathbf{Q}_k^{-1} \cdot \mathbf{S}_k \tag{15.7}$$

This is a set of twelve equations in three unknowns, but the orthogonality condition on the rotational part of the transformation matrices decreases the number of independent equations to six. Then, while searching the solutions of this equation, at least three constraints on the assigned position and orientation must also be found.

Observing that the matrix $\mathbf{S}_k$ can also be expressed using the unit vectors $\mathbf{x}_{0k}$, $\mathbf{y}_{0k}$ and $\mathbf{z}_{0k}$ of the auxiliary body reference frame axes as

$$\mathbf{S}_k = \begin{bmatrix} \mathbf{x}_{0k} & \mathbf{y}_{0k} & \mathbf{z}_{0k} & d_0\mathbf{z}_{0k} \\ 0 & 0 & 0 & 1 \end{bmatrix} \tag{15.8}$$

the right term of Eq. (15.7) after some manipulations can be expressed by

$$\mathbf{E} \cdot \mathbf{Q}_k^{-1} \cdot \mathbf{S}_k = \begin{bmatrix} \mathbf{s}_k \cdot \mathbf{x}_{0k} & \mathbf{s}_k \cdot \mathbf{y}_{0k} & \mathbf{s}_k \cdot \mathbf{z}_{0k} & \mathbf{s}_k \cdot (d_0\mathbf{z}_{0k} - \mathbf{p}_k) \\ -\mathbf{n}_k \cdot \mathbf{x}_{0k} & -\mathbf{n}_k \cdot \mathbf{y}_{0k} & -\mathbf{n}_k \cdot \mathbf{z}_{0k} & -\mathbf{n}_k \cdot (d_0\mathbf{z}_{0k} - \mathbf{p}_k) \\ \mathbf{a}_k \cdot \mathbf{x}_{0k} & \mathbf{a}_k \cdot \mathbf{y}_{0k} & \mathbf{a}_k \cdot \mathbf{z}_{0k} & \mathbf{a}_k \cdot (d_0\mathbf{z}_{0k} - \mathbf{p}_k) + d_4 \\ 0 & 0 & 0 & 1 \end{bmatrix} \tag{15.9}$$

while the left term of the same equation can be expressed by

$$\mathbf{T}_3^{-1}(\vartheta_{1k}, \vartheta_{2k}, \vartheta_{3k}) = \begin{bmatrix} c_{1k}c_k & s_{1k}c_k & -s_k & -a_3 - a_2c_{3k} \\ s_{1k} & -c_{1k} & 0 & 0 \\ -c_{1k}s_k & -s_{1k}s_k & -c_k & a_2s_{3k} \\ 0 & 0 & 0 & 1 \end{bmatrix} \tag{15.10}$$

By equating the correspondent elements in the rotational part of the matrices of these two equations, it is possible to obtain the five expressions

$$\mathbf{n}_k \cdot \mathbf{z}_{0k} = 0 \quad \begin{cases} c_{1k} = \mathbf{n}_k \cdot \mathbf{y}_{0k} \\ s_{1k} = -\mathbf{n}_k \cdot \mathbf{x}_{0k} \end{cases} \quad \begin{cases} c_k = -\mathbf{a}_k \cdot \mathbf{z}_{0k} \\ s_k = -\mathbf{s}_k \cdot \mathbf{z}_{0k} \end{cases} \tag{15.11}$$

The first expression is one of the cited constrains, while the others allow to obtain univocally the angle ϑ_{1k} and the sum of the angles ϑ_{2k} and ϑ_{3k}.

It must be observed that, under the found constraint, the expressions that bring the angles are well posed; as example, the trigonometric identity for the angle ϑ_{1k} can be expressed as

$$c_{1k}^2 + s_{1k}^2 = (\mathbf{n}_k \cdot \mathbf{y}_{0k})^2 + (\mathbf{n}_k \cdot \mathbf{x}_{0k})^2 = |\mathbf{n}_k|^2 - (\mathbf{n}_k \cdot \mathbf{z}_{0k})^2 = 1 \tag{15.12}$$

By equating the correspondent elements in the translational part of the matrices of Eqs. (15.9) and (15.10), it is possible to obtain the three expressions

$$\mathbf{n}_k \cdot (d_0\mathbf{z}_{0k} - \mathbf{p}_k) = 0 \begin{cases} a_2c_{3k} = -\mathbf{s}_k \cdot (d_0\mathbf{z}_{0k} - \mathbf{p}_k) - a_3 \\ a_2s_{3k} = \mathbf{a}_k \cdot (d_0\mathbf{z}_{0k} - \mathbf{p}_k) + d_4 \end{cases} \tag{15.13}$$

The first is another of the mentioned constrains, that can be reduced to $\mathbf{n}_k \cdot \mathbf{p}_k = 0$, while the others allows to obtain univocally the angle ϑ_{3k} and then the angle ϑ_{2k} from their sum; however, the two found constraints do not ensure that the trigonometric identity is verified for the angle ϑ_{3k} and then it must imposed, obtaining in this way the third constraint.

For this reason the expressions in Eq. 15.13 are rewritten as

$$\begin{cases} (\mathbf{p}_k - d_0\mathbf{z}_{0k} - a_3\mathbf{s}_k d_4\mathbf{a}_k) \cdot \mathbf{n}_k = 0 \\ (\mathbf{p}_k - d_0\mathbf{z}_{0k} - a_3\mathbf{s}_k - d_4\mathbf{a}_k) \cdot \mathbf{s}_k = a_2c_{3k} \\ (\mathbf{p}_k - d_0\mathbf{z}_{0k} - a_3\mathbf{s}_k - d_4\mathbf{a}_k) \cdot \mathbf{a}_k = -a_2s_{3k} \end{cases} \tag{15.14}$$

and the third constraint is then expressed by

$$|\mathbf{p}_k - d_0\mathbf{z}_{0k} - a_3\mathbf{s}_k - d_4\mathbf{a}_k| = a_2 \tag{15.15}$$

An obtained solution must also be further examined for its feasibility, it is possible that it may lead a link to close to the central body (or even inside). For this reason, safety acceptance limits have been defined: safe values for the angle ϑ_{1k} can range from −180 to 180°, for the angle ϑ_{2k} from −150 to 30° and for the sum $\vartheta_{2k} + \vartheta_{3k}$ from −120 to 60°.

15.5 Conclusions

The forward and the inverse kinematics of a single TORVEastro robot arm have been fully investigated; the obtained results will be the premise for one of the next developments of the TORVEastro research project, the coordination of two or all the three arms in the fulfillment of a task.

Acknowledgements The TORVEastro project activities reported in this paper were funded by Regione Lazio [Project 36517 POR FESR LAZIO 2014-20].

References

1. Samani, F., Ceccarelli, M.: Prototype design and testing of TORVEastro, cable-driven astronaut robot. In: Advances in Service and Industrial Robotics–Results of RAAD2020, pp. 448–455. Springer Nature Switzerland AG2020, Cham (2020)
2. Araque-Isidro, J.E., Cafolla, D., Ceccarelli, M.: Problems and requirements for outer space astronaut service robot. In: Advances in Italian Mechanism Science. IFToMM Italy 2022. Mechanisms and Machine Science, vol. 122, pp. 603–611. Springer, Cham (2022)
3. Ceccarelli, M.: Fundamentals of Mechanics of Robotic Manipulation–2nd edn (Series: Mechanisms and Machine Science). Springer (2022)

Chapter 16
Evaluation by Numerical Simulation of Friction Forces in Spherical Joints of a 6-DOF Parallel Topology Robot

C.-O. Miclosina, V. Cojocaru, and Z.-I. Korka

Abstract Friction forces occur in all operating mechanisms and machines. Usually, it is hard to determine the friction forces in an experimental way or in an analytical way. Friction forces values between the machine components may vary during operation, depending on the specific motion parameters and on the applied loads. An easier solution is to determine the friction forces by numerical simulation. Thus, by using a specialized software, the friction forces are determined as a result of the motion simulation of the mechanism's geometrical model. The paper presents the geometrical model of a parallel topology robot having the guiding device mechanism of $FP_3 + 6{\cdot}S\underline{P}S + MP_3$ (or MSSM) type. The mechanism components are modelled and kinematical joints are defined. For certain strokes of driving kinematical joints, the motion simulation is accomplished. As results of the simulation process, the friction forces values between the components of certain spherical joints are obtained. In order to achieve the geometrical model and the motion simulation, the SolidWorks software is used.

Keywords Geometrical model · Motion simulation · Friction forces · Solid · Works

16.1 Introduction

Friction forces influence the operation of mechanical systems. For operation at the desired parameters, it is recommended to take into account the friction forces in conditions as close as possible to the real ones.

C.-O. Miclosina · V. Cojocaru (✉) · Z.-I. Korka
Department of Engineering Science, Faculty of Engineering, UBB University Center in Reşiţa, Babeş-Bolyai University, Cluj-Napoca, Romania
e-mail: vasile.cojocaru@ubbcluj.ro

C.-O. Miclosina
e-mail: calin.miclosina@ubbcluj.ro

Z.-I. Korka
e-mail: zoltan.korka@ubbcluj.ro

I. Doroftei et al. (eds.), *Proceedings of SYROM 2022 & ROBOTICS 2022*, Mechanisms and Machine Science 127, https://doi.org/10.1007/978-3-031-25655-4_16

Friction is analyzed for both categories of robots from the link rigidity point of view–for robots with rigid links [1–5] and for robots with flexible links [6, 8, 20]. Different types of kinematical joints and mechanical transmissions [9–11] are geometrically modeled and analyzed.

Another friction modeling approach, applied to robot joints, is by using neural networks [7, 13].

Friction compensation is a closed-loop control method that improves the robot operation. For example, friction compensation is studied experimentally for the high-precision tracking control of heavy-duty parallel robots in [14], and for the case of a sensorless human–robot collaborative assembly in [15].

In certain cases, the friction forces must have high values, for example in the case of walking robots, between the leg and the ground, especially in wet conditions. Thus, in order to maximize the friction forces of robotic legs on an unknown surface, a granular media friction pad is developed, containing a thin elastic membrane which encases loosely filled granular material [12].

This paper presents a method to evaluate by numerical simulation the values of friction forces that occur in certain spherical kinematical joints of a parallel topology robot.

16.2 The Geometrical Model

The structure of the parallel topology guiding device mechanism is $FP_3 + 6\ S\underline{P}S + MP_3$ [16–18], equivalent to MSSM [19].

The mobile platform MP_3 is linked to the fixed one FP_3 by 6 kinematical chains SPS, each one having 2 binary links, 1 prismatic joint *P* and 2 spherical joints *S*. The centers of the spherical joints that belong to different kinematical chains SPS are coincident 2 by 2, thus composing double spherical joints. The mechanism has 6 degrees of freedom (DOF), and the driving kinematical joints are the prismatic ones $\underline{P}$, underlined in the symbolical notation of the robot's structure [16–18].

In order to obtain the assembly geometrical model, *concentric mates* were used for relative positioning the kinematical joints links.

The geometrical model, accomplished in SolidWorks software, is presented in Fig. 16.1. The model being proposed for a flight simulator, the mobile platform carries the human operator on a seat. Oxyz is the fixed reference system, C_m–the center of mass of the human operator model, and F–the frontal point of the human operator model. The SPS identical kinematical chains are denoted with numbers from 1 to 6.

For the same kinematical chain SPS, the minimum distance between the centers of the 2 spherical joints is 800 [mm] and the maximum distance is 1300 [mm]. Thus, the maximum stroke of the driving prismatic joint is set to 500 [mm], as in previous work [16–18].

The geometrical models of human operator, chair, and mobile platform MP_3 are fixed together.

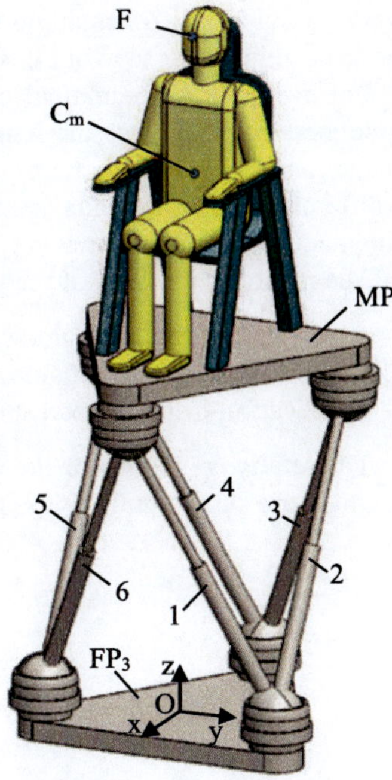

Fig. 16.1 The geometrical model of the parallel topology guiding device, with the mobile platform carrying the human operator [16–18]

The materials of the assembly components are chosen from the SolidWorks library. For the geometrical model of the human operator, which is meant to be an average one–with the height of 1.75 [m] and the mass of 75 [kg], a custom material is defined. This material has the mass density of 1007.27 [kg/m^3] [16–18].

16.3 Motion Simulation

The simulation of robot movement is accomplished in the SolidWorks Motion module, as in previous work [16–18].

The gravitational acceleration is set to the value of 9806.65 [mm/s^2].

The maximum friction forces occur between the surfaces of the components of the lowest spherical joints, due to the fact that they bear the weight of the entire assembly, without the fixed platform. Consequently, the friction forces between the surfaces of spherical shells attached to the fixed platform and the lower nuts of the kinematical chains 1, 3, and 5 are analyzed.

In the previous work [16–18], the friction forces were neglected in the simulation process. In order to determine the friction forces in the spherical joints analyzed, the friction properties in concentric mate properties, and the contacts between the components are defined, as shown in Fig. 16.2a and b respectively. The default friction coefficients and elastic properties of the component's materials, given by software, were used. Figure 16.2c, shows the maximum angle that defines the spherical joint workspace, in order to avoid link collisions.

For every SPS kinematical chain, in order to drive the motion, a linear motor is defined. It establishes the length between the centers of the two spherical joints of the same kinematical chain, and it develops a maximum speed limit of 1 [m/s]. 3 different motions are analyzed, based on the initial and final lengths of the SPS kinematical chains, as shown in Table 16.1.

The resulting motions of the mobile platform are as follows:

1 an advance to the front side combined with the elevation of the back side;
2 an elevation of the right side—a roll from right to left;
3 an elevation from the lowest to the uppermost position.

The relative positions of the parallel topology robot components for motions 1, 2, and 3, are shown in the Figs. 16.3, 16.4, and 16.5, respectively.

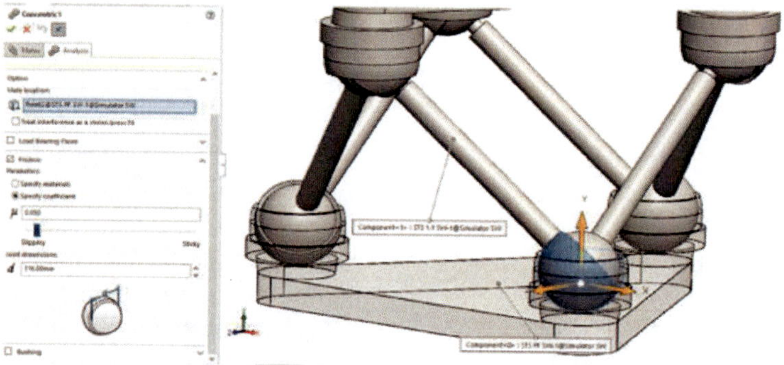

(a) Concentric mate properties

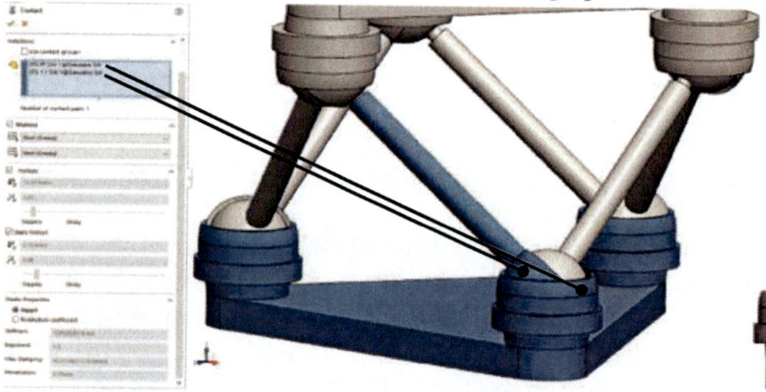

(b) Contact properties

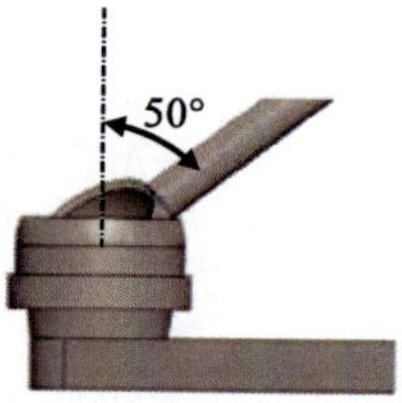

(c) Maximum angle

Fig. 16.2 Setting properties for the spherical joints analyzed

Table 16.1 Variation of lengths $L_1 \div L_6$, between the centers of the spherical joints for the 3 motions

Length	Motion 1		Motion 2		Motion 3	
	Initial position (t = 0 s)	Final position (t = 0.5 s)	Initial position (t = 0 s)	Final position (t = 0.5 s)	Initial position (t = 0 s)	Final position (t = 0.5 s)
L_1 [mm]	1300	1300	800	800	800	1300
L_2 [mm]	1000	1300	800	800	800	1300
L_3 [mm]	900	1300	800	800	800	1300
L_4 [mm]	900	1300	800	1000	800	1300
L_5 [mm]	1000	1300	800	1000	800	1300
L_6 [mm]	1300	1300	800	1000	800	1300

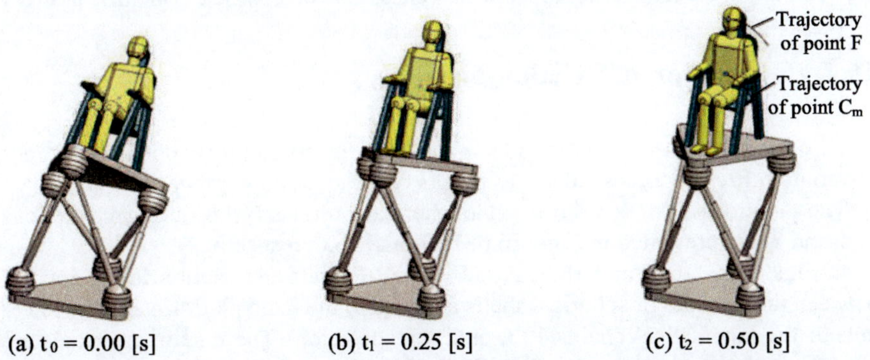

Fig. 16.3 Initial, intermediate and final positions-orientations of the mobile platform in motion 1

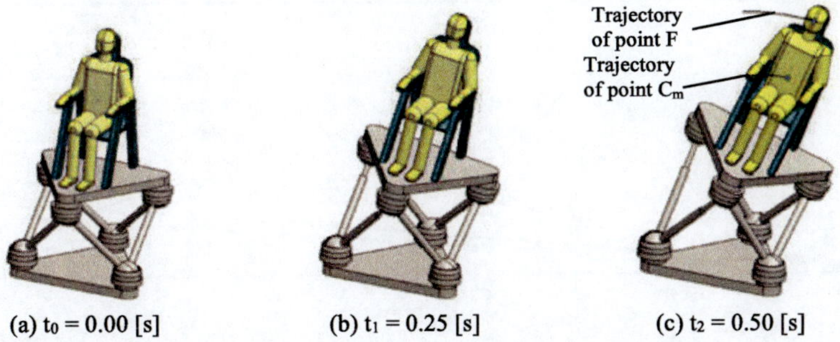

Fig. 16.4 Initial, intermediate and final positions-orientations of the mobile platform in motion 2

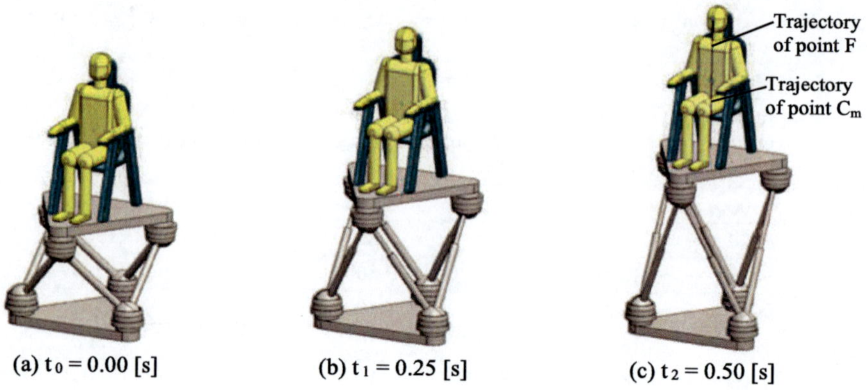

Fig. 16.5 Initial, intermediate and final positions-orientations of the mobile platform in motion 3

16.4 Evaluation of Friction Forces

The Motion module of SolidWorks software offers the possibility to obtain the graphs of variation friction forces values, as results of the simulation process.

The variations of friction forces in the spherical joints analyzed, during the motions 1, 2, and 3, are presented in Figs. 16.6, 16.7, and 16.8, respectively.

In Figs. 16.6, 16.7, and 16.8, F_{fs1}, F_{fs3}, and F_{fs5} are the friction forces moduli between the surfaces of spherical shells attached to the fixed platform and the lower nuts of the kinematical chains 1, 3, and 5, respectively. The maximum values are taken from the results provided SolidWorks Motion module.

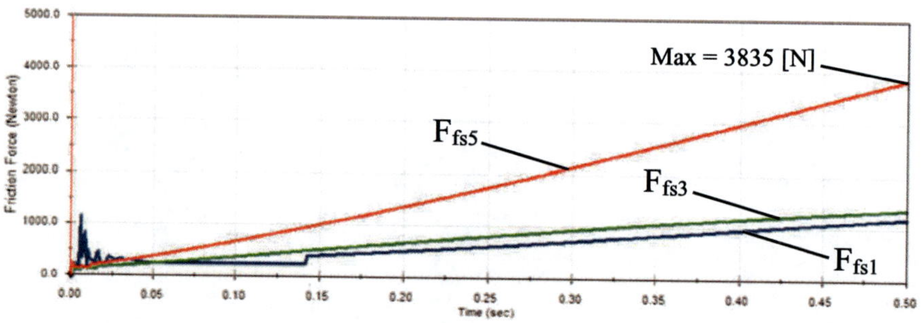

Fig. 16.6 The variations of friction forces modulus between the surfaces of components of spherical joints analyzed, during motion 1

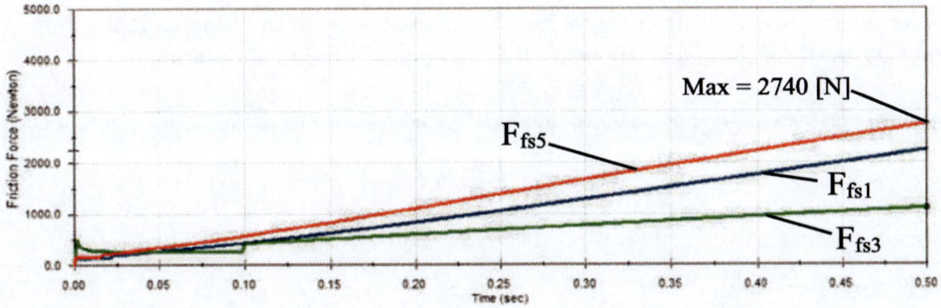

Fig. 16.7 The variations of friction forces modulus between the surfaces of components of spherical joints analyzed, during motion 2

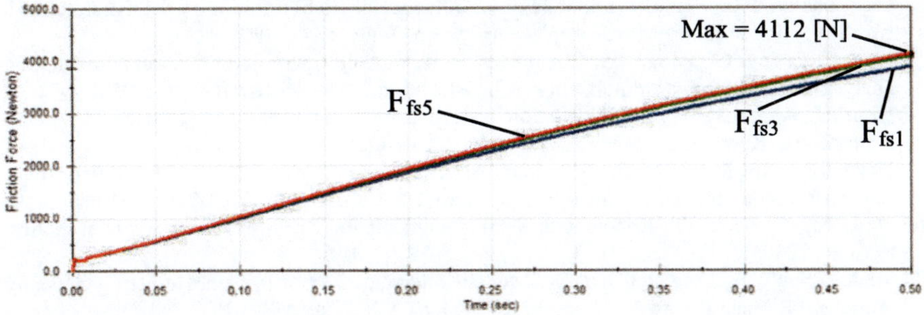

Fig. 16.8 The variations of friction forces modulus between the surfaces of components of spherical joints analyzed, during motion 3

16.5 Conclusions

The paper proposes a method of evaluation of friction forces in spherical joints during motion, by using SolidWorks software.

In the simulation process, interferences between the components of spherical kinematical joints with defined contacts are observed. This is the reason why the concentric mates are superposed to the contacts. When only the concentric mates are defined, there are no friction forces values provided as results.

It can be observed that the maximum values of friction force moduli occur in the lowest spherical joint of the kinematical chain no. 5, in all motions analyzed.

A possible explanation for the peaks of the graph F_{fs1} in Fig. 16.6 is the fact that the spherical shell may turn around the longitudinal axis of the links of kinematical chain no. 5, and thus the edge of the partial nut in contact with the spherical shell may influence the friction smoothness.

As further research, friction forces can be evaluated in the kinematical joints of other types of mechanisms, by using this manner of motion simulation.

Acknowledgements The publication of this article was supported by the 2022 Development Fund of the UBB.

References

1. Aarts, R.G.K.M., Jonker, B.J.B., Waiboer, R.R.: Realistic dynamic simulation of an industrial robot with joint friction. In: Informatics in Control Automation and Robotics I, pp. 187–194. (2006)
2. Chernous'ko, F.L., Shunderyuk, M.M.: The influence of friction forces on the dynamics of a two-link mobile robot. PMM J. Appl. Math. Mech. **74**(1), 13–23 (2010)
3. Guo, F., Cheng, G., Pang, Y.S.: Explicit dynamic modeling with joint friction and coupling analysis of a 5-DOF hybrid polishing robot. Mech Mach Theory 167:104509 (2022)
4. Kim, T.J., Ahn, K.H., Song, J.B.: Friction model of a robot joint considering torque and moment loads. Trans. Korean Soc. Mech. Eng. A 45(1), 27–34 (2021)
5. Nevmerzhitskiy, M.N., Notkin, B.S., Vara, A.V., Zmeu K.V.: Friction model of industrial robot joint with temperature correction by example of KUKA KR10. J. Robot. **2019**, 6931563 (2019)
6. De Luca, A., Book, W.: Robots with flexible elements. In: Springer Handbook of Robotics. Springer (2008)
7. Tu, X., Zhou, Y.F., Zhao, P., Cheng, X.: Modeling the static friction in a robot joint by genetically optimized BP neural network. J. Intell. Robot. Syst. **94**(1), 29–41 (2019)
8. Wollhaf, K., Buxbaum, H.-J.: Simulation of robots with flexible links. In: Robot Control 1991–IFAC Symposia Series, 3rd IFAC/IFIP/IMACS Symposium, pp. 327–332 (1992)
9. Cojocaru, V., Miclosina, C.-O.: Numerical analysis of the influence of clearance on stress state and contact pressure in plain bearings. Robot. Manag. **22**(2), 4–7 (2017)
10. Miclosina, C.-O., Cojocaru, V., Vela, D.-G.: Friction forces in numerical simulations of kinematical joints of mechanical systems. Robot. Manag. **26**(1), 9–12 (2021)
11. Duan, L., An, Z., Yang, R., Fu, Z.: Mechanical model of coupling rolling and sliding friction in real-time non-clearance precision ball transmission. Tribol. Int. **103**, 218–227 (2016)
12. Tramsen, H.T., Heepe, L., Gorb, S.N.: Granular media friction pad for robot shoes-hexagon patterning enhances friction on wet surfaces. Appl. Sci. Basel **11**(23), 11287 (2021)
13. Suchy, J., Skrinarova, J., Trhan, P.: Neural networks modelling of friction in robot joints. In: Proceedings of the Eighth International Symposium on Measurement and Control in Robotics, pp.157–162 (1998)
14. Sancak, K.V., Bayraktaroglu, Z.Y.: Observer-based friction compensation in heavy-duty parallel robot control. J. Mech. Sci. Technol. 35(8), 3693–3704 (2021)
15. Xiao, J.L., Dou, S.X., Zhao, W., Liu, H.T.: Sensorless human-robot collaborative assembly considering load and friction compensation. IEEE Robot. Automat. Lett. **6**(3), 5945–5952 (2021)
16. Miclosina, C.-O., Cojocaru, V., Korka Z.-I.: Dynamic simulation of a parallel topology robot operation. Appl. Mech. Mater. **762**, 107–112 (2015)
17. Miclosina, C.-O., Cojocaru, V., Korka, Z.: Analysis of a 6-DOF parallel robot motion simulation. In: International Conference on Applied Sciences 2019, Journal of Physics: Conference Series, vol. 1426, p. 012003 (2020)

18. Miclosina, C.-O., Korka, Z.-I., Cojocaru, V.: Evaluation by simulation of reaction forces that occur in spherical joints of parallel topology robots. In: Joint International Conference of the International Conference on Mechanisms and Mechanical Transmissions and the International Conference on Robotics–MTM & Robotics 2020, published in Mechanisms and Machine Science book series, vol. 88, pp. 226–234 (2021)
19. Merlet, J.-P.: Parallel robots (Series: Solid Mechanics and Its Applications). Springer (2006)
20. Zhang, Q., Liu, X.F., Cai, G.P.: Dynamics and control of a flexible-link flexible-joint space robot with joint friction. Int. J. Aeronaut. Space Sci. **22**(2), 415–432 (2021)

Chapter 17
Comparative Analysis of Four Solutions for Command and Control Applied to Anthropomorphic Grippers for Robots

I. Staretu and C. Moldovan

Abstract This paper proposes a solution for interconnecting the human hand with a virtual hand, from which the data is transmitted to a real anthropomorphic gripper, which performs accurately and safely gripping operations of various objects in shapes and sizes. We present four solutions for capturing the configurations of the human hand, namely the use of: a data glove made using bending sensors, a webcam, the Kinect sensor and the Motion Leap sensor. For each solution, some specific essential aspects are briefly indicated. Then a comparative study of the four solutions is presented, based on a test to identify eight different configurations of the human hand, to determine the best solution, which turned out to be the use of the Motion Leap sensor.

Keywords Data gloves · Bending sensors · Kinect sensor · Motion leap sensor · Webcam · Comparative analysis

17.1 Introduction

Command and control of an anthropomorphic gripper (with fingers) is a complex problem of great importance because performances of the gripper and obviously gripping safety depend on that. In the present, existing methods of an anthropomorphic gripper control can be grouped in two classes: methods based on motion capture using capture devices of the data gloves type and methods based on motion capture using digital imaging processing algorithms. Approaches based on usage of data gloves can digitize movements of human hand and fingers in the input parameters for a virtual reality system of. Approaches based on the use of image processing algorithms, involve image capture or depth data capture devices.

I. Staretu (✉) · C. Moldovan
Transilvania University of Brașov, Brașov, Romania
e-mail: staretu@unitbv.ro

I. Staretu
Romanian Academy of Technical Sciences, Bucharest, Romania

I. Doroftei et al. (eds.), *Proceedings of SYROM 2022 & ROBOTICS 2022*, Mechanisms and Machine Science 127, https://doi.org/10.1007/978-3-031-25655-4_17

The disadvantage of approaches based on image processing algorithms on accuracy is that it is getting outdated with the evolution of image capture devices. In the development of such a system's architecture, we took into account a minimum of requirements which a software system of human hand gesture recognition must comply with in order to be considered good. These requirements [1] are divided in: *functional requirements*: they represent a collection of services which a system must offer: detection and tracking; *non–functional requirements*–they represent the minimum quality of a service: latency; resolution and stability. The methods presented in this paper are considered successful because they meet both functional and non–functional requirements.

17.2 General Structure of the System Proposed

The structure of the control and command system of the anthropomorphic gripper proposed, first solve the control problem of a virtual anthropomorphic gripper from which all the data are transferred to the actual gripper. As a result, it was established that this system consist of: data capture equipment, which may be: Digital Glove data glove, web camera, Microsoft Kinect device or Motion Leap sensor, two data processing software applications: Interface Controller and Hand Commander that transmit data in the virtual hand, practically a virtual simulator (HandSIM), according to Fig. 17.1.

The Interface Controller application processes the movements of the human hand for each phalanx as well as actions of gripping and releasing an object. Hand Commander transmits data to the serial port and facilitates data to be read from the serial port and works with ViewComponent–a visual component that uses the GraspIT application [2] that allows the reception of signals and their transformation into commands for a virtual gripper, from which data is transmitted to the real gripper. The following are the analyzed modalities for the detection of the human

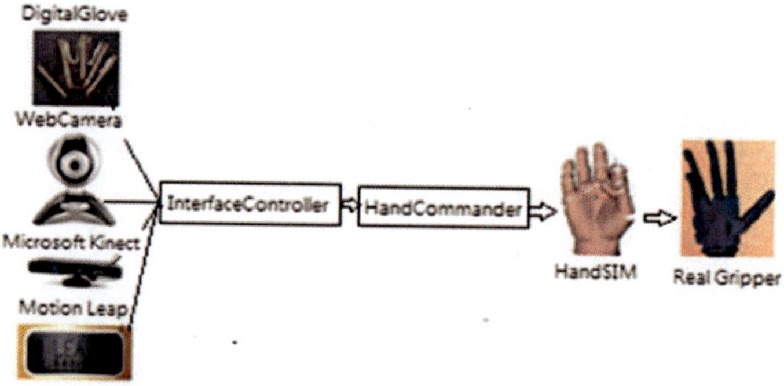

Fig. 17.1 The structure of the control and command system

hand, its movements and the recognition of gestures and the comparative analysis of the analyzed solutions.

17.3 Command and Control with Data Gloves

In the manufacture data gloves, Cyber Glove Systems is the market leader. They provide the most sophisticated and full solutions for capturing hand motion and its implementation in the three-dimensional environment. The products offered by Cyber Glove Systems [3] allow detailed capture of hand and fingers movement.

A variant of the data gloves is the one with bending sensors (Fig. 17.2a) [4]. The five bending sensors are used for each finger with the purpose of bending detection. At the moment of each finger bending, the sensors are bent in the same angle as the finger. All the 5 sensors are connected to the microcontroller monitoring each port separately continuously. Once the signal received by the microcontroller, it is analyzed and transmitted to the computer serial port. The computer will receive the signal used in the virtual environment for the motion of a virtual anthropomorphic gripper's phalanges.

The system implemented for capturing data using bending sensors has four main components (Fig. 17.2b): data glove with five sensors–DigitalGlov: it allows capturing movement data from the human hand and their transfer to the plate Arduino Leonardo; ATmega32u4 microcontroller (using Arduino Leonardo implementation): it allows analyze and data transferring to a computer; the Hand Commander component: transforms data received by a data glove, interprets and transmits them to the visualization component; the ViewComponent: it allows visualisation of human hand movements through GraspIT application [2].

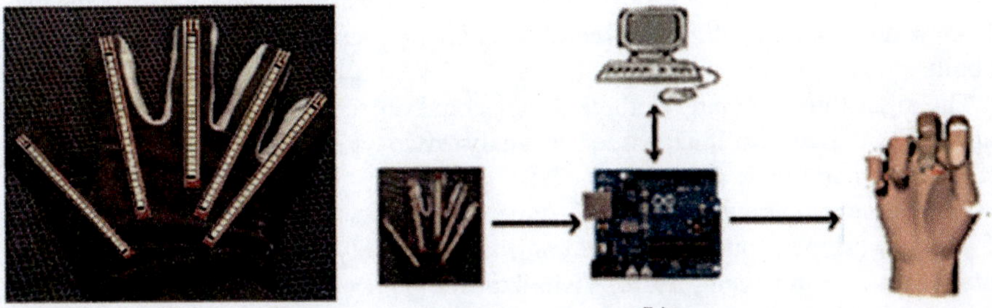

(a) (b)

Fig. 17.2 Glove with bending sensors (**a**) and control diagram with data glove control (**b**)

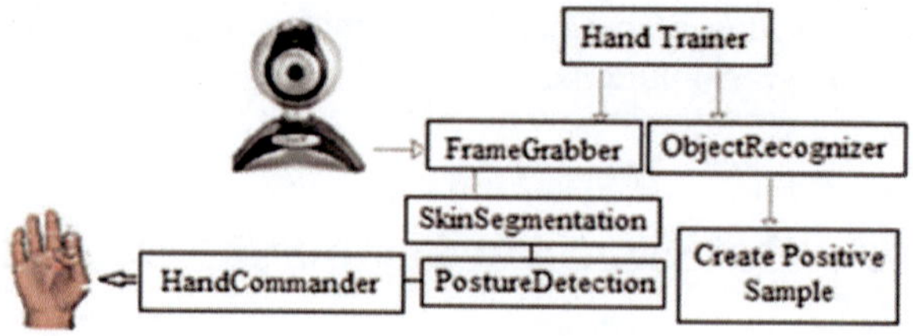

Fig. 17.3 Webcam virtual hand control system components

17.4 Webcam Command and Control

Approaches based on data capture and analysis using image capture sensors were investigated in detail. Initially, direct techniques for detection of skin color pixels in pictures were elaborated, if the images have very large resolution [5]. As analyze of very large resolution images requires very high processing power, in time, other solutions have been developed. One of these solutions that presents promising results, on images with relatively average resolution, is represented by segmentation techniques, which consist in the the establishment of an initial color limit [6]. Using this technique to implement the human hand posture detection system, also using the data presented in [7, 8], an architecture based on functional components was proposed. Figure 17.3 shows the general scheme of the system architecture. The system consists of five modules, namely: **Frame Grabber**; **Skin Segmentation**; **Posture Detection**; **Control Hand**; **Display Module** [9].

17.5 Command and Control with Kinect Sensor

The Microsoft Kinect sensor [10] combines a set of hardware and software mechanisms through which a digital, three-dimensional representation of a physical scene is built.

The algorithms created for three-dimensional object recognition, based on data captured by the Microsoft Kinect sensor, analyze two types of data in parallel, namely: RGB image and image depth data.

For the development of the three-dimensional data capture and control system of a virtual gripper, four steps were considered, namely: **Initialization**; **Detection**; **Interpretation and recognition**; **Visualization** [11].

Simulation of gestures: closed hand and open hand as well as the action of moving the hand corresponding to the actions of gripping, releasing and moving the object are visible in Fig. 17.4.

In Fig. 17.4 we can visualize depth image, RGB image, user's right and left hand. Before the system initialization, all important boolean values for parameters hand closed, hand open and value that says if hand is in motion, are set on *false*.

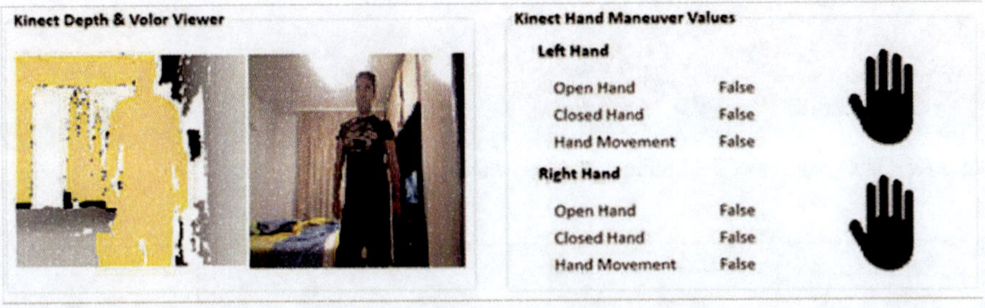

Fig. 17.4 The visual interface of the application for detecting the movement and gestures of the human hand

17.6 Command and Control with Motion Leap Sensor

Compared to the Microsoft Kinect sensor which, with decreasing the 1 m distance between sensors and user makes the recognition of gestures unstable, MotionLeap sensor provides a new hardware system much more advanced.

Information about a user's hand, fingers and gestures, can be recognized if the human hand is between 25 and 500 mm above center of the sensor (Fig. 17.5a). Compared to the accuracy of Microsoft Kinect sensor, the accuracy with which the MotionLeap sensor is endowed, is of submillimeter order, as specified in the MotionLeap documentation, more precisely 0.01 mm [12]. One of the great advantages MotionLeap sensor is endowed with, compared to its predecessors, is that it can distinguish each finger separately. The positional data, which can be captured for each object separately, are given with a submillimeter accuracy on each coordinate axis, X, Y, Z, with respect to the center of the device (Fig. 17.5b).

For hand recognition, more data sequences captured with MotionLeap device were analyzed. These sequences were captured in the differently lighting conditions (naturally or artificial light) and from different viewpoints. In Fig. 17.6 we can observe the gestures used in the recognition test.

When one of the gestures has been detected, the gesture detection module notifies the **Hand Commander** functional simulation component for transmitting the gesture execution data in the virtual environment. As results of the system execution, the gestures detected in Fig. 17.6 are used to control a virtual hand (Fig. 17.7) [13].

Fig. 17.5 Device center and coordinate axes (**a**); Interaction cube (**b**) [10]

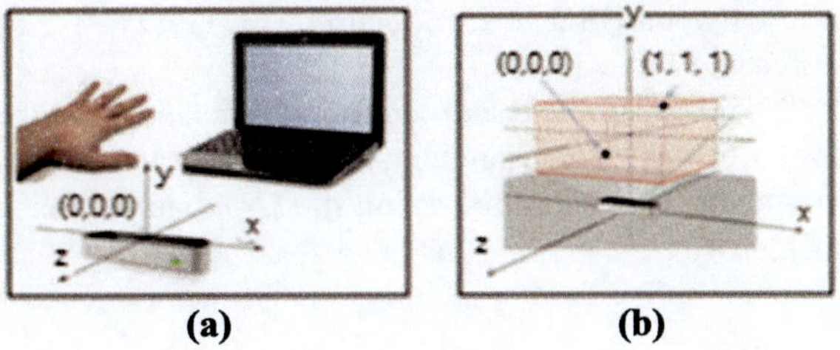

Fig. 17.6 The gestures of the human hand for which recognition is sought

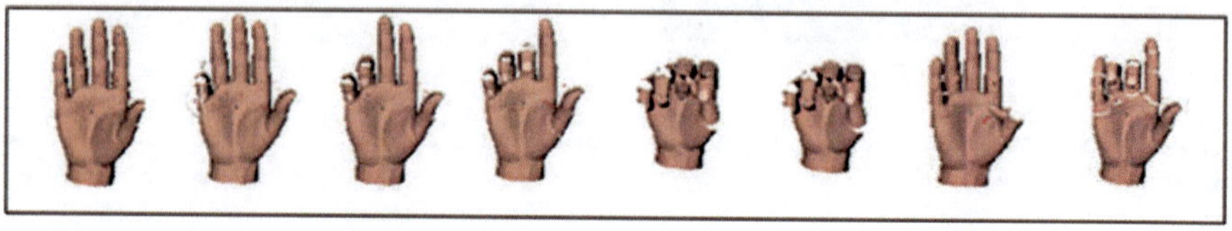

Fig. 17.7 Virtual hand gestures

17.7 Critical Comparative Evaluation of the Proposed Solutions for Command and Control

To evaluate the ability of the equipment to recognize the same gestures repeatedly and with high accuracy, the following test is performed: the human hand will move cyclically through the gestures shown in Fig. 17.6, time between a gesture execution and shifting to the other gesture being 2 s. After each posture is executed, the user's hand will be moved in the open hand position, which is considered initial position. The test is considered final when all positions are carried out and processed by the system. The measurement process is executed in a loop of 20 cycles, the human hand will take a particular position p[i], and the measurement process is considered ended when all P positions are measured, using the four solutions proposed, similar to sensors: DataGlove, webcamera, Microsoft Kinect and MotionLeap. These sensors are used to capture a certain framework and to detect a particular position p[i]. The outcome the measurement process is represented in a set of tuplets T, where each tuplet t has the following attributes: the image of the human hand captured using an RGB camera and the outcome obtained through the transmission of algorithm capture data to the virtual environment. Using these tuplets, a visual comparison can be made taking into account the positions executed.

The detection accuracy of a position p [i] is calculated using the ratio [ISO 5725]: *accuracy* = (TP + TN)/(TP + FP + FN + TN), and the accuracy is given by the ratio between [ISO 5725]: *accuracy* = TP/(TP + FP), where TP, TN, FP and FN represent (Table 17.1).

The results of the test execution are presented in Fig. 17.8a–d. For each hand position in the virtual environment, in addition to accuracy and precision, the repeatability coefficient was measured using the Cronbach alpha internal consistency indicator [14].

Table 17.1 Possible test results

	Positive test	Negative test
Position is known to the measurement algorithm	TP–position is known to the recognition system and the result is what the user expects–the virtual hand is moved to the position of the human hand	FN–the position is known to the recognition algorithm but it is unable to recognize the gesture–the virtual hand position is set to the initial position
The position is unknown to the measurement algorithm	FP–position is unknown to the recognition algorithm, but the algorithm determines however a moving position of the hand–the position of the virtual hand differs from the position of the user's hand	TN–position is unknown to the system and the result is what the user expects–virtual hand is moved in the the initial position

For a system or algorithm to be considered efficient, the accuracy, precision and repeatability must be very high, accuracy >0.9, precision >0.8 and repeatability must be = Excellent.

Following the test performed, the following observations can be made on how to capture human hand gestures using the different methods used for capture.

Using **the DataGlove** (see Fig. 17.8a): accuracy and precision are very high and the algorithm performs very well in terms of repeatability.

Using **CameraWeb** (Fig. 17.8b): in terms of accuracy and precision, the algorithm cannot effectively find the proposed gestures for recognition, and in addition, the repeatability coefficient is very small, which leads to the conclusion that the algorithm is not stable and cannot be used for control a virtual gripper. Using the **Microsoft Kinect device** (Fig. 17.8c): in terms of precision and accuracy, the proposed algorithm repeatedly finds two of the 8 proposed gestures, so it can be concluded that the Kinect sensor can be used exclusively for capturing type 1 and type 6 gestures. Using the **MotionLeap device** (Fig. 17.8d): in terms of accuracy and precision, the device and software application can recognize gestures efficiently and repeatedly.

17.8 Conclusions

Following the experimental results and the tests performed, the following conclusions were drawn:

– for a hand tracking, the user must be at least 1 m away from the Microsoft Kinect sensor, but the application, based on the API provided by Microsoft, cannot effectively detect the fingers, so it cannot be used in recognition complex gestures due to the low resolution of the Microsoft Kinect sensor, which, however, works very

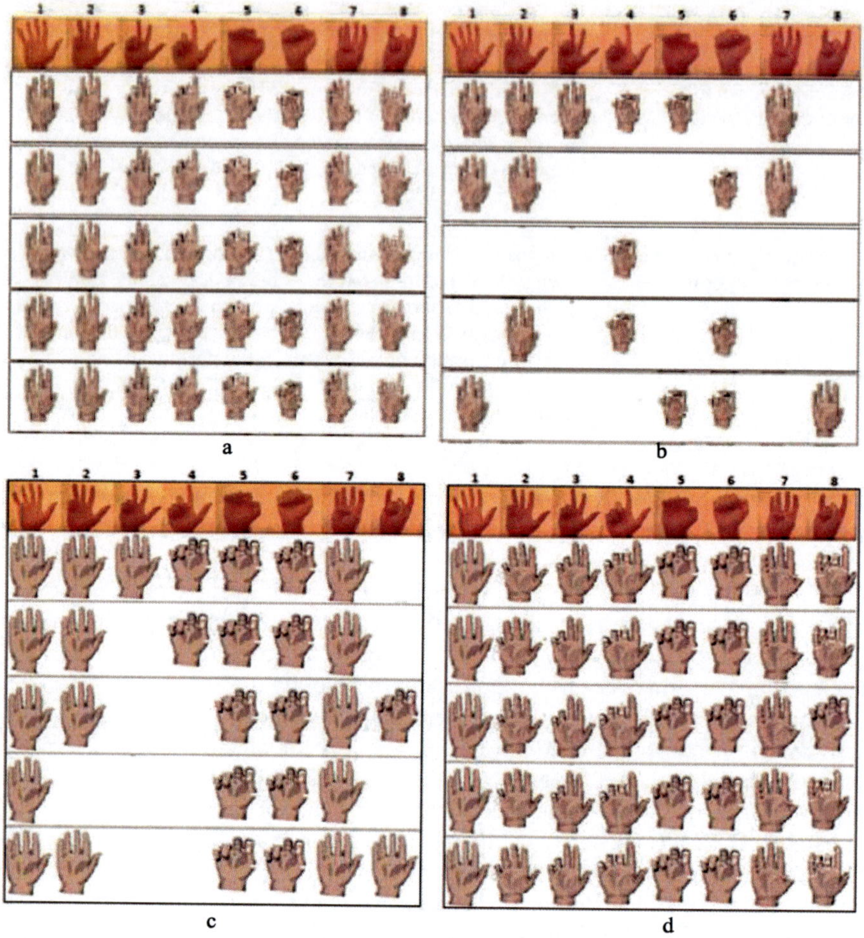

Fig. 17.8 System execution results using: data glove (**a**), WEB camera (**b**), Microsoft Kinect device (**c**), MotionLeap device (**d**)

well if you want to track the human body or a robotic arm or if you want to see the hand closed/open;

- with the help of the Microsoft Kinect sensor, you can still perform a power gripping operation, using the whole hand;
- the MotionLeap device allows a natural interaction between the user and the virtual reality at the level of gestures, this being comparable in accuracy with the interaction based on data gloves.

References

1. Hardenberg Von, C., Berard, F.F.: Bare hand human computer interaction. In: Proceedings of the 2001 Workshop on Perceptive user interfaces, pp.1–8 (2001)
2. Miller, A.T., Allen, P.K.: GraspIt!: a versatile simulator for grasp analysis. IEEE Robot. Autom. Mag. **11**(4), 110–122 (2004)
3. Information on: http://www.cyberglovesystems.com/. Accessed June 2022
4. Staretu, I., Moldovan, C.: Bending sensor used to command a virtual anthropomorphic gripper for robots. Acta Electrotehn. Special Issue **60**(1), 157–162 (2018)
5. Vezhnevets, V., Sazonov, V., Andreeva, A.: A survey on pixel-based skin color detection techniques. In: Proc. Graphicon 2003, Moscow, Russia, pp. 85–92 (2003)
6. Gu, J., et al.: Compressive structured light for recovering inhomogeneous participating media. IEEE Trans. Pattern Anal. Mach. Intell. **35**(3), 1–1 (2013)
7. He, Z., Tan, T., Sun, Z.: Topology modeling for adaboost-cascade based object detection. J. Pattern Recogn. Lett. **31**(9), 912–919 (2010)
8. Yogarajah, P., Condell, J., Curran, K., Cheddad, A.A., McKevitt, P.: A dynamic threshold approach for skin segmentation in color images. In: Proceedings of the 2010 IEEE International Conference on Image Processing, pp. 2225–2228 (2010)
9. Moldovan, C.C., Staretu, I.: Capturing human hand movements with a webcam to control an anthropomorphic gripper. Proc. Manuf. **22**, 519–526 (2018)
10. Information on: https://news.microsoft.com/2010/06/13/kinect-for-xbox-360-is-official-name-of-microsofts-controller-free-game-device/. Accessed June 2022
11. Moldovan, C.C., Staretu, I.: Analysis of the accuracy and robustness of kinect sensor used with the scope of robot manipulation. Appl. Mech. Mater. **555**, 170–177 (2014)
12. Weichert, F., Bachmann, D., Rudak, B., Fisseler, D.: Analysis of the accuracy and robustness of the leap motion controller. Sensors (Basel) **5**(5), 6380–6393 (2013)
13. Moldovan, C.C., Staretu, I.: An anthropomorphic hand with five fingers controlled by a motion leap device. Proc. Eng. **181**, 575–582 (2017)
14. Cronbach, L.J.: Coefficient alpha and the internal structure of tests. Psychometrika **16**, 297–334 (1951)

Chapter 18
Model-Free Reaching of a 2-DOF Robotic Arm Using Neural Networks

A. I. Călin, C. A. Popa, C. Moldovan, and A. Davidescu

Abstract In this paper, we present our results when using a Regression Deep Neural Network in an attempt to position the end-effector of a 2 Degrees of Freedom robotic arm to reach the target. We first train the DNN to understand the correspondence between the target position and the joint angles, and then we use the trained neural network to predict the joint angles such that the target position is reached. This method is an alternative to the classic Direct and Inverse Kinematic approach and comes with some benefits but also drawbacks which are discussed in the paper's Conclusion.

Keywords Model-free · Robotic arm · Position · Deep neural networks · Regression

18.1 Introduction

Robotics is an integral part of the industry and slowly taking over also our daily lives. There is an increasing demand and pressure on robotic tasks that, to be executed, need flexibility and adaptability, this drives the advancement of sensors, actuators, and also control methods. From the theoretical point of view, in industrial robotics, the robot performs in a very precise structured environment, whereas in emerging fields (collaborative or service robotics), the robot has to perform in an unstructured and even dynamic environment where target positions vary depending on the state of the world. In this new, unstructured environment, it has been proven that classic

A. I. Călin (✉) · C. A. Popa · C. Moldovan · A. Davidescu
Politehnica University Timișoara, Timișoara, Romania
e-mail: albert.calin@student.upt.ro

C. A. Popa
e-mail: calin.popa@cs.upt.ro

C. Moldovan
e-mail: cristian.moldovan@upt.ro

A. Davidescu
e-mail: arjana.davidescu@upt.ro

I. Doroftei et al. (eds.), *Proceedings of SYROM 2022 & ROBOTICS 2022*, Mechanisms and Machine Science 127, https://doi.org/10.1007/978-3-031-25655-4_18

robotic control theory is sometimes sub-optimal and new control strategies need to be explored [7].

In papers [1, 2] is presented a 2-DOF arm controlled using video feedback and path-planning algorithms. In [3] is presented a method of control using motion primitives for a 2-DOF arm without prior kinematic knowledge regarding the link lengths and joint positions and also without encoder feedback. Motion primitives were first introduced by Bizzi in [5] where he shows that a frog limb has some preferred states and depending on stimuli, it makes a certain motion. This idea of motion primitives is also explored in [4] where Pahic builds a Neural Network (NN) to reproduce this behavior. Several studied papers refer to the learning of inverse kinematics (IKs) using various types of Neural Networks architectures [7]. In [6] the learning of the inverse kinematics of a humanoid robot is accomplished by using Locally Weighted Projection Regression (LWPR). The paper in [8] surveys the body representations in biology from a functional or computational perspective to present the concept of body schema in robotics. Another approach is done by Sigaud et al. [9] where they present a survey of multiple regression algorithms applied in robotics.

In [10] are presented the Classical robotic control theories based on Direct and Inverse Kinematics. For the Direct Kinematic (DK) problem, the arm's joint coordinates q and link lengths L are known and the end-effector position p can be determined.

Additional control methods are based on Path planning [12] but can integrate Direct and Inverse Kinematics or Neural Networks [11, 14].

In [13] Artificial Neural Network (ANN) was used in a 6-DOF manipulator to output the angular position and velocities of the joints and to solve the singularities issue. Duka also applied Neural Network to estimate the inverse kinematics of a robotic arm [18]. In [15] three Neural Network architectures are used to estimate the Inverse Kinematics of an industrial parallel robot: Multi-Layer Perceptron (MLP), Long-Short Term Memory (LSTM), and Gated Recurrent Unit (GRU). The authors compared the models and concluded that the GRU model had the best accuracy during the test phase and also the estimation of the IK was the fastest of the three. In [19] the estimation of inverse kinematics is made with MLP and Recurrent Neural Network (RNN). In [20] the applications of deep learning in robotics are presented.

There is also a wide spectrum of research covering deep reinforcement learning for robotic manipulation, which also does not rely on direct and inverse kinematics and an analytical predefined model.

Kumar et al. [16] applied Proximal Policy Optimization (PPO) algorithm to control the Baxter robot to reach different targets and also to avoid obstacles. We refer the reader to [17, 21] where a more in-depth of the applications of neural networks and deep reinforcement learning algorithms in robotics are presented.

In our paper, we implement a regression neural network that outputs the joint angles of a 2-DOF robotic arm. Because the whole setup is simulated, we train the network on a dataset generated using direct kinematics relationships for the 2-DOF arm. This dataset gives us the correspondence between q, the joint angles, and p, the cartesian coordinates of the end-effector. After the network is trained on this dataset, we use it to estimate the joint angles given the cartesian coordinates of the end

Table 18.1 Performance of different architectures

No	Hidden layers	Nodes	Average error [mm]
1	3	256, 256, 256	6.78
2	3	256, 256, 128	5.79
3	2	1024, 1024	4.11
4	2	1024, 512	4.72
5	2	512, 512	4.84
6	2	512, 256	5.4
7	2	512, 128	5.08
8	2	512, 64	4.76
9	2	256, 256	3.19
10	2	256, 128	4.01
11	2	256, 64	5.63

effector. Our strategy is to use the kinematic model of a 2DOF robotic arm presented in Chap. 2.1 from the current paper to simulate a sequence of positions and record the joint and corresponding cartesian coordinates, so creating the dataset. We also present the results of various configurations of neural networks in Table 18.1.

18.2 System Architecture

18.2.1 The Simulated 2-DOF Robotic Arm Direct Kinematic Model for Data Generation

The kinematic model of the robotic arm is used for data generation and the testing phase. The error is measured between the precise analytical and the estimated position by the Neural Network. This is similar to the situation when a newborn is starting to explore the space his hand can reach, but in our situation, the arm is virtual. A simple kinematic schema for the arm is presented in Fig. 18.1.

Fig. 18.1 Model of a 2-DOF robotic arm

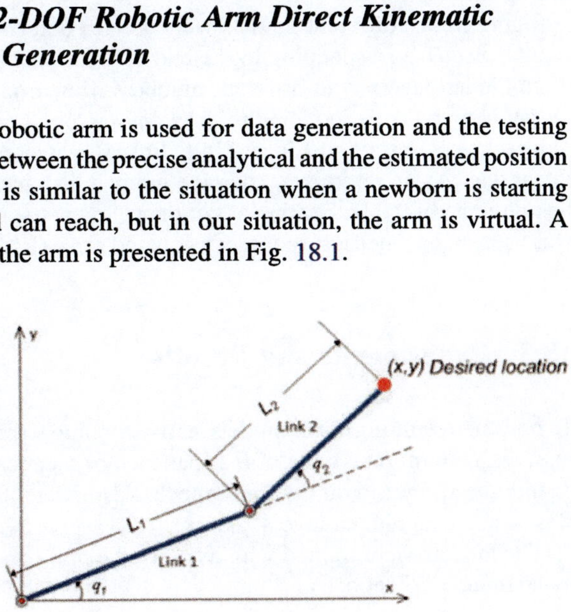

To derive the position (x,y) of the end-point we use relationships presented in Eq. 18.1:

$$x = L_1 \cos(q_1) + L_2 \cos(q_1 + q_2) \tag{18.1}$$

$$y = L_1 \sin(q_1) + L_2 \sin(q_1 + q_2) \tag{18.2}$$

18.2.2 The Position Dataset and Neural Network Architecture

In this implementation, both links of the 2-DOF robot have the same size and both joints have the same motion range:

$$L_1 = L_2 = 100\,mm \tag{18.3}$$

$$q_1, q_2 \in [0, \pi] \tag{18.4}$$

To train the neural network, q_1 and q_2 angles were randomly generated, and the positions of the end effector were calculated using Eq. 18.1. 8100 positions were used with their corresponding joint angles for the training dataset and 8100 positions for the evaluation dataset. Also, 20% of the training dataset was reserved for validation during training.

Using Keras, different models were created and compared. Each model has two input neurons corresponding to the end effector position in cartesian space, two or three hidden layers with different numbers of neurons, and the output layer has 2 output neurons corresponding to the joint angles. As an activation function for the hidden layers, Rectified Linear Unit (ReLU) was used, and for the output layer, a linear one. As an optimizer, adaptive moment estimation (Adam) was used with a learning rate of 0.001. For the loss of the neural network, mean squared error (MSE) was used and as metrics, root mean squared error (RMSE).

18.3 Experiments and Results

To prevent overfitting of the models, early stopping was used to monitor the validation loss, with a minimum delta of 0.1, patience of 5 epochs, and also the possibility to restore the weights from the best epoch. Also, weight decay with a value of 0.001 was used for all hidden layers.

The models were trained with a batch size of 32 and the training process ran for a maximum of 50 epochs.

The models were evaluated on the positions from the evaluation set and Eq. 18.3 was used to calculate the positions from the angles that the models predicted. Each model was trained and evaluated five times and in Table 18.1, each model's architecture and its average error are presented. The root mean squared error for each model at each run was calculated between the positions from the evaluation set and the calculated positions from the angles which the model predicted.

In Fig. 18.2, the architecture of the best model with an average error of 3.19 mm is presented.

In Fig. 18.3, the values of the root mean squared error of the training set and validation set during the training phase of the model are presented. The training ran for 42 epochs, but the weights from epoch 37 were restored.

In Table 18.2 is presented 9 randomly chosen angle predictions and compared with the actual angles and in Fig. 18.4, the actual and predicted position of the end-point from the last example is presented.

Fig. 18.2 The architecture of the model

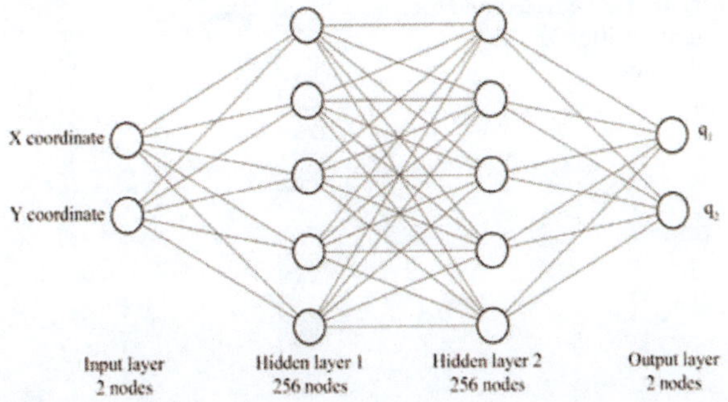

Fig. 18.3 Training root mean squared error

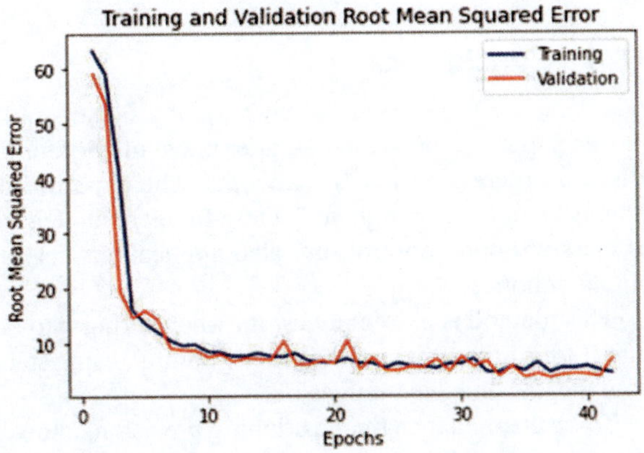

Table 18.2 Predictions made by the best model

No	Actual		Predicted	
	q_1 [°]	q_2 [°]	q_1 [°]	q_2 [°]
1	29.94	16.61	30.59	14.91
2	36.77	27.82	39.64	22.46
3	105.06	24.17	104.23	26.04
4	76.14	87.94	75.26	88.55
5	153.49	98.21	155.13	95.41
6	126.64	133.42	127.7	130.93
7	128.91	27.82	129.03	29.6
8	172.94	53.89	177.09	48.16
9	71.24	49.02	71.21	48.69

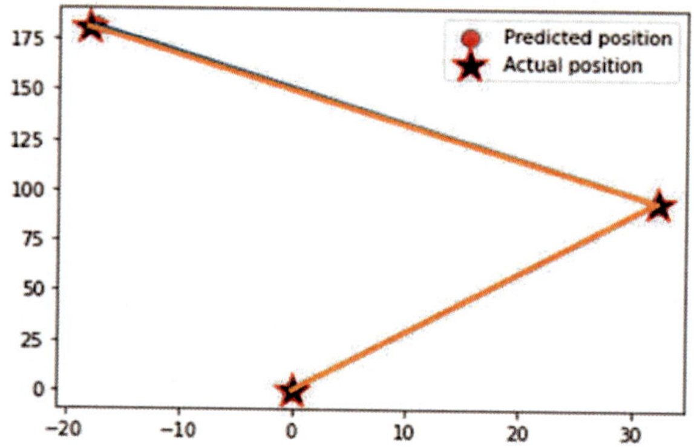

Fig. 18.4 Predicted and the actual position of the end-point

18.4 Conclusions

In this paper, we presented an alternative method to position a 2DOF robotic arm, based on regression neural networks. The experiments have shown that the positioning error is quite high, at 3.2 mm for element lengths of 100 mm each, compared to classic robotic control, but also applications for this method are different than classic robotics.

This method is extremely useful when taking into consideration compliant structures (arms or legs) where the precise control, using standard Control theory becomes complicated and even unfeasible.

To further enhance the positioning precision, a local planner can be used, mainly if employing additional sensors, similar to the situation when a human needs to do precision work, he uses the touch sense and visual servoing.

Future work in developing this project is to make the transition from a simulated environment to a real one. The system proposed to be used is presented in [1, 2], it uses video feedback to track the end-effector position. Our future target is to test the feasibility of the trained neural network in a simulated environment and to adapt it to a real situation.

In a real robotic system, the positioning error depends on actuator and sensor precision. The setup presented in this paper does not take into account real-life errors, being only a simulation and a comparison between two methods that allow the positioning of the end effector.

References

1. Moldovan, C., Ciupe, V., Dolga, V., Lovasz, E.C., Maniu, I.: Design and novel control solution for a modular mechatronic demonstrator with video feedback used in research and education. ICMV (2019)
2. Moldovan, C., Ciupe, V., Filipescu, H., Kristof, R., Dolga, V.: Model-free continuous to discrete workspace transformation and path planning of a 2dof serial arm for visual obstacle avoidance, new advances in mechanisms, mechanical transmissions and robotics. MTM&Robotics (2020)
3. Kormushev, P., Demiris, Y., Caldwell, D.G.: Kinematic-free position control of a 2-DOF planar robot arm. In: 2015 IEEE/RSJ International Conference on Intelligent Robots and Systems (IROS), Hamburg. (2015)
4. Ridge, B., Gams, A., Morimoto, J., Ude, A.: Training of deep neural networks for the generation of dynamic movement primitives. Neural Networks **127** (2020)
5. Bizzi, E., Mussa-Ivaldi, F.A., Giszter, S.: Computations underlying the execution of movement: a biological perspective. American Association for the Advancement of Science. Sci. J. (1991)
6. D'Souza, A., Vijayakumar, S., Schaal, S.: Learning inverse kinematics. In: International Conference on Intelligence in Robotics and Autonomous Systems (IROS 2001). Maui, Hawaii, 2001
7. Sünderhauf, N., Brock, O., Scheirer, W., Hadsell, R., Fox, D., Leitner, J., Upcroft, B., Abbeel, P., Burgard, W., Milford, M., Corke, P.: The limits and potentials of deep learning for robotics. Int. J. Robot. Res. (2018)
8. Hoffmann, M., Marques, H., Arieta, A., Sumioka, H., Lungarella, M., Pfeifer, R.: Body schema in robotics: a review. IEEE Trans. Auton. Ment. Dev. **2**(4), 304–324 (2010)
9. Sigaud, O., Salaün, C., Padois, V.: On-line regression algorithms for learning mechanical models of robots: a survey. Robot. Auton. Syst. **59**(12) (2011)
10. Tsai, L.W.: Robot analysis, the mechanics of serial and parallel manipulators. John Wiley & Sons, Inc. (1999)
11. Hasan, A.T., Ismail, N., Hamouda, A.M.S., Aris, I., Marhaban, M.H., Al-Assadi, H.M.: Artificial neural network-based kinematics Jacobian solution for serial manipulator passing through singular configurations. Adv. Eng. Softw. **41** (2010)
12. LaValle, S.: Planning algorithms. Cambridge University Press, Cambridge, MA (2006)
13. Duleba, I., Opalka, M.: A comparison of Jacobian-based methods of inverse kinematics for serial robot manipulators. Int. J. Appl. Math. Comp. Sci. **23**(2) (2013)
14. Takahashi, K., Ogata, T., Nakanishi, J., Cheng, G., Sugano, S.: Dynamic motion learning for multi-DOF flexible-joint robots using active–passive motor babbling through deep learning. Adv. Robot. (2017)
15. Toquica, J.S., Oliveira, P.S., Souza, W.S.R., Motta, J.M.S.T., Borges, D.L.: An analytical and a Deep Learning model for solving the inverse kinematic problem of an industrial parallel robot. Comp. Indus. Eng. **151** (2021)
16. Kumar, V., Hoeller, D., Sundaralingam, B., Tremblay, J., Birchfield, S.: Joint space control via deep reinforcement learning. In: 2021 IEEE/RSJ International Conference on Intelligent Robots and Systems (IROS) (2021)

17. Nguyen-Tuong, D., Peters, J.: Model learning for robot control: a survey. Cogn. Process. **12**, 319–340
18. Duka, A.-V.: Neural network based inverse kinematics solution for trajectory tracking of a robotic arm. Proc. Technol. **12**, 20–27 (2014)
19. Bensadoun, R., Gur, S., Blau, N., Shenkar, T., Wolf, L.: Neural inverse kinematics. In: ICML (2022)
20. Pierson, H.A., Gashler, M.S.: Deep learning in robotics: a review of recent research. arXiv https://doi.org/10.48550/ARXIV.1707.07217
21. Liu, R., Nageotte, F., Zanne, P., de Mathelin, M., Dresp-Langley, B.: Deep reinforcement learning for the control of robotic manipulation: a focussed mini-review. CoRR abs/2102.04148 (2021)

Part IV
Exoskeletons and Rehabilitation Systems

Chapter 19
Inverse Dynamic Modeling of a Parallel Robot for Lower Limb Rehabilitation

I. Nadas, B. Gherman, P. Tucan, G. Carbone, A. Pisla, T. Antal, and D. Pisla

Abstract This paper presents the development of the inverse dynamic model for the RECOVER parallel robot, which is used in lower limbs stroke rehabilitation, more precisely in rehabilitation exercises for hip flexion/extension, knee flexion, plantar flexion/dorsiflexion, and inversion/eversion of the ankle. The inverse dynamic model is obtained using the concept of virtual work and the lumped masses approach. The developed model facilitates in-depth analysis of the dynamic capabilities of the RECOVER robot and the development of a precise and effective robotic-assisted procedure for lower limb rehabilitation.

Keywords Parallel robot · Rehabilitation · Dynamic model · Virtual work · Simulation

I. Nadas · B. Gherman (✉) · P. Tucan · A. Pisla · T. Antal · D. Pisla
CESTER, Technical University of Cluj-Napoca, 400114 Cluj-Napoca, Romania
e-mail: bogdangherman@mep.utcluj.ro

I. Nadas
e-mail: iuliu.nadas@itim-cj.ro

P. Tucan
e-mail: paul.tucan@mep.utcluj.ro

A. Pisla
e-mail: adrian.pisla@muri.utcluj.ro

T. Antal
e-mail: tiberiu.alexandru.antal@mep.utcluj.ro

D. Pisla
e-mail: doina.pisla@mep.utcluj.ro

I. Nadas
National Institute for Research and Development of Isotopic and Molecular Technologies, Cluj-Napoca, Romania

G. Carbone
DIMEG, University of Calabria, 87036 Cosenza, Italy
e-mail: giuseppe.carbone@unical.it

I. Doroftei et al. (eds.), *Proceedings of SYROM 2022 & ROBOTICS 2022*, Mechanisms and Machine Science 127, https://doi.org/10.1007/978-3-031-25655-4_19

19.1 Introduction

Stroke victims may have a loss of motor function, which can have a devastating effect on their ability to walk normally [1]. Rehabilitative measures must begin as soon as feasible after a stroke has occurred to lessen the severity of the effects. If possible, the treatment should start within the first few hours after the stroke [2].

The goal of physical rehabilitation is to help persons with physical impairments lead more independent and productive lives by regaining lost motor function and maximizing functional capacity [3]. The rehabilitation sector has seen a substantial increase in the use of robotic technologies over the past few years [4]. Patient-robot interaction is essential for effective physical rehabilitation with the use of medical robots [5].

Robotic assisted rehabilitation devices have been widely explored and utilized in the rehabilitation field because of the repetitive aspect of the operation [6, 7].

The Lambda [8] is a robotic system for lower limb rehabilitation that is specifically designed for use with bedridden patients; the device's parallel design includes two prismatic joints and a rotation joint to allow ankle mobility. The Physioterabot (PT) [9] is a stationary gait system with three degrees of freedom (DOF) that allow it to do knee flexion/extension, hip DOF flexion/extension, and hip DOF adduction/abduction. The RAISE [10] parallel robotic system is a stationary rehabilitation training device designed specifically for bed-ridden patients. It consists of two parallel modules each module having 2 DOF's and it performs the flexion/extension and the abduction/adduction of hip, the flexion/extension of the knee, the inversion/eversion and flexion/dorsiflexion of the ankle.

Applying the dynamic model in the control system may improve the rehabilitation outcomes by increasing the treatment efficiency. Various strategies have been used to build inverse dynamic models. Application of the Lagrange-Euler formalism to a system of parallel robots in [11] demonstrates the algorithm's value for closed-loop chains. The concept of virtual work used by Plitea [12] to calculate the inverse dynamics of a five DOF's modular parallel medical robot. Others [13], have used the Newton–Euler method for the inverse dynamics. To reduce the overall amount of time spent processing, the virtual work method for inverse dynamic model was selected for this article [14]. The virtual work concept, which is applied to calculate the inverse dynamics, is very effective since it does not need the internal forces to calculate the robot torques and forces. In this work, the inverse dynamic model of a parallel robot for lower limb rehabilitation using the virtual work approach is developed. It will be later implemented into the robot control, and thus used for achieving various patient-robot interaction modalities (i.e. assistive, active-assistive, etc.). In addition, in can be further used during the rehabilitation process to improve the treatment procedure safety.

Section 19.2 of paper aims to presents parallel robot for lower limb rehabilitation kinematics and the main elements of the design. Section 19.3 presents the inverse dynamic model of the parallel rehabilitation robot by using virtual work principle and the lumped masses approach. Section 19.4 presents the validation of the inverse dynamics using by comparing the linear forces simulated in Siemens NX 20 CAD and the measured forces simulated in MATLAB programing environment and Sect. 19.5 presents the conclusions of this paper.

19.2 The RECOVER Rehabilitation Parallel Robot

The RECOVER robotic system consists of two modules [15], namely: the hip-knee module (designed for the rehabilitation of the hip and knee) and the ankle module (designed for the rehabilitation of the ankle), [16–19].

The RECOVER hip-knee module has a parallel architecture having two active prismatic joints and five passive revolute joints being able to perform hip flexion/extension and knee flexion. The main components of RECOVER parallel robot hip-knee module Fig. 19.1a and b are: NEMA 23 stepper motors (1) and (2); the rotational joints R_1 and R_2 denoted by (3) and (4) respectively; the links l_1 and l_2 are denoted by (5) and (6) respectively; (7) is the hip revolute joint R_h; (8) is the femoral link L_f situated between the hip joint and knee revolute joint R_k denoted by (9), followed by the tibial link L_t (10) which starts on knee joint. The links (5) and (6) intersect in the same point on the axis (11) where the revolute joint R_3 is placed. Another important element are the thigh support (12) and ankle module (13) which will be described in the next subsection, (14) and (15) are the ball screws. The active prismatic joints q1 and q2 are denoted by (16) and (17) respectively.

The ankle module is a parallel mechanism which performs the rehabilitation of ankle joint having two active prismatic joints q_3 and q_4 denoted by (17_a) and (18_a) respectively, two passive revolute joints R_{a1} and R_{a2} denoted by (12a) and (16a) respectively and four passive spherical joints (7_a), (8_a), (9_a) and (10_a). It performs the plantar flexion/dorsiflexion and ankle eversion/inversion. The main components of RECOVER parallel robot ankle module are shown in Fig. 19.2a and b, which are as follows: (1_a) and (2_a) are the stepper motors NEMA 17 which actuates the trapezoidal

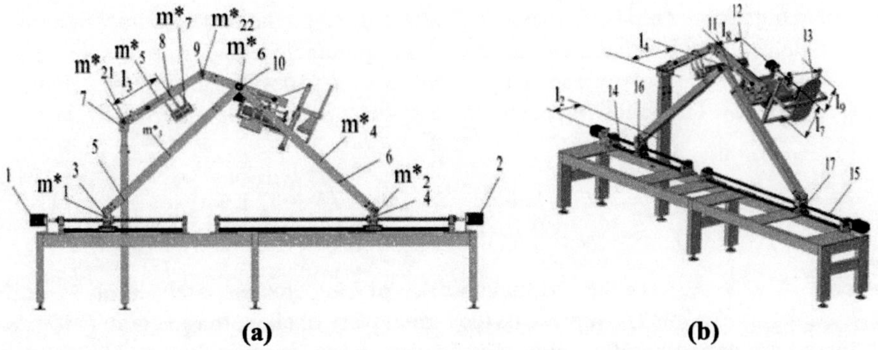

Fig. 19.1 Distribution of equivalent masses of RECOVER rehabilitation robot hip-knee module

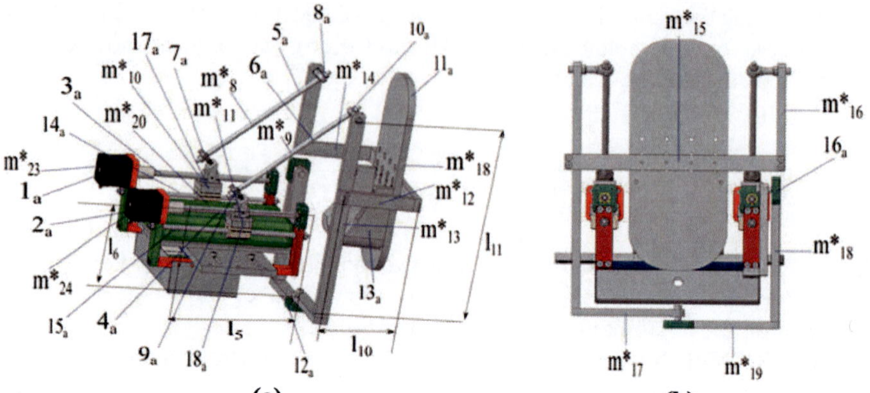

Fig. 19.2 The design of RECOVER rehabilitation robot ankle module

screws (3_a) and (4_a) respectively that moves trapezoidal nuts assemblies (17_a) and (18_a), which are guided by the linear elements (14_a) and (15_a). The translation motions from the two trapezoidal nuts are transmitted to the sole support (11_a) by a pair of connection rods (5_a) and (6_a), both being connected at the nut's assemblies and to the sole support by four spherical joints mentioned above. The heel support (13_a) is placed on the sole support (11_a). Regarding the ankle module, when the stepper motors run in the same direction with the same speed, the ankle module performs plantar flexion/dorsiflexion motions by moving the revolute joint (16_a), when the motors run in opposite direction having the same speed, the ankle module performs ankle eversion/inversion by moving the revolute joint (12_a).

19.3 Inverse Dynamic Model of RECOVER

Motion ranges of the hip flexion/extension, knee flexion, plantar flexion/dorsiflexion and ankle eversion/inversion as well as motion parameters (in terms of speed and acceleration), are considered as input data and are used to develop the inverse dynamic model, [20, 21]. The following equation describes the principle of virtual work:

$$\Upsilon W = \Upsilon q^T \cdot \lambda + \sum_{i=1}^{28} \Upsilon X_{Mi}^T \cdot \left(T_i^{Tr} + T_i^g\right) = 0 \tag{19.1}$$

where: $\Upsilon q^T \cdot \lambda$ is the virtual work of all forces and torques and $\sum_{i=1}^{28} \Upsilon X_{Mi}^T \cdot \left(T_i^{Tr} + T_i^g\right)$ represents the virtual work of all inertia and gravity forces of the rehabilitation parallel robot. The inertia and the gravity forces matrices are expressed as follows:

$$T_i^{Tr} = \begin{bmatrix} m_i^* & 0 & 0 \\ 0 & m_i^* & 0 \\ 0 & 0 & m_i^* \end{bmatrix} \begin{bmatrix} X_i \\ Y_i \\ Z_i \end{bmatrix}; \tag{19.2}$$

;

$$T_i^g = \begin{bmatrix} 0 \\ 0 \\ -m_i^* g_i \end{bmatrix} \tag{19.3}$$

$$T_i = T_i^{T_r} + T_i^g \tag{19.4}$$

The inverse dynamic model is used to simplify principles:

- The friction forces are neglected.
- The lumped masses approach was utilized, the masses of the bodies being concentrated in specific points (and denoted with m_i^*).

The mass of each body (bar) is split into three unique points, one in the middle of the body and the other two at either end of the body. The lumped masses principle is shown graphically in Fig. 19.3. The values of the lumped masses for a linear bar of mass m are calculated as follows:

$$\frac{1}{6}m = m_1^*; \ \frac{2}{3}m = m_2^*; \ \frac{1}{6}m = m_3^* \tag{19.5}$$

The model's calculation requires the following geometric parameters: $D_i, i = 1...9$-diameters of the elements in motion, $l_i, i = 1...11$—the lengths from each mass equivalent to the CSYS (coordinate system), placed on the hip joint on each axis (Ox, Oy and Oz) Fig. 19.1a and b and Fig. 19.2a and b. For the hip-knee module Fig. 19.2a and b m_1 is the mass of q1 joint, m_2 is the mass of q_2 joint, m_3 is the mass of l_1 link, m_4 is the mass of l_2 link, m_5 is the mass of L_f link, m_6 is the mass of L_t link, m_7 is the mass of thigh support. The mass of hip joint is m_{21} and the mass of knee joint is m_{22}. For the ankle module, the CSYS (placed at the intersection of the two orthogonal axes) the masses are noted as follows: m_8 and m_9 are the masses of the rods that connects the two active prismatic joints q_3 and q_4 with their masses m_{10} and m_{11} with a connection element that belongs to sole support having masses m_{12}, m_{13},

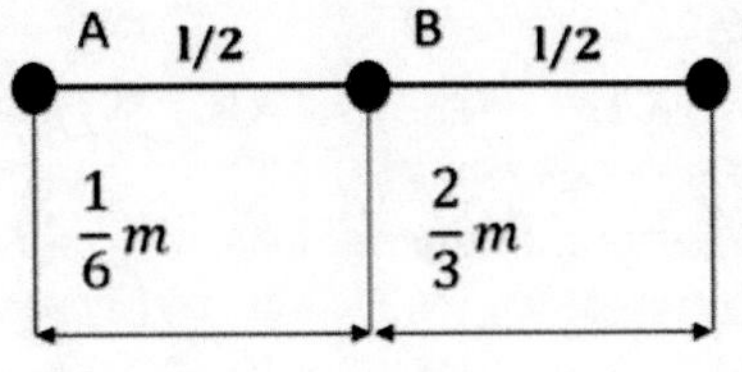

Fig. 19.3 The principle of lumped masses

m_{14}, m_{15}, m_{16}, m_{17} and they being mounted to the sole support having the mass m_{18}. The lateral "L" shape connection element composed by two elements having their masses m_{18} and m_{19}. The mass of lower leg support is m_{20}, m_{23} and m_{24} represent the massed of the NEMA 17 motors that actuates the active prismatic joints q_3 and q_4. The moment of inertia of motors (1) and (2) of hip-knee module Fig. 19.1a and motors (1_a) and (2_a) Fig. 19.2a of ankle module have been obtained from catalogues. The moment of inertia for ball screws (14) and (15) illustrated in Fig. 19.1b of the hip-knee module and the moment of inertia for the trapezoidal screws (3_a) and (4_a) of the ankle module, illustrated in Fig. 19.2a, can be calculated using the following formula.

$$I_{screws} = \left(m_s \cdot D^2\right)/\left(8 \times 10^6\right)\left[\mathrm{kgm}^2\right] \tag{19.6}$$

where m_s is the mass of the screw and D is the diameter of screw. Reduced moment of inertia might be calculated using the formula.

$$I_{red} = I_{motor} + I_{screw} \tag{19.7}$$

The gyration radius determines the position of the dynamic equivalent mass point.

$$i_\Lambda = \sqrt{I_{red}/I_{mot}} \tag{19.8}$$

The equivalent masses are presented in (9). The coordinates of the masses $m_{26}{}^*$–$m_{29}{}^*$ are presented in (11–12), where q_{k0} represents the position of the motor when the motion begins and φ_{k0} is initial rotation angle of the respective lumped mass with respect with the OX axis and P is the screw lead. The rest of the corresponding lumped mass coordinates are presented in (13).

$$\begin{cases} m_1^* = m_1;\, m_2^* = m_2;\, m_3^* = \frac{1}{6}m_1 + \frac{2}{3}m_3 + \frac{1}{6}m_1;\, m_4^* = \frac{1}{6}m_2 + \frac{2}{3}m_4 + \frac{1}{6}m_6 \\ m_5^* = \frac{1}{6}m_{21} + \frac{2}{3}m_5 + \frac{1}{6}m_{22};\, m_6^* = m_6;\, m_7^* = m_7;\, m_8^* = \frac{2}{3}m_8 \\ m_9^* = \frac{2}{3}m_9;\, m_{10}^* = \frac{2}{3}m_{10};\, m_{11}^* = m_{11};\, m_{12}^* = \frac{2}{3}m_{12} + \frac{1}{6}m_{18} \\ m_{13}^* = \frac{1}{6}m_{17} + \frac{2}{3}m_{13};\, m_{14}^* = \frac{1}{6}m_{16} + \frac{2}{3}m_{14};\, m_{15}^* = \frac{1}{6}m_{14} + \frac{2}{3}m_{15} + \frac{1}{6}m_{12} \\ m_{16}^* = \frac{2}{3}m_{16} + \frac{1}{6}m_{16};\, m_{17}^* = \frac{2}{3}m_{17} + \frac{1}{6}m_{13};\, m_{18}^* = \frac{2}{3}m_{18} + \frac{1}{6}m_{19};\, m_{19}^* = \frac{2}{3}m_{19} + \frac{1}{6}m_{18} \\ m_{20}^* = m_{20};\, m_{21}^* = m_{21};\, m_{22}^* = m_{22};\, m_{23}^* = m_{23};\, m_{24}^* = m_{24};\, m_{25}^* = \frac{2}{3}m_{25} + \frac{1}{6}m_{15} \end{cases} \tag{19.9}$$

$$\varphi_k = \varphi_{k0} + (2\pi)/P(q_k - q_{k0}),\ k = 1..4 \tag{19.10}$$

$$\begin{cases} x_{26} = -l_{12} \\ y_{26} = i_{\Delta 26}\cos(\varphi_1); \\ z_{26} = i_{\Delta 26}\sin(\varphi_1) \end{cases} \begin{cases} x_{28} = L_f\cos(u_h) + (L_t - l_5)\cos(u_k - u_h) \\ y_{28} = i_{\Delta 28}\cos(\varphi_3) \\ z_{28} = i_{\Delta 28}\sin(\varphi_3) \end{cases}; \tag{19.11}$$

$$\begin{cases} x_{27} = l_{13} \\ y_{27} = i_{\Delta 27}\cos(\varphi_2) \\ z_{27} = i_{\Delta 27}\sin(\varphi_2) \end{cases}; \begin{cases} x_{29} = L_f\cos(u_h) + (L_t - l_5)\cos(u_k - u_h) \\ y_{29} = i_{\Delta 29}\cos(\varphi_4) \\ z_{28} = -{}^{l_8}/_2 + i_{\Delta 29}\sin(\varphi_4) \end{cases}; \quad (19.12)$$

$$\begin{cases} X_1 = q_1;\ Y_1 = -L_d;\ Z_1 = l_2; \\ X_2 = q_2;\ Y_2 = -L_d;\ Z = l_2; \\ X_3 = \frac{q_2 - q_1}{4};\ Y_3 = -L_d + \sqrt{l_1^2 - \left(\frac{q_2 - q_1}{2}\right)^2};\ Z_3 = l_2; \\ X_4 = 3\frac{q_2 - q_1}{4};\ Y_4 = -L_d + \sqrt{l_1^2 - \left(\frac{q_2 - q_1}{2}\right)^2};\ Z_4 = l_2; \\ X_5 = \frac{L_f}{2}\cdot\cos(u_h);\ Y_5 = \frac{L_f}{2}\cdot\sin(u_h);\ Z_5 = 0; \\ X_6 = L_F\cdot\cos(u_h);\ Y_6 = L_F\cdot\sin(u_h);\ Z_6 = 0; \\ X_7 = l_3\cdot\cos(u_h);\ Y_7 = l_3\cdot\sin(u_h);\ Z_7 = -l_4; \\ \ldots \\ X_{21} = 0;\ Y_{21} = 0;\ Z_{21} = 0; \\ X_{22} = L_f\cdot\cos(u_h) + L_t\cdot\cos(u_k);\ Y_{22} = L_f\cdot\sin(u_h) + L_t\cdot\sin(u_k);\ Z_{22} = 0; \end{cases} \quad (19.13)$$

19.4 Experimental Results

The linear forces of the active prismatic joints q_1, q_2, q_3 and q_4 were obtained from a CAD simulation in Siemens NX 20 [22] and they were compared with the measured linear forces of the active prismatic joints of the RECOVER robotic system simulated in MATLAB. The plots of the simulation performed in Siemens NX have been overlapped with the plots from MATLAB and they are presented in the Fig. 19.4. The green continues line represent forces obtained from NX simulation and the black interrupted line are forces taken from MATLAB simulation.

Figure 19.4a illustrates the forces comparation of q_1 prismatic joint of hip-knee module; in Fig. 19.4b it is shown the compared q_2 prismatic joint of hip-knee module forces values. Fig. 19.4c and d shown the plots representing the simulated linear forces of q_3 and q_4 with the continuous green line and MATLAB simulated linear forces of the prismatic joints q_3 and q_4 represented with the dashed black line. The maximum computed error between the results obtained in Matlab and NX is of 1%, which validates the proposed inverse dynamic model. Figure 19.5 presents the experimental model of RECOVER during initial validation testes in laboratory conditions using healthy subjects.

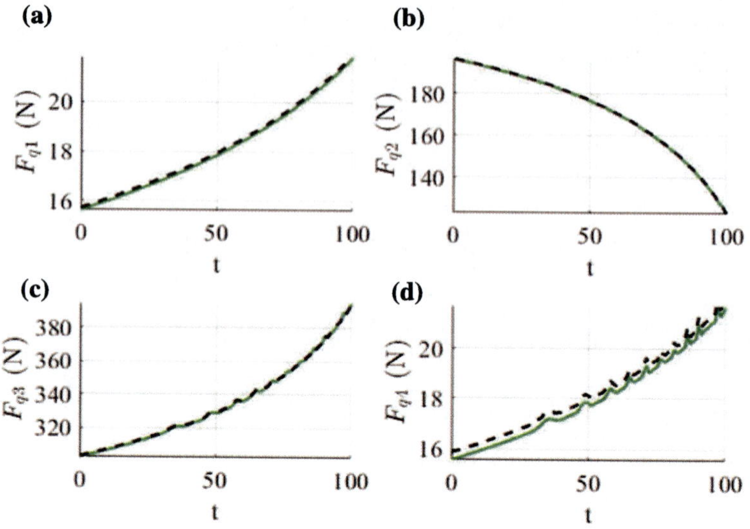

Fig. 19.4 q_1, q_2, q_3 and q_4 linear forces comparison between NX and Matlab simulations

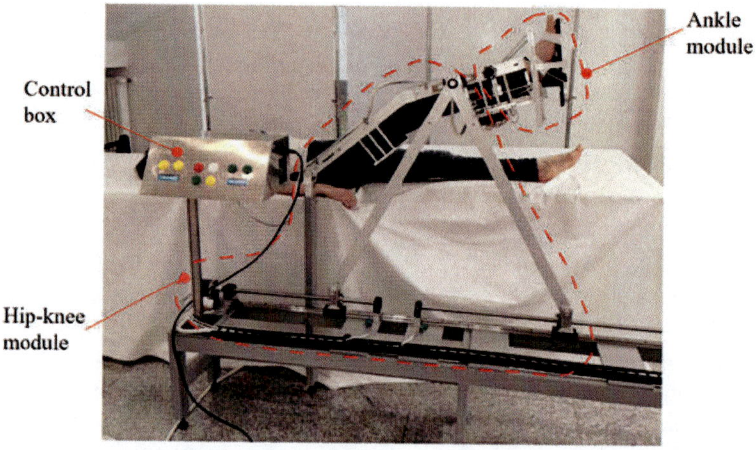

Fig. 19.5 RECOVER experimental model during laboratory tests

19.5 Conclusions

The inverse dynamics of the RECOVER parallel robot for the rehabilitation of the lower limb is presented in this paper. The proposed model uses the principle of virtual work with two simplifying hypotheses: the correct replacement of a given multibody system with equivalent lumped masses and the neglection of friction forces. The inverse dynamic model has been analytically obtained. The results of a Matlab simulation have been compared with a simulation of the same motion using Siemens NX, validating the proposed model. The results prove that the developed inverse dynamics can be implemented into the robot control, leading to multiple advantages. It could be used within a safety algorithm, monitoring the actuation forces during training exercises along various trajectories or for the implementation of different patient–robot interaction modalities (i.e. the active-assisted modality, when the robot intervenes during the rehabilitation training exercise when the patient is unable to completely perform the required motion), thus improving the overall treatment efficiency.

Acknowledgements This work was supported by a grant from the Romanian Ministry of Research and Innovation, CCCDI—UEFISCDI, project number PN-III-P2-2.1-PED-2021-3430/608PED/2022 (Hope2Walk) within PNCDI III.

References

1. Pyo, S., Yoon, J., Oh, M.-K.: A novel robotic knee device with stance control and its kinematic weight optimization for rehabilitation. Robotica **32**, 1245–1263 (2014). https://doi.org/10.1017/S0263574714001453
2. Prieto, P., Auat, F., Escobar, M., Vallejos, R., Maldonado, P., Larrain, C., Serey, M.: Robotic Care: A Low Cost Design to Assist Therapy for Brain Stroke Rehabilitation. Proceedings of the Design Society: International Conference on Engineering Design. **1**, 975–984 (2019). https://doi.org/10.1017/dsi.2019.103
3. Nef, T., Mihelj, M., Riener, R.: ARMin: a robot for patient-cooperative arm therapy. Med Biol Eng Comput. **45**, 887–900 (2007). https://doi.org/10.1007/s11517-007-0226-6
4. Cafolla, D., Russo, M., Carbone, G.: Design of CUBE, a cable-driven device for upper and lower limb exercising. In: Carbone, G., Ceccarelli, M., Pisla, D. (eds.) New trends in medical and service robotics, pp. 255–263. Springer International Publishing, Cham (2019)
5. Banica, A., Gherman, B., Tohanean, N., Antal, T., Pisla, A., Abrudan, C., Carbone, G., Pisla, D.: Inverse dynamic modeling of a parallel elbow rehabilitation robot for spasticity treatment. In: Müller, A., Brandstötter, M. (eds.) Advances in service and industrial robotics, pp. 392–400. Springer International Publishing, Cham (2022)
6. Vaida, C., Birlescu, I., Pisla, A., Ulinici, I., Tarnita, D., Carbone, G., Pisla, D.: Systematic design of a parallel robotic system for lower limb rehabilitation. IEEE Access. **8**, 34522–34537 (2020). https://doi.org/10.1109/ACCESS.2020.2974295
7. Carbone, G., Gherman, B., Ulinici, I., Vaida, C., Pisla, D.: Design issues for an inherently safe robotic rehabilitation device. In: Ferraresi, C., Quaglia, G. (eds.) Advances in service and industrial robotics, pp. 1025–1032. Springer International Publishing, Cham (2018)
8. Bouri, M., Le Gall, B., Clavel, R.: A new concept of parallel robot for rehabilitation and fitness: The Lambda. In: 2009 IEEE International conference on robotics and biomimetics (ROBIO), pp. 2503–2508 (2009)
9. Akdoğan, E., Adli, M.A.: The design and control of a therapeutic exercise robot for lower limb rehabilitation: Physiotherabot. Mechatronics **21**, 509–522 (2011). https://doi.org/10.1016/j.mechatronics.2011.01.005
10. Vaida, C., Birlescu, I., Pisla, A., Carbone, G., Plitea, N., Ulinici, I., Gherman, B., Puskas, F., Tucan, P., Pisla, D.: RAISE—An innovative parallel robotic system for lower limb rehabilitation. In: Carbone, G., Ceccarelli, M., Pisla, D. (eds.) New trends in medical and service robotics, pp. 293–302. Springer International Publishing, Cham (2019)

11. Abdellatif, H., Heimann, B.: Computational efficient inverse dynamics of 6-DOF fully parallel manipulators by using the Lagrangian formalism. Mech. Mach. Theory **44**, 192–207 (2009). https://doi.org/10.1016/j.mechmachtheory.2008.02.003
12. Plitea, N., Szilaghyi, A., Cocorean, D., Vaida, C., Pisla, D.: Inverse dynamics and simulation of A 5-DOF modular parallel robot used in brachytherapy. Proceedings of the Romanian academy series A—Mathematics physics technical sciences information science. **17**, 67–75 (2016)
13. Dasgupta, B., Mruthyunjaya, T.S.: Closed-form dynamic equations of the general stewart platform through the newton-euler approach. Mech. Mach. Theory **33**, 993–1012 (1998). https://doi.org/10.1016/S0094-114X(97)00087-6
14. Li, Y., Xu, Q.: Dynamic analysis of a modified DELTA parallel robot for cardiopulmonary resuscitation. In: 2005 IEEE/RSJ International conference on intelligent robots and systems, pp. 233–238 (2005)
15. Gherman, B., Birlescu, I., Plitea, N., Carbone, G., Tarnita, D., Pisla, D.: On the singularity-free workspace of a parallel robot for lower-limb rehabilitation. Proceedings of The Romanian Academy, Series A. **20**, 383–391 (2019)
16. Nadas, I., Gherman, B., Albert, S., Surducan, V., Pop, N., Carbone, G., Pisla, D.: Innovative development of a parallel robotic system for lower limb rehabilitation. Acta Tech Napocensis—Series: Appl Math, Mech, Eng. **64**(1-S2), (2021)
17. Pisla, D., Nadas, I., Tucan, P., Albert, S., Carbone, G., Antal, T., Banica, A., Gherman, B.: Development of a control system and functional validation of a parallel robot for lower limb rehabilitation. Actuators. **10**, 277 (2021). https://doi.org/10.3390/act10100277
18. Nadas, I., Tucan, P., Gherman, B., Banica, A., Rednic, V., Carbone, G., Pisla, D.: On the design and validation of a parallel robot for lower limb rehabilitation. Rom. J. Tech. Sciences. Appl. Mechanics. **67**, 187–187 (2022)
19. Carbone, G., Plitea, N., Gherman, B.G., Vaida, C., Pisla, D., Birlescu, I., Craciun, F., Nadas, I., Tucan, P., Pop, N.: Parallel Robot for Medical Recovery of Lower Limbs. Patent Number: RO133815B1.Publication Date: 2020–02–28 Technical University of Cluj Napoca.
20. Plitea, N., Gherman, B., Carbone, G., Ceccarelli, M., Vaida, C., Banica, A., Pisla, D., Pisla, A.: Kinematic analysis of an exoskeleton-based robot for elbow and wrist rehabilitation. In: Carvalho, J.C.M., Martins, D., Simoni, R., Simas, H. (eds.) Multibody mechatronic systems, pp. 424–433. Springer International Publishing, Cham (2018)
21. Gherman, B., Pisla, D., Vaida, C., Plitea, N.: Development of inverse dynamic model for a surgical hybrid parallel robot with equivalent lumped masses. Robot. Comput.-Integr. Manuf. **28**, 402–415 (2012). https://doi.org/10.1016/j.rcim.2011.11.003
22. NX | Siemens Software, https://www.plm.automation.siemens.com/global/en/products/nx/

Chapter 20
Kinematic and Dynamic Analysis of an Exoskeleton Robotic System

I. Geonea, D. Tarnita, and L. Grigorie

Abstract This paper presents the kinematic and dynamic modelling of an exoskeleton robotic system for assisting people with locomotor disabilities. In the first part of the paper a kinematic analysis of the proposed mechanism as a constructive solution is presented. In the second part of the paper a dynamic analysis of the robotic system is performed with the help of the software for dynamic analysis of mobile mechanical systems: ADAMS. The obtained results are presented and interpreted, the most important being the laws of variation of rotation angles in kinematic joints, the connecting forces in kinematic joints, and the ground reaction forces. In the last part of the paper, the solution of easy manufacture of the proposed robotic system by 3D printing method is presented. The optimal design shape of the elements is considered, in order to give the system modularity and structural strength.

Keywords Kinematic analysis · Dynamics · Robot exoskeleton · Locomotor rehabilitation

20.1 Introduction

Around the world, medical robotics is actively developing in three areas: rehabilitation, service and clinical [1, 2]. Rehabilitation robots are designed to solve the problems of restoring lost limb function and life support for bed-bound disabled people (with visual impairments, musculoskeletal system and other serious diseases) [3–5]. Medical robots for service purposes are designed to solve the problems of transporting moving patients, various loads, and caring for bed-bound patients. Clinical robotics provides full or partial automation of diagnostic, therapeutic and surgical treatment of various diseases [6–8].

In this paper, a new structural solution of a mechanism designed for the leg of an exoskeleton for the rehabilitation of people with locomotor disabilities is presented. The purpose of the exoskeleton is to provide locomotor assistance to people with

I. Geonea (✉) · D. Tarnita · L. Grigorie
University of Craiova, Craiova, Romania
e-mail: ionut.geonea@edu.ucv.ro

I. Doroftei et al. (eds.), *Proceedings of SYROM 2022 & ROBOTICS 2022*, Mechanisms and Machine Science 127, https://doi.org/10.1007/978-3-031-25655-4_20

disabilities. The presented solution is different from existing solutions on the market in that it uses a kinematic chain for the legs of the exoskeleton driven by a single motor. The paper is structured in three parts. After the introductory part, a kinematic analysis of the exoskeleton leg mechanism is presented, followed by a dynamic modelling in ADAMS.

The contribution of this work consists in proposing an original solution of the kinematic structure of an exoskeleton robot, for which the kinematics and dynamics during use by a virtual patient, materialized in the form of a manikin, are analyzed. Compared to previous achievements, presented in [5–8], this solution is new and structurally different.

20.2 Structural and Kinematic Analysis of the Exoskeleton Leg Mechanism

As shown in Fig. 20.1, the proposed construction solution for the exoskeleton leg consists of a desmodromic kinematic chain with 9 kinematic elements and 14 rotational kinematic joints. The kinematic chain reproduces the human foot through elements 6 and 9 (LM segment). The F coupling of the mechanism is equivalent to the hip joint and the L coupling of the mechanism is equivalent to the knee joint. Based on this structural solution a virtual model is made, as shown in Fig. 20.1. The virtual model is worn by a virtual human dummy. This virtual prototyping is useful in two essential aspects, namely: kinematic simulation with multibody mechanical systems analysis software and additive manufacturing.

The role of the exoskeleton is to provide additional strength to the disabled patient. To this purpose, the exoskeleton elements are attached to the human foot by means of ergonomic fasteners. The robotic exoskeleton system is worn by the patient, and is not intended to take any of the patient's weight. The elements attached to the human foot are the FL segment of the exoskeleton, which performs the structural role of the human femur, and the LM segment of the exoskeleton which is attached to the tibia of the human foot.

Among the advantages of this exoskeleton solution are the fact that the actuation is performed with a single drive motor, the modular construction, which allows adjustment on patients with different anthropometric sizes. Another major advantage of this exoskeleton solution is the safety in operation, i.e. due to the fact that it is driven by a single motor operating a desmodromic kinematic chain (with determined motions), the possibility of executing wrong motions is excluded, as can happen with exoskeleton solutions with motors placed in joints, where angular position sensors can provide erroneous information. Because of this major advantage, this robotic system will be well accepted by patients with locomotor disabilities.

For the kinematic analysis of the mechanism, we know from the virtual model designed in Solid Works: the lengths of the elements, the coordinates of the kinematic rotational torques at base A, F and E. The equations describing the kinematic model

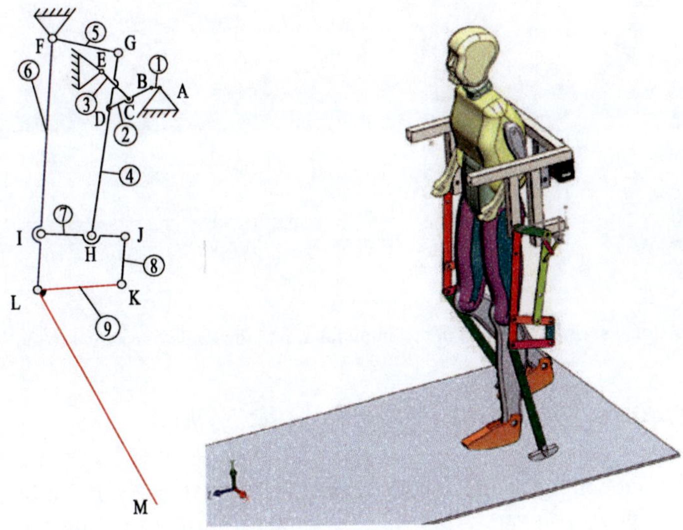

Fig. 20.1 Kinematic scheme and CAD model of the exoskeleton robotic system

of the mechanism, are developed based on the vector contour method, as presented in the paper [5]. In the present paper the kinematic modelling has been carried out using the ADAMS program. The mathematical equations for the mechanism kinematics allow us to optimize the angles of variation in the kinematic joints of the mechanism. The graphical results obtained for the rotation angles in the knee and hip joint are shown in Fig. 20.2.

The robotic system can work successfully if in the ankle joints of the two legs we define rotational couplings, which are not actuated, and in these couplings pre-tensioned rotational springs are defined, which allow rotation in the coupling only within certain limits. The final construction of the robotic system will be realized with this passive rotation torques, provided with torsional springs to limit oscillations. In the Fig. 20.3 shows the rotation angle calculated in the ADAMS simulation.

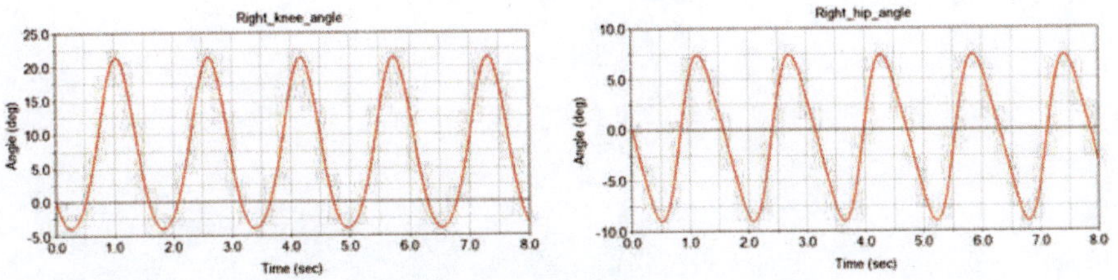

Fig. 20.2 Laws of variation over time for the angle of rotation in the knee and hip joint

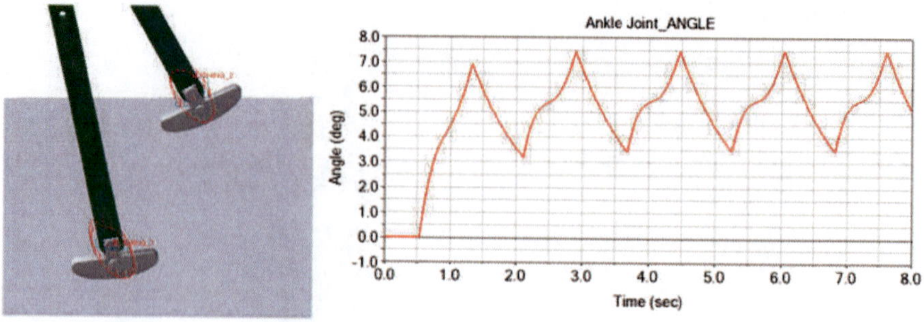

Fig. 20.3 Torsional springs mounted in the ankle joint, and the angle of rotation in the ankle joint

20.3 Dynamic Modelling of the Robotic System

The dynamic modelling of the robotic system will be done with the help of the dynamic simulation software ADAMS. The steps followed to build the dynamic model are characterized by key points: definition of the materials from which the robot elements are made, definition of the kinematic joints and the law of motion of the driving element (4 rad/sec), and definition of the contact between the foot and the ground. In order to make a correct model, the most important step is to define the contact between the foot and the ground. The type of contact formulated is by the impact method, and the specified parameters, namely: stiffness $= 1.0\text{E} + 07$ N/m, force exponent = 1.3, damping 7000 N·s/m and penetration depth 1.0E-05 m. With these considerations, we simulated the motion of the robotic system, in the situation when it steps on the ground. In Fig. 20.4 shows the trajectory described by point M, for both legs.

This analysis gives us the forces in the joints of the exoskeleton leg mechanism. We present the forces obtained for the hip and knee joints of the exoskeleton. For the knee and hip joints, the resulting forces are shown in Fig. 20.5.

Fig. 20.4 Trajectory of the exoskeleton legs when walking on the ground

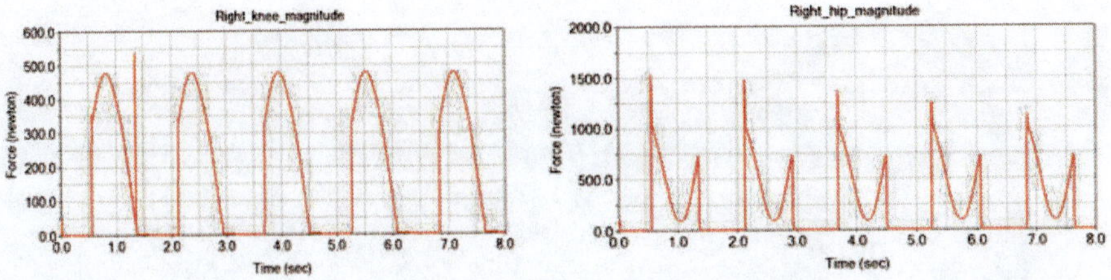

Fig. 20.5 Connecting forces in the hip and knee joints, for the right leg

In Fig. 20.6, the ground reaction forces calculated by simulation in ADAMS are shown and in Fig. 20.7 the motor torque and power consumed by the exoskeleton during the run. It can be seen that due to the rigid foot the reaction force with the ground has a step-like variation around 1000 N. According to Fig. 20.5 it can be seen that the hip joint of the exoskeleton is the most loaded, the resultant force being around 1000 N. The sharp jumps that occur in the variation are due to the impact between the rigid sole of the exoskeleton and the ground considered rigid. In reality these spikes do not occur, because the exoskeleton sole will be fitted with a cushioning rubber. The knee joint of the exoskeleton is less loaded, the maximum connection force is less than 500 N. To realize the physical prototype of the exoskeleton, the components will be made in modular construction, as shown in Fig. 20.8, in order to achieve the adjustment of the exoskeleton to the patient's dimensions.

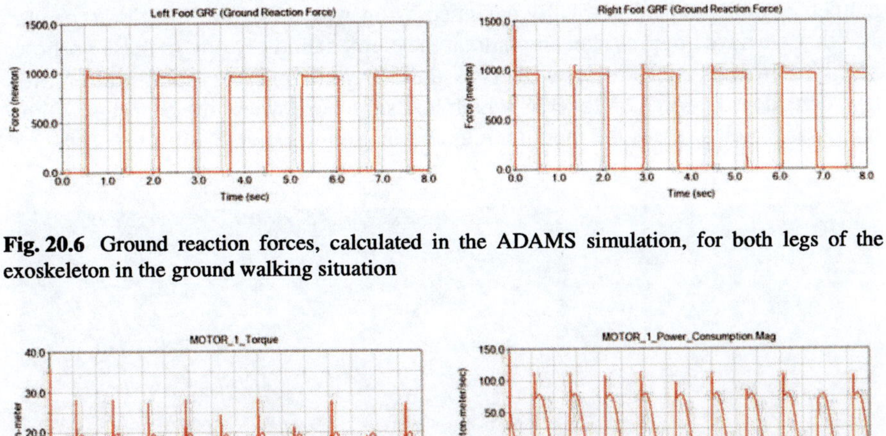

Fig. 20.6 Ground reaction forces, calculated in the ADAMS simulation, for both legs of the exoskeleton in the ground walking situation

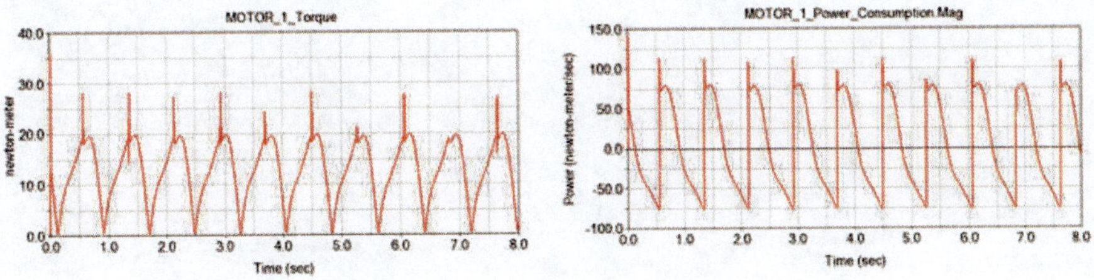

Fig. 20.7 Calculated motor torque and power consumed by the exoskeleton

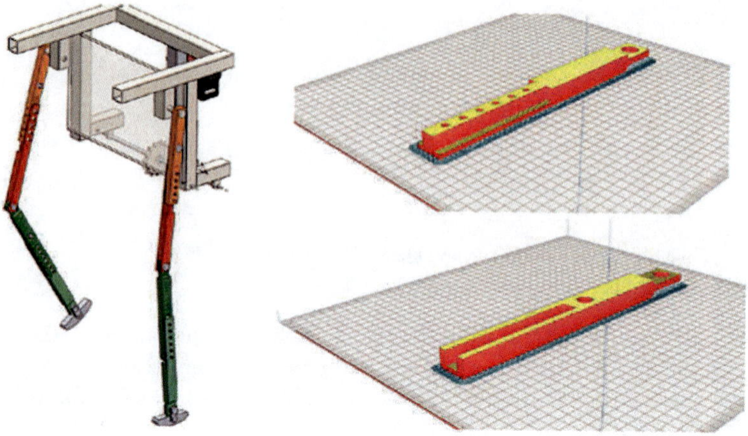

Fig. 20.8 Modular element design solution for structural optimization in 3D printing

20.4 Structural Optimization of Modular Exoskeleton Components

In this section we will present the structural optimization of the modular elements of the exoskeleton using ANSYS.

The initial designed shape of the modular elements is as shown in Fig. 20.9. The geometric model is parametrically designed using Ansys Design Modeler. It can be seen that we have not provided any radii at the element end cut-out. In these portions the highest mechanical stresses will occur, these areas being stress concentrators. We have considered a force of 1500 N, which loads the bore where the bearings will be mounted, which stresses the modular element in compression.

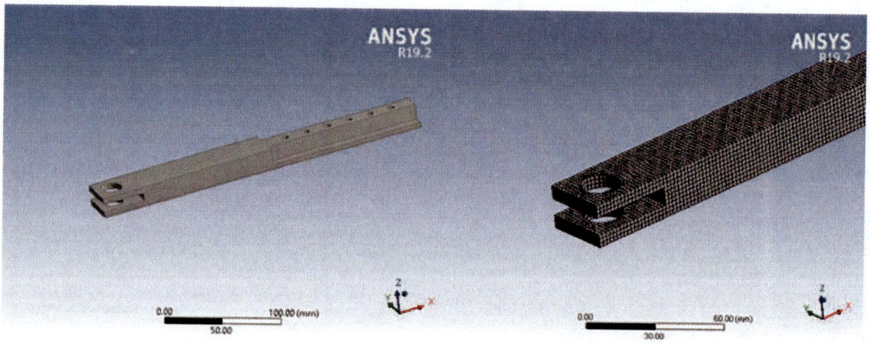

Fig. 20.9 ANSYS parametric model design and mesh with hexahedral elements

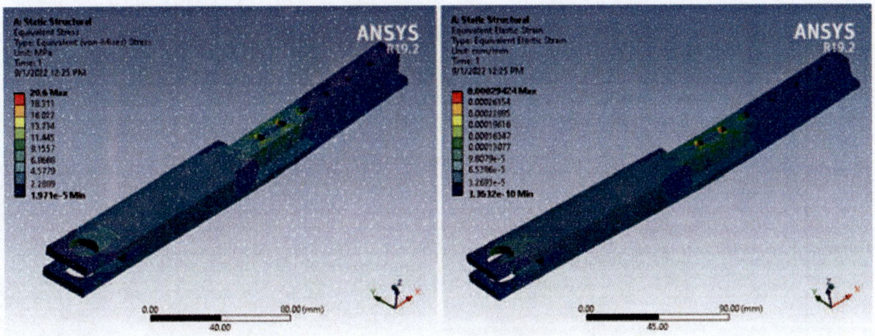

Fig. 20.10 Distribution maps for the equivalent (von-Mises) stress and equivalent elastic strain

With these loads we obtained values (for the non-optimized construction) for the equivalent stress value of 20.6 MPa and for the specific deformations value of 0.0002942 mm/mm. These results, in the form of distribution maps, for specific strains and equivalent mechanical stresses according to the von Mises method, obtained for the initial design are visible in Fig. 20.10.

As input parameters for the optimization process, we consider the connection radius R1(denoted FDBlend1) and the size (width) of the channel cut in the end of the element. The output parameters considered are the mass of the modular element (which is made of Al) as well as the specific strain. After running the optimization, we obtained the response surfaces shown in Fig. 20.20. 20.11, were we seen that the minimum stress is recorded for 8 mm channel width and 4 mm radius Fig. 20.12.

As a result, we obtained the optimal solution (Fig. 20.13), i.e. the radius value is 4 mm and the channel width is 8 mm, for which the stress distribution and strain is as in Fig. 20.14.

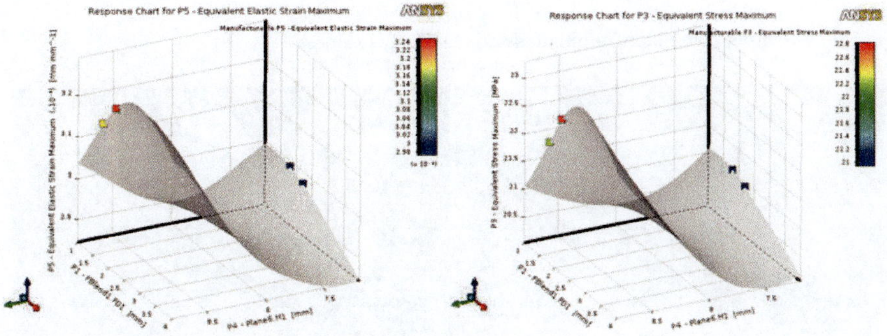

Fig. 20.11 Response surfacs for output parameters

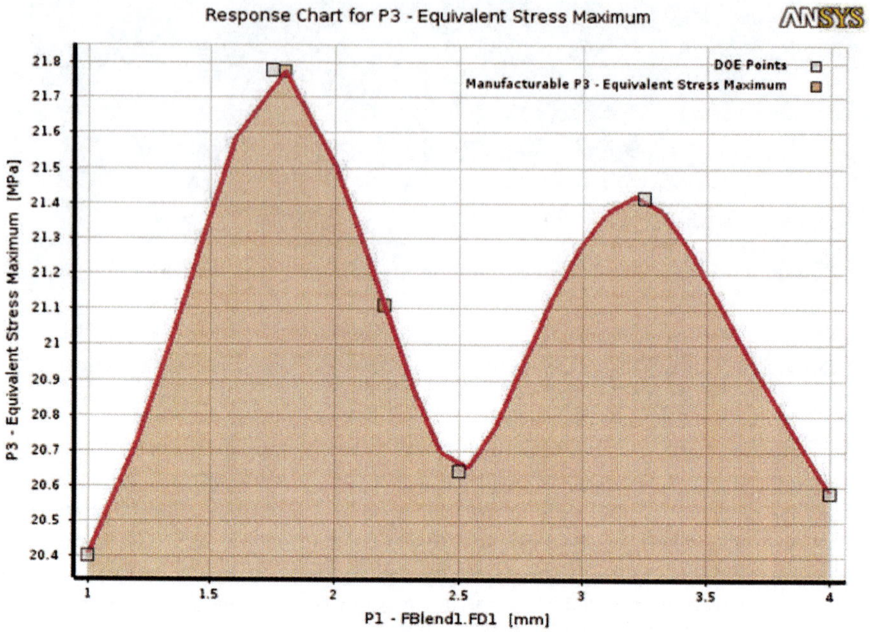

Fig. 20.12 Response chart for equivalent stress

	A	B	C	D	E
1	Name	P1 - FBlend1.FD1 (mm)	P2 - Solid Mass (kg)	P3 - Equivalent Stress Maximum (MPa)	P4 - Equivalent Elastic Strain Maximum (mm mm^-1)
2	Output Parameter Minimums				
3	P2 - Solid Mass	1.8	**0.36834**	21.777	0.00030831
4	P3 - Equivalent Stress Maximum	2.2	0.3684	**21.112**	0.000299
5	P4 - Equivalent Elastic Strain Maximum	2.2	0.3684	21.112	**0.000299**
6	Output Parameter Maximums				
7	P2 - Solid Mass	2.2	**0.3684**	21.112	0.000299
8	P3 - Equivalent Stress Maximum	1.8	0.36834	**21.777**	0.00030831
9	P4 - Equivalent Elastic Strain Maximum	1.8	0.36834	21.777	**0.00030831**

Fig. 20.13 Output parameters minimums and maximums values

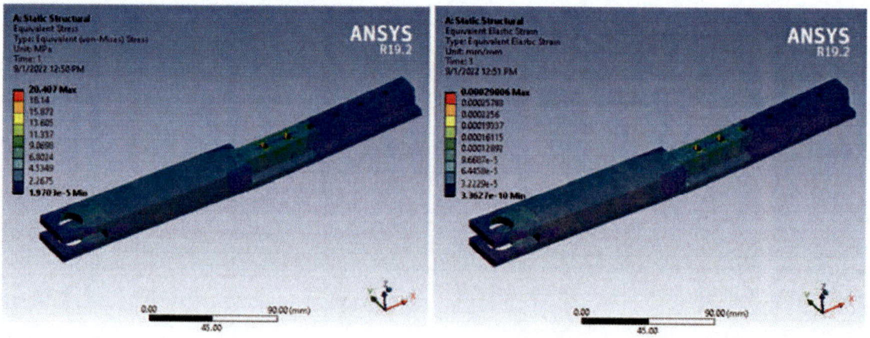

Fig. 20.14 The equivalent (von-Mises) stress and equivalent elastic strain for optimized design

Fig. 20.15 Aspects of 3D printing of exoskeleton elements

In order to realize the elements of the robot we used 3D printing technology. We used a Tronxy printer, and a filament made of PLA-F material. To create the G-Code file needed for printing we used CURA Ultimaker software. The models obtained by 3D printing are shown in Fig. 20.15, as well as an aspect of the CURA software. The next step is to assemble the exoskeleton and test it.

20.5 Conclusions

In this paper we have presented an original structural solution for the leg of an exoskeleton designed to assist people with locomotor disabilities. The developed solution, is a single mobility solution, and is designed to provide assistance to the hip and knee joints, only in the sagital plane. Through the kinematic simulation performed we obtained the variation angles in the knee and hip, which should be similar in variation and angular amplitude as that of a healthy human subject. The solution developed and presented in this paper achieves smaller angular amplitudes, but this is not an impediment to use it in gait assistance. Regarding future studies, we propose to analyze a comparative study between the robotic systems previously realized and presented in the papers [5–8], which will highlight the optimal solution from a kinematic, dynamic and ergonomic point of view.

References

1. Wang, D.: Design and development of a portable exoskeleton for hand rehabilitation. IEEE Trans. Neural Syst. Rehabil. Eng. **26**(12), 2376–2386 (2018)
2. Rehmat, N.: Upper limb rehabilitation using robotic exoskeleton systems: A systematic review. Int. J. Intell. Robot. Appl. **2**(3), 283–295 (2018)
3. Shi, D.: A review on lower limb rehabilitation exoskeleton robots. Chin. J. Mech. Eng. **32**(1), 1–11 (2019)
4. Wang, X.: Design and experimental verification of a parallel hip Exoskeleton system for full-gait-cycle rehabilitation. J. Mech. Robot.,1–17 (2022)

5. Geonea, I.: Design and motion analysis of an exoskeleton robot for assisting human locomotion. In: International workshop on medical and service robots. Springer, Cham, pp. 44–52 (2020)
6. Geonea I.: Motion assistance with an exoskeleton for stair climb. In: 2018 IEEE International conference on automation, quality and testing, robotics (AQTR). IEEE, (2018)
7. Geonea, I.: Design and evaluation of a new exoskeleton for gait rehabilitation. Mech. Sci. **8**(2), 307–321 (2017)
8. RO134430 Patent, Exoskeleton type mechatronic system for assisting walking of locomotor disabled persons. Inventors: Geonea I, Dumitru N, Dumitru S, Copilusi C, Ciurezu Gh

Chapter 21
Design and Operation of a Cable-Driven Robot for Lower-Limb Rehabilitation

G. Ferrise, E. M. Curcio, F. Lago, and G. Carbone

Abstract This paper discusses the design and implementation of a cable robotic system as a lower limb rehabilitation assistance device for hip surgery patients in the immediate postoperative period. The idea stems from the desire to implement an inexpensive and reliable robot adaptable for home use, to avoid overcrowding in hospital environments and prevent the spread of Covid 19 infection. The design logic has four parts: an estimation of competitors is made, the product specification table is compiled, optimization of the structure is carried out using kinematic and dynamic analyses, and experimental tests are carried out on a prototype.

Keywords Cable-driven robot · Lower limb · Rehabilitation

21.1 Introduction

The existing mechatronic devices for lower limb rehabilitation can be classified into three macro-categories: conventional systems, robotic/exoskeleton systems, and cable systems. Each category has different technical and operational characteristics. Cable robotic systems prove to be a good compromise between cost-effectiveness and reliability as reported for example in [1–4]. However, certain issues must be considered during design. For example, [5] addresses the issue related with gravity compensation to achieve balance during handling. References [6, 7] address the kinematic modelling where [7], particularly focuses at the role of singularities within the workspace of parallel kinematic architectures. Reference [8] outlines the role of stiffness modelling for a proper human–robot interaction. Reference [9] addresses the role of compliance within elastodynamic multibody models. Reference [10] introduces the significance of parameter identification for a suitable dynamic modelling.

This paper addresses the design and implementation of a novel cable-based robotic system for lower limb rehabilitation characterized by a crane structure, with the intention of using the functionality of an industrially used machine to adapt it for

G. Ferrise · E. M. Curcio · F. Lago · G. Carbone (✉)
DIMEG, University of Calabria, Calabria, Italy
e-mail: giuseppe.carbone@unical.it

I. Doroftei et al. (eds.), *Proceedings of SYROM 2022 & ROBOTICS 2022*, Mechanisms and Machine Science 127, https://doi.org/10.1007/978-3-031-25655-4_21

domestic rehabilitative use, considering the above-mentioned design issues to achieve a feasible design solution for rehabilitation. Specific simulations and a preliminary experimental validation have been carried out to demonstrate the soundness and engineering feasibility of the proposed design concept after carefully outlining the design requirements and product specifications.

21.2 Design Requirements and Specifications

A product specifications table (PDS) is described in Table 21.1 summarizing the main requirements and constraints for a cable driven lower limb rehabilitation device. One should note that it refers to the anthropometric measurements of an Italian adult male in [11] by considering the 90% percentile with a body weight of 120 kg, a thigh length of 60 cm, the tibia parts of 40 cm and a leg weight of 21 kg.

21.3 Conceptual Design

In this paper, we propose a new concept as outlined in Fig.21.1. It is composed of two modules having the same control logic. Namely, it is necessary to monitor the angle of the thigh to provide feedback from the Arduino closed-loop control to perform the leg lifting and lowering strokes. In this regard, the use of an IMU sensor was chosen, in this case the GY-521 MPU 6050 module. As described in Fig. 21.1, the sensor is placed on the appropriate support on the patient's thigh, and is connected to the Arduino board, from which it receives power and exchanges data via the SCL and SDA ports, using the I2C (Inter Integrated Circuit) protocol.

In Fig. 21.1, the proposed concept aims to perform the flexion–extension and adduction in the frontal plane of the lower limb in a modular manner. Note that the minimum height necessary to avoid the jamming condition is obtained as equal to 89 cm by modelling the leg with a push handle. In Fig. 21.2, a 3D CAD modeling of the device components was conducted. Numerical analysis of the robot's kinematics is carried out with Working Model. Figure 21.3 shows the results obtained by modeling the leg in accordance with the anthropometric study, with reference to the lower limb flexion–extension exercise with results within the requirements given by the PDS. Figure 21.4 shows the results obtained for the adduction motion in the frontal plane. From these results it can be seen that the maximum values obtained are consistent with the PDS.

Table 21.1 Product design specifications (PDS)–(D: defined; W: user wish)

Specifications for: Robotic cable system as an assistive device for the rehabilitation of the lower limb			
1		*Function (s)*	Robot for the accompanied movement of the hip for rehabilitation purposes, to perform exercises such as flexion–extension and adduction on the frontal plane of the lower limb
2		*Performance/size*	
	D D D	Length: 2000–2500 mm Width: 50–150 mm Height: 1500–2500 mm	Dimensions like devices and such as to allow the device [12] to be placed in a lateral position with respect to a bed
	W	Weight: 30–50 kg	Weight comparable with that of devices [12]
	D	Velocity of movement: 0,001–0,05 m/s	The handling speed is estimated according to the guidelines [13]
	D	Maximum load: 40–50 kg	It comes from anthropometric study [12]
3		*Operation*	
	D	Procedure of use: independently of the user upon medical prescription	With start and stop and emergency buttons, as a device and for CEI 64–8 standard [13]
	D	Maintenance: periodic inspection cables/pulleys	As per UNI EN 62,353
4		*Manufacture*	
	D	Quantity to be realized: not defined, realization	It comes from an analysis of competitors, (Average price 2000–7000 €)- Design intent: economic device
	W	Target selling price: 1000–2000€	
5		*Operating environment*	
	D D	Ambient temperature: 20 °C Avoid exposure and use on the device of alcohol, solvents, and flammables	As per CEI EN 60,601, CEI 62-File 1276 G [12]
6		*Material properties for declared parts*	
	W W W	Portal: Metal alloy (aluminum or steel) Pulley: Nylon the Teflon Trolley wheels for handling: Steel and Nylon	

21.4 Dimensional Synthesis

21.4.1 Dynamic Simulations

Dynamic analysis of the device under consideration was carried out with MATLAB software; specifically, Simulink's Simscape library was used with reference to the Simulink model of flexion–extension: a force switch was generated to assist the lifting

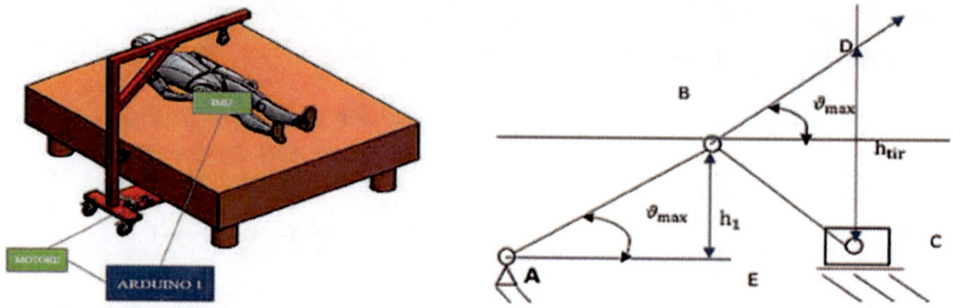

Fig. 21.1 **a** Concept; **b** Leg modeling with thrust crank

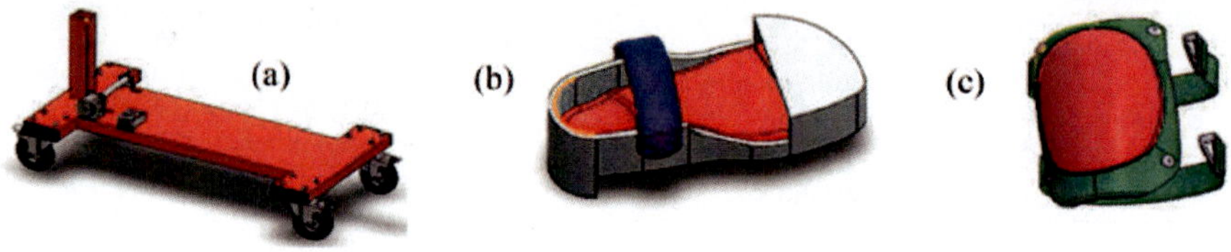

Fig. 21.2 **a** Trolley with engine block; **b** foot end effector; **c** knee end effector

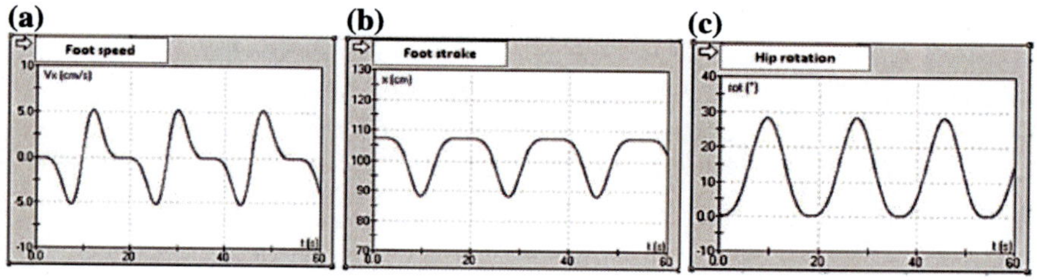

Fig. 21.3 Flexion–extension exercise: **a** foot speed; **b** foot stroke; **c** hip rotation

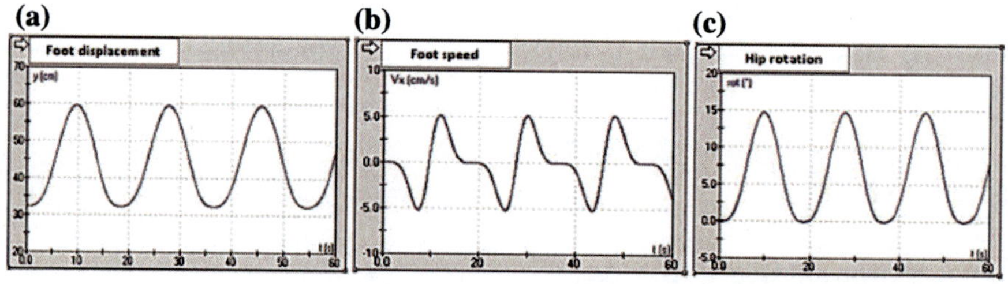

Fig. 21.4 Adduction in the frontal plane: **a** foot displacement; **b** foot speed; **c** hip rotation

and lowering phase of the leg. As described in the Simulink model in Fig. 21.5, 100 N are required for the leg lift. Similarly, Fig. 21.6 shows the proposed Simulink model of adduction in the frontal plane of the lower limb: the hip was modelled as a hinge, the prismatic guide was omitted. The plots in Fig. 21.7a and b show the calculated angular displacement of the hip for the case for flexion–extension and abduction, respectively. One can note that both are comparable with the results obtained in kinematic simulations. Analysing the force switch shows how about 200 N is required for lifting the limb. With regards to the results obtained in the multibody analyses, the choice of motor is made. The motor torque required for lifting will be equal to the resisting force times the radius of the spool connected to the shaft (subscript 1 denotes the torques required for flex-extension operation, subscript 2 denotes the torques required for adduction operation in the frontal plane). This value should be increased by a safety coefficient of 2.00 as follows.

The choice of motor fell on a Nema 17 stepper motor. Stepper motors are DC motors that move by discrete steps, thanks to windings arranged in a particular sequence and organized in phases. It is then possible to have a motor capable of delivering high torque even at very low rpm, and to keep the motor locked in a desired position. In correlation with the motor, a reducer was chosen, since all categories of stepper motors below NEMA 23 do not have sufficient holding torque to lift the leg; the chosen reducer has a reduction ratio of 20.64. Accordingly, one can calculate the torque at the actuator as

$$\tau_{nom1} = F \cdot r = 120\,\mathrm{N} \cdot 0,01\,\mathrm{m} = 1,2\,\mathrm{Nm} \tag{21.1}$$

$$\tau_{nom2} = F \cdot r = 200\,\mathrm{N} \cdot 0,01\,\mathrm{m} = 2\,\mathrm{Nm} \tag{21.2}$$

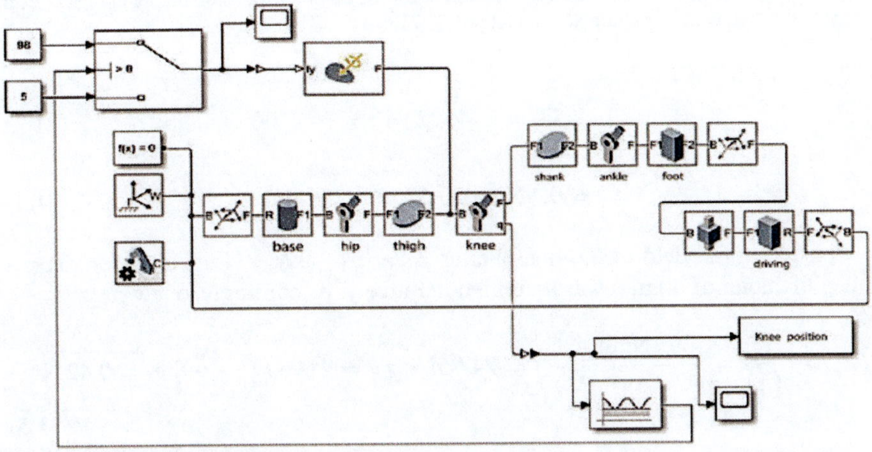

Fig. 21.5 Simulink model for the flexion–extension exercise

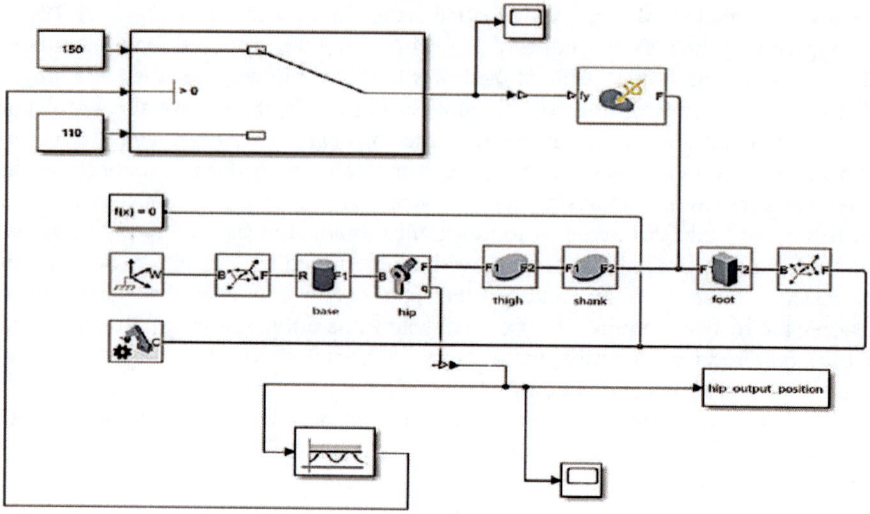

Fig. 21.6 Simulink model for the adduction exercise in the frontal plane

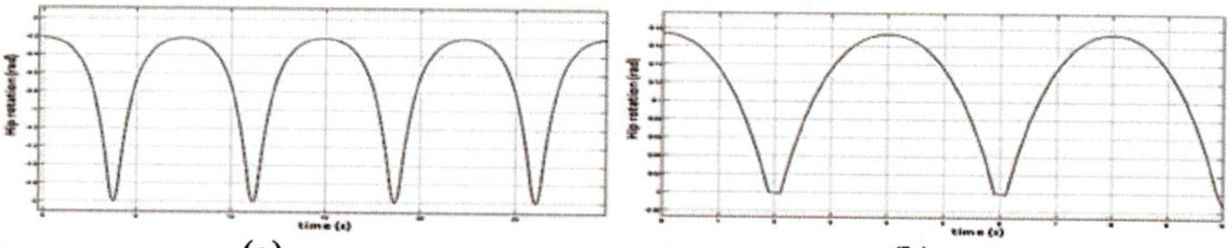

Fig. 21.7 Simulation results for the Hip rotation: **a** for the flexion–extension model in Fig. 21.5; **b** for the adduction in the frontal plane model in Fig. 21.6

$$\tau 1 = \tau_{nom1} \cdot 2,00 = 2,4; \tau 2 = \tau_{nom2} \cdot 2,00 = 4\text{Nm} \tag{21.3}$$

$$\tau_A = 0.4\,\text{Nm} \cdot 20.64 = 8.256\,\text{Nm} \tag{21.4}$$

Considering a yield of 0.6 one obtains $\tau_v = \tau_A \cdot 0.6 = 4.95\text{Nm}$. Moreover, a pulley diameter of 40 mm was identified iteratively. Accordingly, one writes

$$P_{max} = \frac{\tau_v}{\left(\frac{d}{2}\right)} = \frac{4.95\,\text{Nm}}{0.02\,\text{m}} \cdot \left(\frac{1}{g}\right) = 24.75kg;\ \tau_R = 21kg \cdot \left(\frac{0.04}{2}\right)\text{m} = 0.42\,\text{Nm} \tag{21.5}$$

$$\tau_R = 21kg \cdot \left(\frac{0.04}{2}\right)\text{m} = 0.42\,\text{Nm}$$

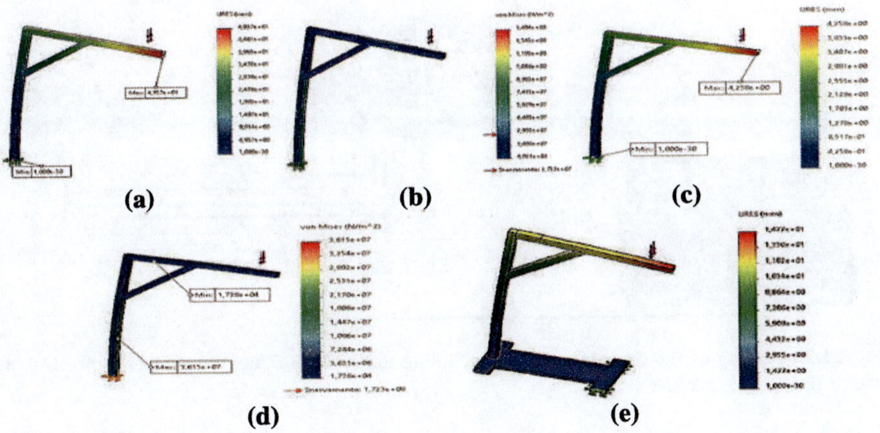

Fig. 21.8 Simulation results: **a** deformations for hollow aluminum profiles 45 × 45 mm; **b** Von Mises stresses for case (**a**); **c** deformations for hollow iron profiles 40 × 40 mm; **d** Von Mises stresses for case (**c**); **e** tipping displacement test for case (**c**)

21.4.2 FEM Simulations

In Fig. 21.8a static analysis was carried out of the portal made with extruded aluminium profiles of 45 × 45 mm with a concentrated load equal to 400 N using the Simulation module of Solidworks. Excessive displacements is shows in Fig. 21.8a and b.

To increase the structural rigidity of the portal, a second analysis was carried out using hollow steel sections with a section of 40 × 40 mm and a thickness of 2 mm; the resulting graphs show limited displacements and stresses. A rollover test was also carried out in Fig. 21.8, constraining the most stressed holes in the carriage with hinges and stressing the structure as before; the following analysis shows no rollover problems with a displacement approximately equal to 0.

21.4.3 Control Architecture

The proposed control architecture is outlined in Fig. 21.9. The robotic device is controlled through an Arduino Uno. The HY-DIV268N driver was chosen to control the motor with an operating voltage of 12–45 V. This driver enables speed control and can directly adjust the resolution of the motor through micro stepping, a technique that allows further subdivision of the pitch (which is initially 1.8°) bringing as advantages better smoothness during rotation, better energy efficiency and achievable speed. A flow-chart of the proposed control loop is shown in Fig. 21.10. A check is made on the angle of inclination of the leg to determine whether the upstroke or the downstroke should be performed. By setting the direction of rotation of the motor via the driver's "dir" pin and sending the HIGH-LOW sequence, a movement step is performed.

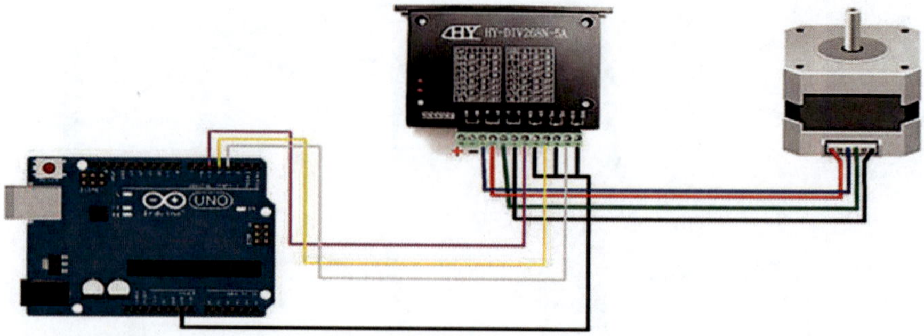

Fig. 21.9 Cabling of the control hardware including the Arduino Uno, the stepper motor Nema and its driver HY-DIV268N-5A

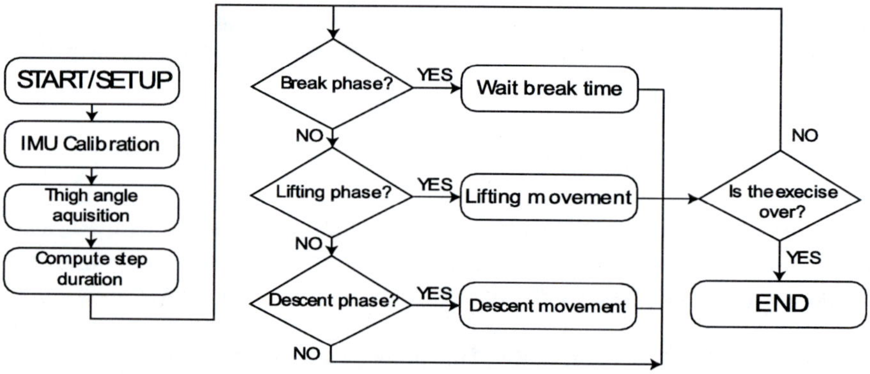

Fig. 21.10 A flow-chart of the proposed control loop

When the iteration is then finished, at the next loop cycle the leg angle is calculated again as a result of the step taken at the previous loop: if the angle has exceeded the maximum angle then it means that the direction of rotation must be reversed to perform the return stroke, vice versa if the angle has fallen below the minimum threshold then it means that the direction must be reversed to perform the upstroke.

21.5 Preliminary Tests

The prototype was made at the MECHLAB laboratory of the University of Calabria. The gantry was made from iron profiles with a cross section of 40 × 40 mm and a thickness of 2 mm as per CAD design. An aluminium plate was used to fix the motor block with motor, gearbox, motor brackets, sprocket, ball bearing. Proper cable housing was made. Figures 21.11 and 21.12 show some image frames taken from a

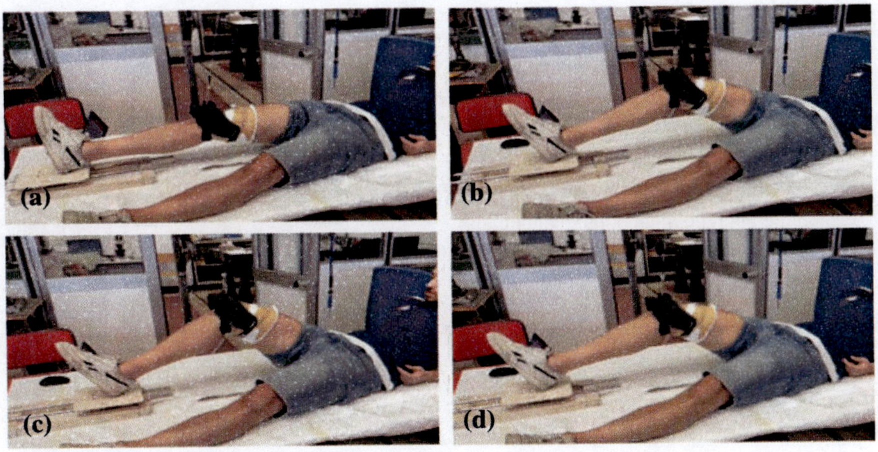

Fig. 21.11 Photo sequence from a flexion–extension exercise

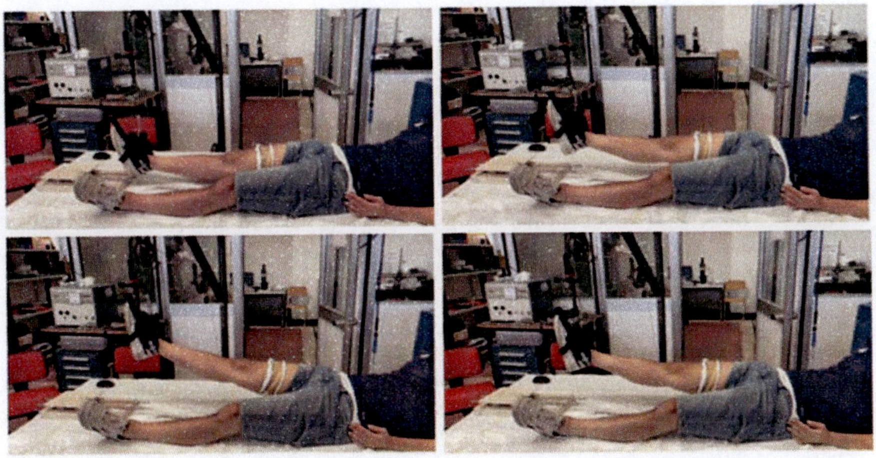

Fig. 21.12 Photo sequence from an exercise of an abduction in the frontal plane

video of the two rehabilitation exercises performed by the robot. Figure 21.13 shows the measured accelerations using the IMU sensor and the computed motor torque.

21.6 Conclusions

This paper described the design methodology of a prototype of cable robotic assistance to lower limb rehabilitation. From the anthropometric study and compilation

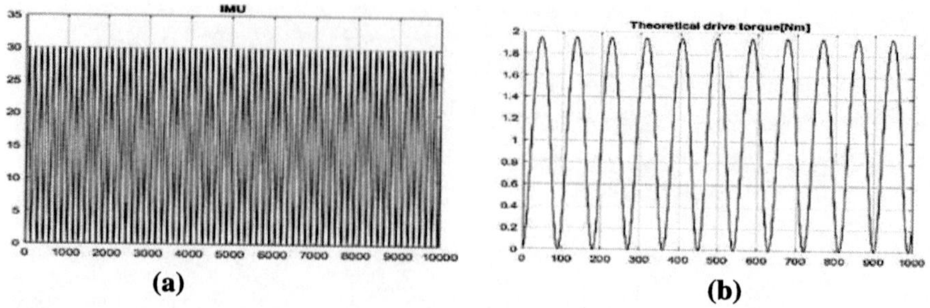

Fig. 21.13 Flexion–extension exercise: **a** IMU sensor test; **b** Computed motor torque

of the PDS, design constraints and guidelines for the design of the robot emerged. Following kinematic sizing, a dynamic analysis was carried out with the intent of figuring out which motor would be suitable for activating the system. 3D CAD modelling of the device components was also optimized. The control architecture was developed using an Arduino Uno. The system is capable of correctly performing lower limb movements within the ranges of motion that were defined in the design phase.

References

1. Vaida, C., Birlescu, I., Pisla, A., Ulinici, I., Tarnita, D., Carbone, G., Pisla, D Systematic design of a parallel robotic system for lower limb rehabilitation. IEEE Access **8**, 34522-34537 (2020)
2. Cafolla, D., Russo, M., Carbone, G.: CUBE, a cable-driven device for limb rehabilitation. J. Bionic. Eng. **16**(3), 492–502 (2019). https://doi.org/10.1007/s42235-019-0040-5
3. Hernández-Martínez, E.E., Ceccarelli, M., Carbone, G., López-Cajún, C.S., Jáuregui-Correa, J.C.: Characterization of a cable-based parallel mechanism for measurement purposes#. Mech. Based Des. Struct. Mach. **38**(1), 25–49 (2010)
4. Laribi, M.A., Carbone, G., Zeghloul, S.: On the optimal design of cable driven parallel robot with a prescribed workspace for upper limb rehabilitation tasks. J. Bionic Eng. **16**(3), 503–513 (2019)
5. Aldanmaz, A., Ayit, O., Kiper, G., Dede, M.: Gravity compensation of a 2R1T mechanism with remote center of motion for minimally invasive transnasal surgery applications. Robotica, 1–14 (2022)
6. Cammarata, A., Lacagnina, M., Sinatra, R.: Closed-form solutions for the inverse kinematics of the Agile Eye with constraint errors on the revolute joint axes. In: IEEE International conference on intelligent robots and systems. Art. No. 7759073. pp. 317–322, (2016)
7. Gherman, B., Birlescu, I., Plitea, N., Carbone, G., Tarnita, D., Pisla, D.: On the singularity-free workspace of a parallel robot for lower-limb rehabilitation. In: Proceedings of the Romanian academy series A—Mathematics physics technical sciences information science, vol. 20 issue no.4. pp. 383–391 (2019)
8. Sanjeevi, N., Vashista, V.: Stiffness modulation of a cable-driven leg exoskeleton for effective human–robot interaction. Robotica **39**(12), 2172–2192 (2021)
9. Tanev, T.K., Cammarata, A., Marano, D., Sinatra, R.: Elastostatic model of a new hybrid minimally-invasive-surgery robot. In: IFToMM world congress proceedings, IFToMM, (2015)

10. Zhang, P., Zhang, J.: Lower limb exoskeleton robots' dynamics parameters identification based on improved beetle swarm optimization algorithm. Robotica **40**(8), 2716–2731 (2022)
11. Molenbroek, J.F.M., Albin, T.J., Vink, P.: Thirty years of anthropometric changes relevant to the width and depth of transportation seating spaces, present and future. Appl Ergon **65**, 130–138 (2017). https://doi.org/10.1016/j.apergo.2017.06.003
12. Jull, G., Moore, A., Falla, D., Lewis, J., McCarthy, C., Sterling, M. (eds.): Grieve's modern musculoskeletal physiotherapy, 4th edition. Elsevier, Edinburgh (2015)
13. Di Monaco, M., Vallero, F., Tappero, R., Cavanna, A.: Rehabilitation after total hip arthroplasty: a systematic review of controlled trials on physical exercise programs. Eur. J. Phys. Rehab. Med. **45**(3),303–17 (2009)

Chapter 22
A State of Art Overwiew on Wrist Rehabilitation Exoskeletons

N. G. Crăciun, E. C. Lovasz, M. C. Mateaş, and C. E. Moldovan

Abstract Exoskeleton-type robotic systems have been of great interest in scientific research in recent years, due to the many applications in which they can be used. They can be found in industry to enhance the physical abilities of the human operator such as speed and strength, in the military, where in addition to enhancing physical abilities, it also provides protection for the operator and in the medical field, in rehabilitation applications where it provides support. and assistance in performing various movements. In this paper, the focus is on exoskeleton robotic devices used to rehabilitate the upper limb for patients recovering from a stroke. There will be a review in the literature for the latest devices developed, they will be classified according to the number of degrees of freedom rendered and respectively the type of drive system used.

Keywords Upper-limb exoskeleton · Exoskeleton review · Rehabilitation after stroke · Assistive technology · Human–robot interaction

22.1 Introduction

An older definition given by the World Health Organization (WHO) describes stroke as a clinical syndrome described by "rapidly developing clinical signs of focal or global disturbance of cerebral function, lasting more than 24 h or leading to death, with no apparent cause apart that of vascular origin" [1]. However, this definition is considered to be outdated, because since its issuance, important discoveries and considerable progress have been made with the development of technology. Stroke can be defined by a more recent definition as a neurological deficit attributed to an acute focal lesion of the central nervous system by a vascular cause, including cerebral infarction, intracerebral hemorrhage, and subarachnoid hemorrhage [1].

N. G. Crăciun · E. C. Lovasz (✉) · M. C. Mateaș · C. E. Moldovan
Polytechnic University of Timisoara, Timisoara, Romania
e-mail: erwin.lovasz@upt.ro

C. E. Moldovan
e-mail: cristian.moldovan@upt.ro

I. Doroftei et al. (eds.), *Proceedings of SYROM 2022 & ROBOTICS 2022*, Mechanisms and Machine Science 127, https://doi.org/10.1007/978-3-031-25655-4_22

In recent decades, stroke has become a common disease in most countries around the world. In the United States, Europe, Japan, China, New Zealand, Canada, but also in India, the number of people suffering a stroke is increasing [2–8]. In our times, stroke is the second most common cause of illness worldwide and is also one of the most common causes of locomotor disabilities. In 2019, approximately 80 million people were registered who suffered a stroke [9].

Another important effect of stroke is the economic one. In 2017, at Oxford University in England, a study was conducted that generated some worrying data. That year, in all 32 European countries under study, 1.5 million people suffered a stroke [10]. The total cost of stroke in these 32 countries was 60 billion euros. Most of these costs are generated by the need for permanent patient care. Because of this, the need to develop devices that allow patients to perform daily activities on their own, without permanent assistance is huge [10, 11]. A possible solution to this problem may be the use of exoskeleton devices for rehabilitation applications.

An exoskeleton used for medical purpose can be defined as a robotic/mechatronic external skeletal device that can be worn by a person and which has the role to provide support and assistance in performing different types of movements during rehabilitation sessions [12–15].

The idea of using exoskeleton devices in industrial and medical applications dates back to the 1960s, when the Hardiman device was developed. This was the first exoskeleton for the entire human body. It was originally designed for military applications and was powered by hydraulic actuators [16]. Over the years, many types of exoskeletons used to rehabilitate the upper limb have been developed. In general, they can be divided into three categories in terms of the anatomical area for which they are built, so we have: exoskeletons for the upper limbs, for the lower limbs and exoskeletons for the whole body. And from a structural point of view, they are divided into final effector devices and robotic exoskeletons.

This paper aims to provide a brief summary of the robotic exoskeletons used to rehabilitate the upper limb, especially for the rehabilitation of the wrist joint for patients who are recovering from a stroke. Devices developed in the last 15 years will be presented, divided according to the number of possible degrees of freedom and will be analyzed from a structural point of view, the mode of action and the control system. There will also be devices that, in addition to the wrist joint, are also designed for other joints of the upper limb.

22.2 Mechanical Design

The mechanical design as well as the actuation methods of the exoskeleton/orthosis devices used for the rehabilitation of the wrist joint are presented below. The devices are classified according to the number of possible degrees of freedom (DOFs). The devices that, in addition to the two degrees of freedom of the wrist joint, also have another degree of freedom for the pronation/supination movement of the forearm, will also be presented.

22.2.1 Wrist Exoskeletons with One Degree of Freedom

An exoskeleton with one degree of freedom for the flexion/extension movement of the wrist joint used for bilateral training in patients who have suffered a stroke is presented in [17]. The device is operated with the help of a brushless electric motor that has a high power density and high-gear-ratio gearheads. The torque is transferred by means of a cable drive which is lighter than a gear, and most of the components in the mechanical structure are made of aluminum.

In [18] is presented a device used to rehabilitate the wrist joint in a compliant way. The device has a parallel mechanism created for the wrist joint and is operated using artificial pneumatic muscles (PMA) (Fig. 22.1a). A mechanical hinge is placed co-axially with the wrist joint, which leads to the need to connect the pneumatic muscles to the hinge pulleys through a redirected steel wire. Hinge friction is reduced by the use of ball bearings. For the control of PMAs a controller with an open-loop trajectory control with force feedback and a closed-loop pressure control is used.

Another device used for wrist and finger joint rehabilitation applications is SCRIPT Active Orthosis Iteration 3 (SAO-i3) [19]. The device is based on the concept of passive exoskeleton called SCRIPT passive orthosis (SPO) [20] previously developed. The SPO finger joint mechanism was reused, but the wrist self-alignment mechanism was replaced with a fixed mechanism (Fig. 22.1b). The device al- lows only the flexion/extension movement for the wrist, and the actuation is done with a single electric actuator by means of a cable both for the finger joints and for the wrist joint.

SCRIPT SPO-F [21] is also an exoskeleton device used for wrist rehabilitation applications performed at the patient's home (Fig. 22.1). The device is based on previous exoskeleton models from SCRIPT, namely SPO and SAO-i1 to SAO-i3, having a single DOF for flexion/extension movement. The electric motors used in SAO- I1 to SAO-i3 have been removed, the drive being made by an extension spring that can be adjusted manually. This makes the device a passive exoskeleton. The double parallelogram mechanism from the SPO model is preserved, it decouples the rotational translations and prevents the misalignment of the device.

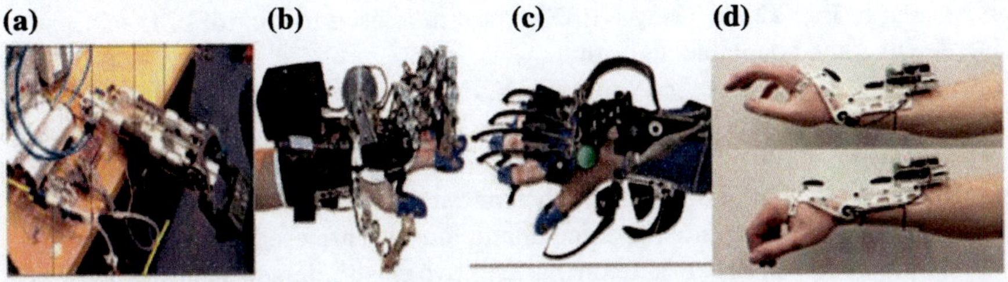

Fig. 22.1 **a** Exosckeleton actionated by artificial pneumatic muscles [18]. **b** SAO-i3 exoskeleton [19]. **c** SPO-F exoskeleton [21]. **d** eWrist exoskeleton [22]

A prototype exoskeleton device was developed at Seoul National University (SNU) to rehabilitate the flexion/extension movement of the wrist. The device is operated by brushless DC motors and sensors are used to determine the reaction torque between the motors and the patient's arm [22, 23].

eWrist [24] is an exoskeleton with one DOF for the flexion/extension movement of the wrist joint specially created to be used independently by the patient at home, with the possibility to monitor his progress. The device has been developed to be as comfortable as possible for the patient to use, so most of the components are 3D printed, in order to reduce the weight as much as possible (Fig. 22.1). The actuation is done by the DC motors and Myo [25] armband which has surface electromyography (sEMG) sensors to detect the patient's intention to move.

A new version of the eWrist device is presented in [26]. The device is an updated version of eWrist, the emphasis being on ergonomics and on the attachment of the device as easily as possible by the patient. The device has three modules, namely, the forearm module, the upper arm module and the Myo armband. To make it easier to attach the device, a one-handed spider-type fastening system has been developed. The device also has a single DOF for flexion/extension of the wrist, and the operation is also done also with a DC motor.

22.2.2 Wrist Exoskeletons with Two Degrees of Freedom

In [27] is presented an exoskeleton with two DOFs for the flexion/extension movements, respectively abduction/adduction of the wrist. From a structural point of view, the device can be divided into three modules, namely: a forearm brace, an actuator module and a hand brace. A geared motor is used for actuation, mounted in the middle of the actuator module. The device is controlled by a microcontroller and has two control modes, namely, position and torque control.

EXOWRIST [27] is an exoskeleton that uses artificial pneumatic muscles (PMA) to perform wrist movements. The device is mostly made of soft materials, to be as comfortable as possible to use. It consists of a neoprene glove that covers the forearm and wrist, over which are mounted plastic modules that fix the four PMAs as it can be seen in Fig. 22.2a. The ANPID algorithm is used to control PMAs, because it is a flexible and adaptable algorithm.

Another exoskeleton device with two DOFs that uses PMAs is presented in [28]. The device consists of a glove that covers the palm and wrist, to which are attached five PMAs (Fig. 22.2b). The flexion movement is performed by the contraction of 2 muscles, the extension of the wrist by the contraction of a single muscle, and for the abduction/adduction movements 2 pneumatic muscles are used.

In [29] is presented an exoskeleton that uses two elastic elements that are deformed during the movement to transmit the torque generated by two linear motors. When the two spring blades move in the same direction, it generates the flexion/extension movement, and when the two move in different directions, the abduction/adduction movement is generated.

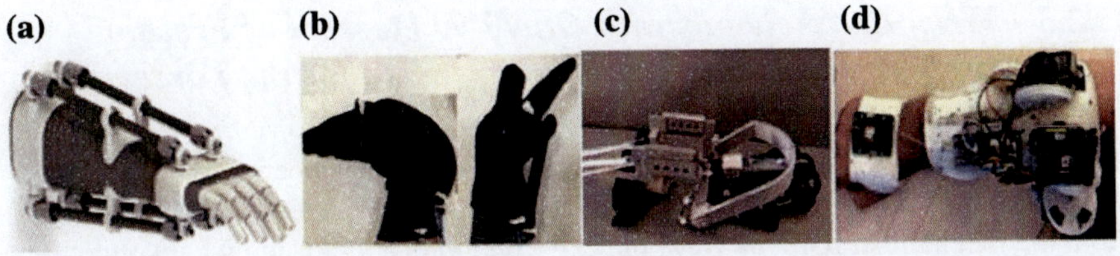

Fig. 22.2 **a** EXOWRIST prototype [27]. **b** Exoskeleton with PMAs [28]. **c** SMA based exoskeleton [31]. **d** Tendon-actuated exosckeelton [32]

An exoskeleton used for rehabilitation with parallel structure is presented in [30]. The device has an over-actuated structure to extend the range of motion (ROM) and to eliminate the singularities during operation.

A 2DOFs exoskeleton that is actuated using Shape Memory Alloy (SMA) based actuators is presented in Fig. 22.2c [31]. The device is highlighted by its low weight and noise-free operation. SMA is a metal alloy that has the property of deforming and returning to its original state after being heated above the transformation temperature between the martensite-austenite phases. The device is attached to the patient's hand by means of a glove that makes the use of the device pleasant, also the structure of the device is symmetrical.

A device, easy to use created both for use in daily activities and for long-term rehabilitation applications is presented in [32]. From a mechanical point of view, the device consists of two modules, a module that is fixed on the hand and a module fixed on the forearm presented in Fig. 22.3d. The two modules are connected to each other by means of three tendons driven by three electric motors. The device is controlled via a microcontroller, and a Bluetooth module communicates the data of a PC application that allows the user to select and track the desired applications.

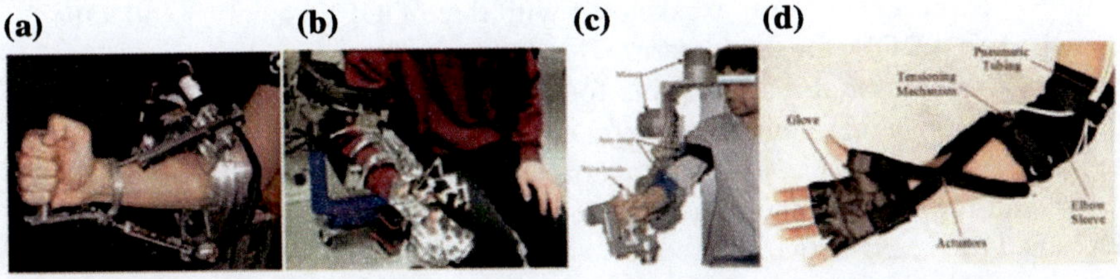

Fig. 22.3 **a** Rice wrist exoskeelton [35]. **b** BRAVO exoskeleton [36]. **c** EST-MARSE exoskeelton.[36] **d** Soft exoskeleton [37]

22.2.3 Wrist Exoskeletons with One/Two Degrees of Freedom for Wrist and One Degree of Freedom for the Forearm

At Rice University, a device called the Rice wrist [33] has been developed for the wrist joint and also reproduces the pronation/supination movement of the forearm. From the mechanical point of view, the exoskeleton is composed by a revolute joint at the forearm and a 3 revolute-prismatic-spherical serial in parallel wrist. This par allel mechanism was chosen over a serial mechanism due to the compactness of the parallel mechanism. For the actuation of the forearm, a frameless brushless DC motor is used and for the wrist actuation high torque rotary electric motors are used. The exoskeleton is presented in Fig. 22.3a.

IntelliArm is an exoskeleton device that has six DOFs, four active and two passive. For the wrist joint, a single degree of freedom is provided for the flexion/extension movement and a degree is provided for the forearm joint. The actuation of the device is performed with the help of DC motors, and the torque is transmitted in several ways, either by cable-driven transmission, or by connecting the motor shaft to gear transmissions. A central digital controller is used to servo control the 4 DOFs [34].

The BRAVO exoskeleton presented in [35] is a device that allows the flexion/extension movement of the wrist and the pronation/supination movement of the forearm and is presented in Fig. 22.3b. The device has a kinematic chain in which the pronation/supination movement is the first active joint, followed by the joint for flexion/extension. The device is driven by two DC geared motors.

ETS-MARSE [36] is an exoskeleton for the upper limb with seven DOFs, of which two DOFs for the wrist joint and a DOFs for the forearm joint (Fig. 22.3c). The device has been designed to be attached to the side of the arm to increase comfort and naturalness during use. From a mechanical point of view, the device can be divided into 3 modules, namely, shoulder mode, elbow and forearm mode and wrist mode. Electric motors and potentiometers are used to operate the device.

In [37] is presented a soft exoskeleton with three DOFs, namely, two DOFs for the wrist and a DOFs for the forearm. The device is made of lightweight materials to make it as comfortable as possible for the patient to use it as it can be seen in Fig. 22.3d. From a mechanical point of view, it consists of a textile glove that covers the hand, an elbow sleeve, and an actuator tensioning mechanism that have the role of transmitting the movement generated by the device to the user. The action is performed by means of two crossed pairs of PMAs, one pair being positioned on the upper part of the forearm, and the other pair on the lower part of the forearm. The device is controlled by an Arduino Mega 2560 and a pressure sensor monitors the state of each actuator.

The WRES [38] is an exoskeleton for the wrist joint that presents a novel capstan-based tendon driven solution to actuate a differential transmission. Using this horse-power transmission, the actuators that generate the motion to perform the flexion/extension and abduction/adduction movements can be positioned closer to the base to reduce the moving mass. The WRES device can also be mounted on an exoskeleton intended for arm rehabilitation and can also be combined with a hand exoskeleton so that it can be used as a rehabilitation device for the entire upper limb. This paper [38] shows the installation of the WRES device with the ALEx elbow exoskeleton and with the hand exoskeleton device named PERCRO.

In this paper [39] is presented an exoskeleton with three DOFs, two DOFs for the two movements of the wrist respectively 1 DOF for the movement of the forearm. The device is made up of 3D printed segments to lighten its total mass which with- out the batteries is only 248 g. The device is operated by means of two linear actuators. They push or pull two sliding springs to make all possible movements of the joint of the hand and forearm.

A device with three DOFs for the two movements of the wrist and for the movement of the forearm joint is presented in [40]. The device is mostly made of nylon by 3d printing. To make the device safe to use, two safety modes have been configured. One is used for protection during the training process when either the preset movement range is exceeded or the tension in the cables is too high. The other mod, is used to get a quick release of the device (Table 22.1).

22.3 Conclusions

Over the time, many types of exoskeletons have been used for upper limb rehabilitation. Due to the increase in the number of people suffering a stroke worldwide, the interest for such devices is enormous, this can be seen by developing more and more suitable devices to be used by patients at home. Devices for the wrist joint generally have one or two degrees of freedom, because the joint has only two degrees of freedom, but for a device to be as comfortable as possible to use and as efficient as possible in wrist rehabilitation applications, it is useful to or include a degree of freedom for the forearm joint. Electric motors are generally used to drive the devices, but other drive methods such as SMA and PMAs have been used due to their low weight. A trend used in recent years is the creation of plastic devices by 3D printing, because it drastically reduces the mass of the device. One thing is clear, with the advancement of technology, this field will be improved and adapted more and more for the required needs.

Table 22.1 Summary of wrist rehabilitation exoskeletons

Device name	*DOFs for wrist reha bilitation*	*Active/Passive device*	*Actuation type*
ULERD	1	Active	DC motor
DCRWOJR	1	Active	PMA
SAO-i3	1	Passive	DC motor
SPO-F	1	Active	Spring
BRENRSEW	1	Active	DC motor
eWrist	2	Active	DC motor
TDPWE	2	Active	DC motor
EXOWRIST	2	Active	PMA
WRERPSA	2	Active	PMA
BRAVO	2	Active	DC motor
LPTDFWEDUMSB	2	Active	DC motor
SMAWERT	2	Active	SMA
TAEWRT	2	Active	DC motor
IntelliArm	2	Active	DC motor
WRES	2	Active	DC motor
RiceWris	3	Active	DC motor
ETS-MARSE	3	Active	DC motor
ASROWR	3	Active	PMA
ANWREPMSM	3	Active	DC motor
SEU-WRE	3	Active	DC motor

Abbreviations: DCRWOJR, Design and Control of a Robotic Wrist Orthosis for Joint Rehabilitation; BRENRSEW, Biomechanical Reactions of Exoskeleton Neurorehabilitation Robots in Spastic Elbows and Wrists; TDPWE, Towards the development of a portable wrist exoskeleton; WRERPSA, Wrist Rehabilitation Exoskeleton Robot based on Pneumatic Soft Actuators; LPTDFWEDUMSB, Low Profile Two-Degree-of-Freedom Wrist Exoskeleton Device Using Multiple Spring Blades; SMAWERT, SMA based wrist exoskeleton for rehabilitation therapy; ASROWR, A Soft Robotic Orthosis for Wrist Rehabilitation; ANWREPMSM, A novel wrist rehabilitation exoskeleton using 3D-printedmulti-segment mechanism

References

1. Puthenpurakal, A., Crussell, J.: Stroke 1: definition, burden, risk factors and diagnosis. In: Nursing Times, (2017)
2. Go, A.S., Mozaffarian, D., Roger, V.L., Benjamin, E.J., Berry, J.D., Borden, W.B., et al.: Heart disease and stroke statistics. In: American heart association, (2016)
3. Townsend, K.W.N., Bhatnagar, P., Smolina, K., Nichols, M., Leal, J., Luengo-Fernandez, R. Rayner, M.: Coronary heart disease statistics. In: British heart foundation, (2012)
4. Rehmat, N., Zuo, J., Meng, W. et al.: Upper limb rehabilitation using robotic exoskeleton systems: a systematic review. Int. J. Intell. Robot. Appl., (2018)
5. Chen, J., Zhu, Q., Yu, L., Li, Y., Jia, S., Zhang, Z.: Stroke risk factors of stroke patients in China: A nationwide community-based cross-sectional study. Int. J. Environ. Res. Public Health, (2022)

6. Townsend, N., Nichols, M., Luengo-Fernandez, R., Leal, J., Gray, A., Scarborough, P., Rayner, M.: European cardiovascular disease statistics. Eur. Hear. Netw. AISBL, (2012)
7. Tsugawa, Y., Kumamaru, H., Yasunaga, H., Hashimoto, H., Horiguchi, H., Ayanian, J. Z.: The association of hospital volume with mortality and costs of care for stroke in Japan. In: National Library of Medicine, (2013)
8. http://www.stroke.org.nz/stroke-facts-and-fallacies—Accessed in 20 June 2022.
9. Lindsay, M.P., Norrving, B., Sacco, R.L., Brainin, M., Hacke, W., Martins, S., Pandian, J., Feigin, V.: Global stroke fact sheet. In: National library of Medicine, (2019)
10. https://www.safestroke.eu/wp-content/uploads/2020/10/01-At_What_Cost_EIOS_Factsheet.pdf Accesed in 01 July 2022
11. https://www.socialstyrelsen.se/globalassets/sharepointdokument/artikelkatalog/statistik/2021-12-7644.pdf. Accesed in 01 July 2022
12. Gull, M.A., Bai, S., Bak, T.: A review on design of upper limb exoskeletons. Robotics, (2020)
13. Gopura, R.C., Kiguchi, K., Bandara, S.V.: A brief review on upper extremity robotic exoskeleton systems. In: Industrial and information systems, (2011)
14. Pezent, E., Rose, C., Deshpande, A., Omalley, M.: Design and characterization of the Open-Wrist: A robotic wrist exoskeleton for coordinated hand-wrist rehabilitation. In: International conference on rehabilitations robotics IEEE Xplore, (2017)
15. Esmaeili, M., Dailey, W., Burdet, E., Campolo, D.: Ergonomic design of a wrist exoskeleton and its effects on natural motor strategies during redundant tasks. In: IEEE International conference on robotics and automation, (2013)
16. Kazerooni, H.: Exoskeletons for human performance augmentation. In: Springer hand- book of robotics, (2008)
17. Song, Z., Guo, S.: Design process of exoskeleton rehabilitation device and implemen tation of bilateral upper limb motor movement. J. Med. Biol. Eng., 2011
18. Meng, W., Sheng, B., Klinger, M., Liu, Q., Zhou, Z., Xie, S.Q.: Design and control of a robotic wrist orthosis for joint rehabilitation. In: IEEE/ASME (AIM) International conference on advanced intelligent mechatronics, (2015)
19. Ates, S., Mora-Moreno, I., Wessels, M., Stienen, A.: Combined active wrist and hand orthosis for home use: Lessons learned. In: Conference: Rehabilitation robotics (ICORR), (2015)
20. Ates, S., Lobo-Prat, J., Lammertse, P.,vad der Koji, H., Stenen, A.H.A.: SCRIPT passive orthosis: Design and technical evaluation of the wrist and hand orthosis for rehabilitation training at home. In: IEEE international conference on rehabilitation robotics, (2013)
21. Ates, S., Haarman, C.J., Stenen, A.H.A.: SCRIPT passive orthosis: design of interactive hand and wrist exoskeleton for rehabilitation at home after stroke. Auton. Robot., (2017)
22. Nam, H.S., Koh, S., Kim, Y.J., Beom, J., Lee, W.H., Lee, S.U., Kim, S.: Biomechanical reactions of exoskeleton neurorehabilitation robots in spastic elbows and wrists. In: IEEE Transactions on neural systems and rehabilitation engineering, (2017)
23. Hussain, S., Jamwal, P.K., Van Vliet, P., Ghyesh, M.H.: State of the art robotic devices for wrist rehabilitation: Design and control aspects. In: IEEE Transactions on human-machine systems, (2020)
24. Lambelet, C., Lyu, M., Wenderoth, N., Woolley, D., Gassert, R.: The eWrist—A wearable wrist exoskeleton with sEMG-based force control for stroke rehabilitation. In: IEEE International conference on rehabilitation Robotics, (2017)
25. Visconti, P., Gaetani, F., Zappatore, G.A., Primiceri, P.: Technical features and functionalities of Myo armband: An overview on related literature and advanced appli cations of myoelectric armbands mainly focused on arm prostheses. Int. J. Smart Sens. Intell. Syst., 2018
26. Lambelet, C., Temiraliuly, D., Siegenthaler, M., Wirth, M., Woolley, D.G., Lambercy, O., Gassert, R., Wenderoth, N.: Characterization and wearability evaluation of a fully porta ble wrist exoskeleton for unsupervised training after stroke. J. NeuroEngineering Rehabil., (2020)
27. Xiao, Z.G., Menon, C.: Towards the development of a portable wrist exoskeleton. In: IEEE international conference on robotics and biomimetics, (2011)
28. Andrikopoulos, G., Nikolakopoulos, G., Manesis, S.: Motion control of a novel robotic wrist exoskeleton via pneumatic muscle actuators. In: 20th IEEE international conference on emerging technologies and factory automation (ETFA), (2015)

29. Al-Fahaam, H., Davis, S., Nefti-Meziani, S.: Wrist rehabilitation exoskeleton robot based on pneumatic soft actuators. In: Proceeding of international conference for stu dents on applied engineering (ICSAE), (2017)
30. Higuma, T., Kiguchi, K., Arata, J.: Low-profile two-degree-of-freedom wrist exoskel eton device using multiple spring blades. IEEE Robot. Autom. Lett., (2017)
31. Lee, J., Song, B.W., Yang, W.: Design of exoskeleton-type wrist human–machine inter- face based on over-actuated coaxial spherical parallel mechanism. Adv. Mech. Eng
32. Del Cerro, D.S., Copaci, D., E Moreno, L., Blanco, D.: SMA based wrist exoskeleton for rehabilitation therapy*. IROS, (2018)
33. Dragusanu, M., Lisini, T., Iqbal, M.Z., Prattichizzo, D., Melvezzi, M.: Design, development, and control of a tendon-actuated exoskeleton for wrist rehabilitation and Training. In: IEEE international conference on robotics and automation (ICRA), (2020)
34. Gupta, A., O'Malley, M.K., Patoglu, V., Burgar, C.: Design, control and performance of rice wrist: A force feedback wrist exoskeleton for rehabilitation and training. Int. J. Robot. Res., (2008)
35. Ren, Y., Kang, S.H., Park, H.S., Wu, Y.N., Zhang, L.Q.: Developing a multi-joint upper limb exoskeleton robot for diagnosis, therapy, and outcome evaluation in Neuro rehabilitation. In: IEEE transactions of neural systems and rehabilitation engineer ing, (2013)
36. Troncossi, M., Mozaffari Foumashi, M., Mazzotti, C., Zannoli, D., Castelli, V.P.: Design and manufacturing of a hand-and-wrist exoskeleton prototype for the rehabilitation of post-stroke patients. In: Quaderni del DIEM—GMA. Atti della Sesta Giornata di Studio Ettore Funaioli, (2012)
37. Rahman, M.H., Rahman, M.J., Cristobal,O.L., Saad, M., Kenne, J.P, Archambault, P.S.: Development of a whole arm wearable robotic exoskeleton for rehabilitation and to assist upper limb movements. Robotica, (2014)
38. Bartlett, N.W., Lyau, V., Raiford, W.A., Holland, D., Gafford, J.B., Ellis, T.D., Walsh, C.J.: A soft robotic orthosis for wrist rehabilitation. In: J. Med. Devices, (2015)
39. Buongiorno, D., Sotgiu, E., Leonardis, D., Marchschi, S., Solassi, M., Frisoli, A.: WRES: A novel 3 DoF wrist exoskeleton with tendon-driven differential transmission for neuro-rehabilitation and teleoperation. In: IEEE robotics and automation letters, (2018)
40. Yang, S., Li, M., Wang, J., Wang, T., Liang, Z., He, B., Xie, J., Xu, G.: A novel wrist reha- bilitation exoskeleton using 3D-printed multi-segment mechanism. In: 43rd Annual international conference of the IEEE engineering in medicine & biology society (EMBC), (2021)
41. Shi, K., Song, A., Li, Y., Li H., Chen, D., Zhu, L.: A cable-driven three-DOF wrist rehabilitation exoskeleton with improved performance. Front. Neuro Robot., (2021)

Chapter 23
Design of an Exoskeleton for Rehabilitation Ankle Joint

Z. Nursultan, M. Ceccarelli, and G. Balbayev

Abstract This paper presents a low-cost design of a device to assist in ankle rehabilitation. The article introduces the mechanical design of the proposed device, covering the ankle joint of the lower body. The size of the exoskeleton, referring to the segments of the limbs, is limited to a certain range of proportions of the object. Each joint is motorized and equipped with a specific mechanism, such as a linear actuator and a four-bar mechanism efficiency in terms of speed and power transmission help a patient to regain the ability to move. The exoskeleton is made specifically for outdoor use so that the lower body can move as well as support the weight of the assisted person.

Keywords Medical devices · Biomechanics · Motion assisting devices · Ankle exoskeleton · Design

23.1 Introduction

The ankle joint is a complex joint that forms a kinematic connection between the lower limb and the foot, which allows you to perform everyday tasks of walking [1]. Unfortunately, this joint is very susceptible to acute and long-term injury in physically active people. Thus, rehabilitation is necessary to ensure full recovery of the injured joint.

Physical therapy and motor exercises are required to treat these cases. With the help of robotic devices, the training of people's movements can be improved with controlled exercises. Many existing solutions for rehabilitation exercises have

Z. Nursultan (✉)
Satbayev University, Almaty, Kazakhstan
e-mail: n.zhetenbayev@gmail.com

M. Ceccarelli
LARM2 Laboratory of Robot Mechatronics, University of Rome Tor Vergata, Rome, Italy
e-mail: marco.ceccarelli@uniroma2.it

G. Balbayev
Eurasian National University, Nur-Sultan, Kazakhstan

I. Doroftei et al. (eds.), *Proceedings of SYROM 2022 & ROBOTICS 2022*, Mechanisms and Machine Science 127, https://doi.org/10.1007/978-3-031-25655-4_23

revealed the need to develop inexpensive devices that allow performing rehabilitation exercises for patients with damaged limbs. An exoskeleton with artificial muscles can help people restore the function of individual joints of the lower and upper extremities [2].

Effective rehabilitation is a long process that also requires a to qualified physiotherapist who will help restore joint mobility. Rehabilitation training is usually carried out by a physiotherapist in individual sessions with a patient since it requires the full attention of a physiotherapist [3]. Consequently, in this regard, robotics can be used in rehabilitation therapy for constant monitoring of the patient and assistance in performing movements [4].

Most of the known robotic structures are based on the construction of a static platform, where a non-portable device requires the patient to keep his foot on a grounded platform (usually sitting) to complete rehabilitation as reported in [5–7].

The proposed design includes a mechanism driven by electric linear actuators that help to run the ankle joint. Such structures have small dimensions and are also not attached to anything. Thus, patients do not need to go to the hospital, where there is a rehabilitation machine. This paper presents the design and analysis of an ankle joint device with electric linear actuators.

23.2 Biomechanics of the Ankle Joint

To design human assisting robots, it is very important to know the biomechanics of the leg-foot joint. It helps to choose the necessary robot parameters for the preparation of the design of rehabilitation robots, which are effective for patients with im-paired motor function, such as walking. The legs are the main part of the human musculoskeletal system, and they can simultaneously have rigid and flexible structures due to various motor features.

As shown in Fig. 23.1a, the combination of the tibia and subtitle joints facilitates movement in the ankle joint [8]. The tibia joint forms a connection between the tibia and the fibula and the upper surface of the talus bone, while the subtitle joint forms a connection between the lower part of the talus bone and the upper surface of the heel bone.

Axes and planes of rotation are used to characterize the spatial configuration of the ankle joint. The foot can be considered divided into sagittal, frontal, and transverse planes (Fig. 23.2b). These planes allow us to find three axes of rotation in the mutual pediculus used to design the movement of the joint (parallel to these planes) [8].

Abduction and Adduction are rotational movements of the heel around its axis parallel to the horizontal plane. With characteristic values of reference listed in Table 23.1.

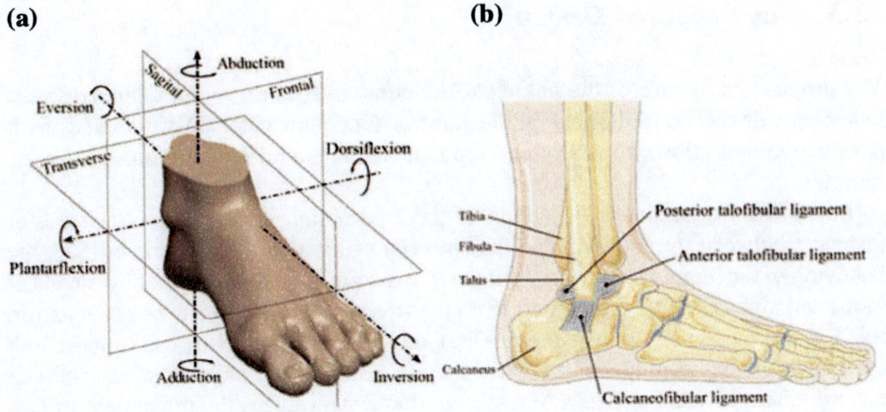

Fig. 23.1 Biomechanical data of the ankle joint: **a** planes and movement **b** bone and muscle structure

Fig. 23.2 The kinematic design of the proposed Ankle Exoskeleton with design parameters

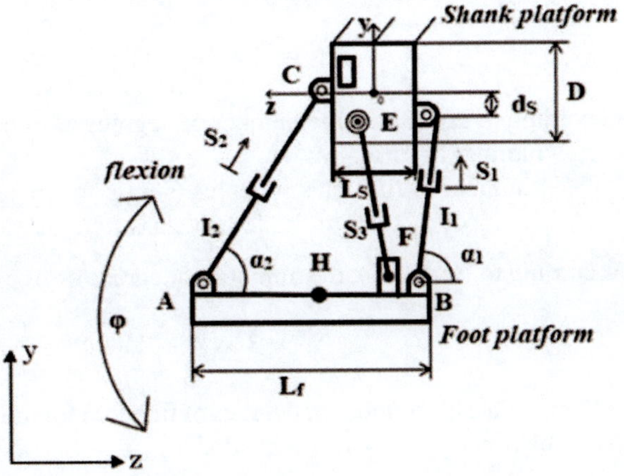

Table 23.1 Characteristic date for movement of the ankle joint, [3]

Motion	Dorsiflexion (deg)	Plantarflexion (deg)	Abduction/Adduction (deg)	Inversion/Eversion (deg)
Range limits	±20	±50	±10	±12

23.3 The Proposed Design

The proposed structure of the ankle exoskeleton consists of a joint connecting the lower part of the body. For the ankle joint, a four-joint mechanism is used, as it is effective and allows to reproduce joint movements and contractions of the leg muscles.

In Fig. 23.2 a passive exoskeleton of the ankle joint, is proposed consisting of fasteners between the foot (A) and the shin (B) using a ball joint element (S3), the fastening of the human foot consists of two parts connected to each other by means of loops and attached to the upper part from the front. (S1) (S2) electric linear actuator, fixed at the back and front of the ankle. Connections E and F of the support link the ball joint is the guide link between the shank and the support platform. The electric linear actuator consists of a system that synchronizes the movement of two interconnected skeletons and moves the ankle together with the exoskeleton.

To determine the number of degrees of freedom, we use P.L. Chebyshev's structural formula for a plane mechanism:

$$W = 3 \times (n - 1) - 2 \times p_5 - p_4 \tag{23.1}$$

According to Eq. (23.1), the number of degrees of freedom of a plane mechanism is:

n − number of links.

P_5 − number of fifth-class couples (single-movement couples);

P_4 − number of pairs of the fourth class (two movable pairs).

According to Eq. (23.1), the number of degrees of freedom of a plane mechanism is:

$$W = 3 \times (8 - 1) - 2 \times 9 = 3$$

Let's calculate the number of degrees of freedom for the plane mechanism in Fig. 23.2 using the Somov–Malyshev formula:

$$W = 6 \times (n - 1) - p_5 - 4 \times p_4 - 3 \times p_3 - 2 \times p_2 - p_1 \tag{23.2}$$

p_i− s the number of kinematic pairs of the−class with $(6 - i)$ degrees of mobility $(i = 1, \ldots 5)$.

The number of degrees of freedom is equal to:

$$W = 6 \times (8 - 1) - 5 \times 9 = -3$$

The mechanism in question has three degrees of freedom. To eliminate redundant connections, we will replace the rotational pairs connected to the output link with spherical ones. In this case, each chain imposes one link. The coupling reaction is directed along the axis of a spherical pair parallel to the z axis. Such three kinematic chains impose three bonds–three forces parallel to the z axis. These three forces create

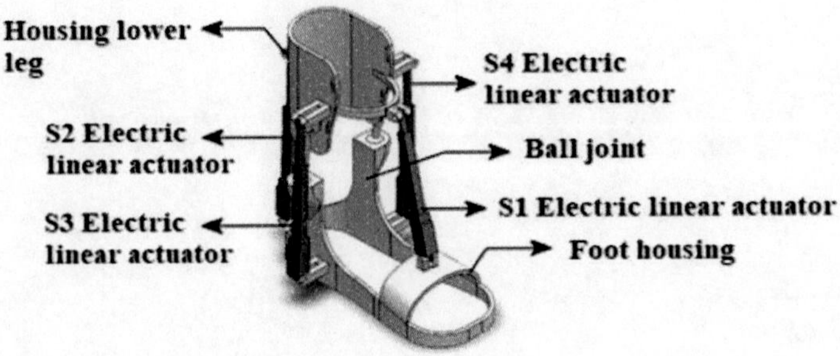

Fig. 23.3 A CAD mechanical design of the design in Fig. 23.2

moments around the x, y axes. Thus, the reactions of the bonds are two moments and a force.

This exoskeleton can be used as an orthosis when it is attached externally to the subject's leg. The frame for the shoe model is prepared in such a way that the shoes can be worn. The four-link connection with which links is designed in such a way that it is in two different planes to avoid collisions between exoskeleton parts.

The CAD design of the exoskeleton model, driven by a linear actuator, is shown in Fig. 23.3, which is equipped with a ball joint mechanism to increase the output torque coefficient.

23.4 Performance Analysis

A 3D modeling and simulation calculations were performed in a virtual environment using the Solidworks Simulation software, and the Motion Simulation package. Using Solidworks Simulation an electric linear actuator input is given, which generates movement of the ankle joint. Figure 23.4 shows simulated Abduction and Abduction obtained using Solidworks Simulation. Adduction the range of motion in this movement will be up 22.0–36°, while the Simulation will bend to 15°. The range of motion in Abduction is from 15.4° to 25.9°, and in Simulation, binding is up to 10°.

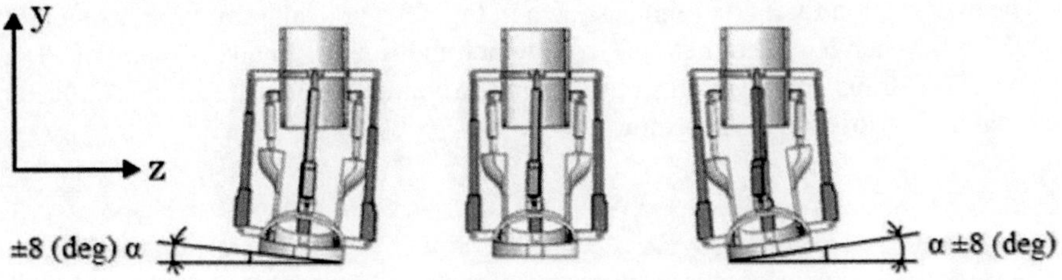

Fig. 23.4 A snapshot of simulated Abduction–Adduction assisted motion

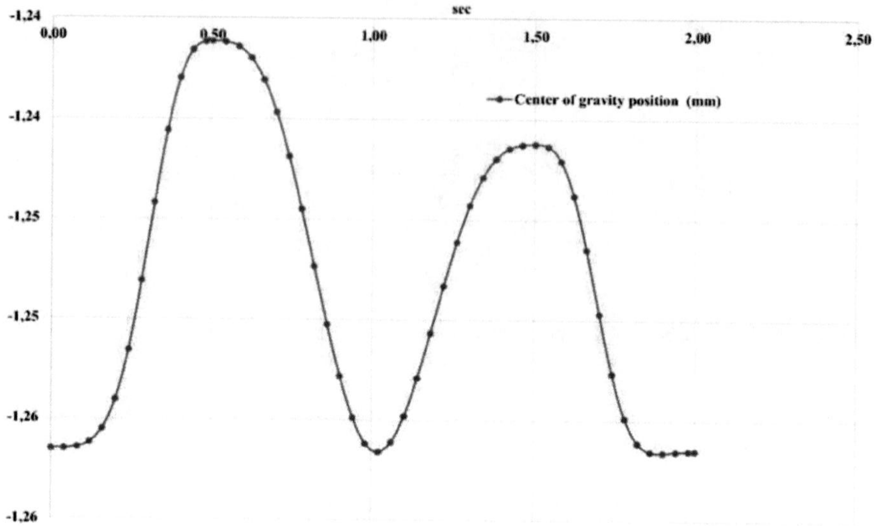

Fig. 23.5 Input data for the simulated motion in Fig. 23.4 in Center of gravity position

Figure 23.5 shows the computed results in of components of Center of gravity position of the platform. The movements of this point along Y axis are equal to −1.23 mm.

The Center of gravity in this study shows the stability of the equilibrium positions of bodies and continuous media under the influence of gravity, namely: in the resistance of materials–when using the Vereshchagin rule.

Figure 23.6 shows components of linear displacement of the linear actuators. From the plot, the motions from all directions are with peaks of maximum values 30 deg/s^2. represented for Y –component and less than 25 deg/s^2 for another components.

The angular acceleration regarding the angle is shown in Fig. 23.7. The largest value of the acceleration, which equals to 33 deg/s^2, is near the top position. Angular acceleration characterizes the intensity of changes in the modulus and direction of angular velocity when the ankle joint moves with an exoskeleton.

This prototype, validated by the reported simulation analysis, will be used in future testing both for checking the feasibility of the design solution and for characterizing its operation on performance in ankle rehabilitation motion.

The exoskeleton for ankle joint is shown in Fig. 23.8 as a lab prototype. Its weight is 205 kg for each leg. The material from which the frame is made is made of PLA plastic. The straps for fastening the leg and drive are also made of PLA plastic material and printed on a 3D printer.

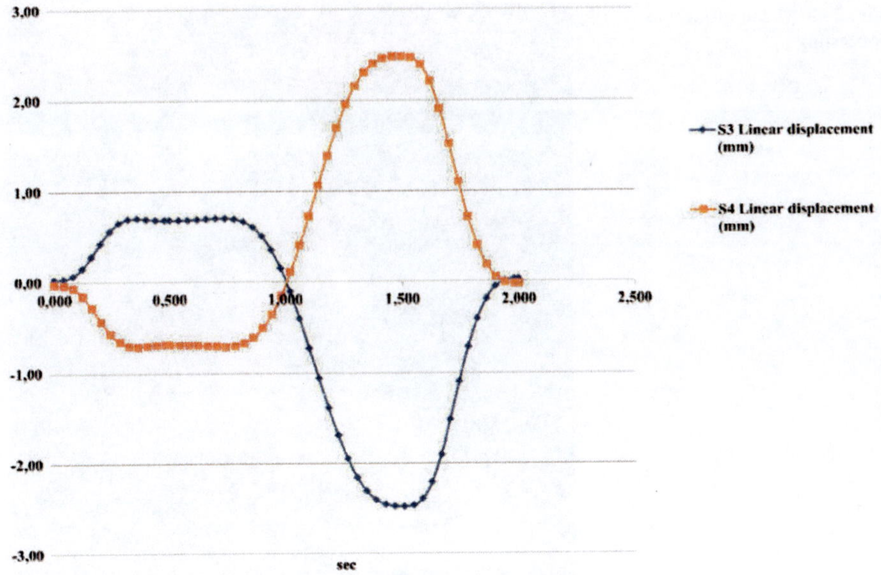

Fig. 23.6 Computed simulated motion in Fig. 23.4 in terms of displacement of linear actuators

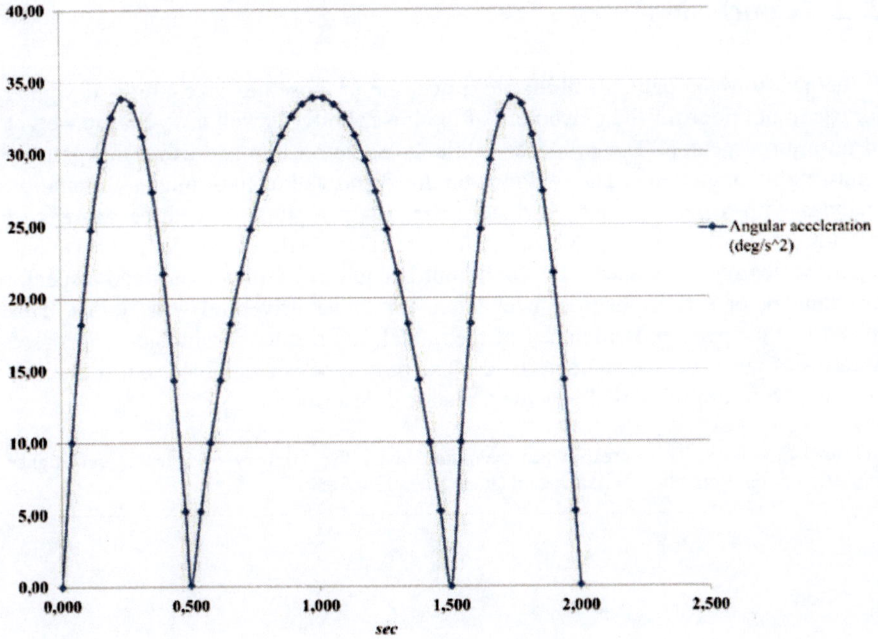

Fig. 23.7 Computed results of the simulated motion in Fig. 23.4 in terms of angle of the foot platform

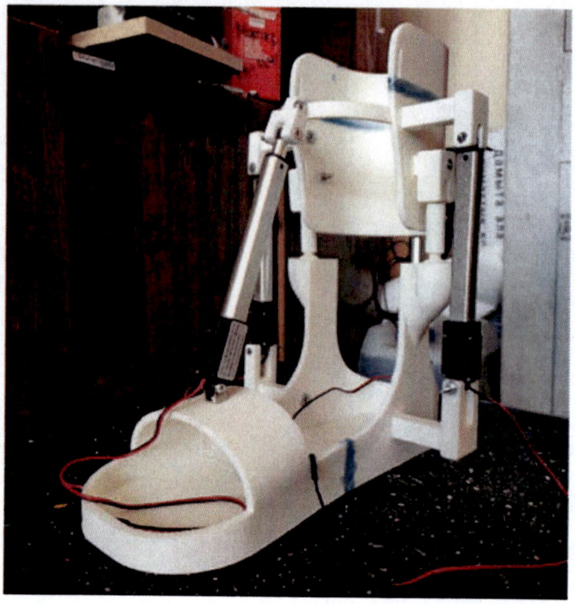

Fig. 23.8 A lab prototype for testing

23.5 Conclusion

To help restore the patient's ability to move, the proposed exoskeleton is designed for use in out door for the movement of the lower body, as well as to lift the weight of an injured person. The proposed ankle exoskeleton is made using inexpensive lightweight components. The design is finalized and validated through a Simulation analysis with a performance characterization. A prototype with three degrees of freedom of rotation around a virtual stationary center is developed for intensive and repetitive leg recovery exercises. Each joint is equipped with a motor and a special mechanism, of a four-position mechanism for linear drive and wrist joints. This increases the speed and efficiency of the model, the results of which are discussed in terms of performance characteristics. The prototype is assembled to test the new design, and future work will provide a practical description.

Acknowledgements This research has been funded by the Ministry of Science and Higher Education of the Republic of Kazakhstan, Grant № AP13268857.

References

1. Alvarez-Perez, M.G., Garcia-Murillo, M.A., Cervantes-Sánchez, J.J.: Robot-assisted ankle rehabilitation: A review. Disabil. Rehabil. Assist. Technol. **15**, 394–408 (2020)
2. Idumudi, V.S.P., Ceccarelli, M., Russo M.: Control design for cable ankle, a cable driven manipulator for ankle motion assistance. In: MDPI Actuators, **11**(2), 63 (2022). https://doi.org/10.3390/act11020063
3. Zhetenbayev, N., Titov, A., Ceccarelli, M., Balbayev, G.: Design and performance of a motion—assisting device for ankle. In: Proceedings mechanisms and machine science book series (Mechan. Machine Science, vol. 113) ASIAN MMS 2021, MMS 113, pp. 659–668(2021)
4. Russo, M., Ceccarelli, M.: Analysis of a wearable robotic system for ankle rehabilitation MDPI. Machines, **8**(3), 48 (2020). https://doi.org/10.3390/machines8030048
5. Yoon, J., Ryu, J., Lim, K.-B.: Reconfigurable ankle rehabilitation robot for various exercises. J. Robot. Syst. **22**, S15–S33 (2006)
6. Roy, A., Krebs, H.I., Williams, D.J., Bever, C.T., Forrester, L.W., Macko, R.M., Hogan, N.: Robot-aided neurorehabilitation: A novel robot for ankle rehabilitation. IEEE Trans. Robot. **25**, 569–582 (2009)
7. Rodríguez-León, J.F., Chaparro-Rico, B.D.M., Russo, M., Cafolla, D.: An autotuning cable-driven device for home rehabilitation. J. Health. Eng, 6680762, (2021)
8. Siegler, S., Chen, J., Schneck, C.: The three-dimensional kinematics and flexibility characteristics of the human ankle and subtalar joint's part i: kinematics. J Biomech Eng. **110**, 364–373 (1988)

Chapter 24
Requirements and Characteristics of Finger Motion Assistance

K.-E. Lovasz, A. Olah, and M. Ceccarelli

Abstract Lab experiences are presented in characterizing human finger motion and grasping with comparison to the operation of ExoFinger, finger exoskeleton, with the aim of reformulating design and operation requirements for a new finger exoskeleton. Motion assistance of index finger is specifically addressed in order to look for exoskeleton features in wearability, comfort, ease of operation, lightweight and compact design.

Keywords Biomechanics · Motion assistance · Finger analysis · Finger exoskeleton · Testing

24.1 Introduction

Hands are among the most used parts of the human body, with an important role in day-to-day activities. Fingers can be easily injured, which decreases the capabilities of the human hand. Motion assistance of finger function is achieved by manual operation of physiotherapists with long complex operations. Alternatively, for a couple of decades, finger exoskeletons have been used to assist finger motion, even in other daily applications. Examples of existing finger exoskeletons with positive experiences can be referred to the cases in [2–5]. In [2] a lightweight prototype of a pneumatically actuated glove-hand exoskeleton is presented with 3D printed flexible thermoplastic glove giving a limited performance in the index finger with obstructions for the thumb. Reference [3] presents an eight-link mechanism actuated by two

K.-E. Lovasz (✉) · A. Olah
Politehnica University Timisoara, 300222 Timisoara, Romania
e-mail: karla.lovasz@student.upt.ro

A. Olah
e-mail: andrada.olah@student.upt.ro

M. Ceccarelli
Department of Industrial Engineering, University of Rome Tor Vergata, 00133 Roma, Italy
e-mail: marco.ceccarelli@uniroma2.it

I. Doroftei et al. (eds.), *Proceedings of SYROM 2022 & ROBOTICS 2022*, Mechanisms and Machine Science 127, https://doi.org/10.1007/978-3-031-25655-4_24

cables that allow 3 degrees of freedom for the finger adduction with a low resistance against finger movement and easily adjustable. The solution in [4] is a cable driven design with 5 degrees of freedom with poor feature of portability because of its heavy design. The glove-based patent in [5] is an exoskeleton for the index and thumb using pneumatic actuators to move the fingers fluidly, but the linkage design limits the movement. Previous designs and experiences were carried out at LARM2 in Rome as reported in [6–8]. Reference [6] shows a 2 DOF linkage design to assist the phalanx motion independently and in a coordinated way. An experimental characterization is reported in [7] by also looking at the physiological effects on the finger tissues. The novel solution is patented, [8].

This paper presents recent work to revise requirements and characteristics of the human finger motion assistance with experimental lab results to look for new solutions of the LARM2 index finger exoskeleton.

24.2 Anatomy and Functioning of Human Finger

The index finger contains tendons and bones that enable its movement, as shown in Fig. 24.1a. Figure 24.1b shows a kinematic functional scheme with three joints: φ for flexion/extension of the MCP (metacarpophalangeal joint), θ for flexion/extension of the PIP (proximal interphalangeal joints), and ψ for the flexion/extension of the DIP (distal interphalangeal joints). FT stands for "fingertip". The considered average link sizes for an adult index finger are considered as, [1], L1 = 41 mm from the MCP to PIP, L2 = 25 mm from the PIP to DIP, and L3 = 18 mm from the DIP to FT, giving an overall length of 88 mm. Expected ranges for the angular motion are $\varphi = 90°$, $\theta = 90°$, and $\psi = 90°$.

In Fig. 24.2 and Table 24.1 considered general requirements are summarized for design and operation of finger motion assistance using exoskeleton mechanisms for motion and force transmission features.

In Fig. 24.2 requirements are outlined in term of main aspects, considering size constraints to fit on the finger size of a user, expected motion ranges in the articulation of the assisted phalanxes, smooth finger assisted motion with limited continuous

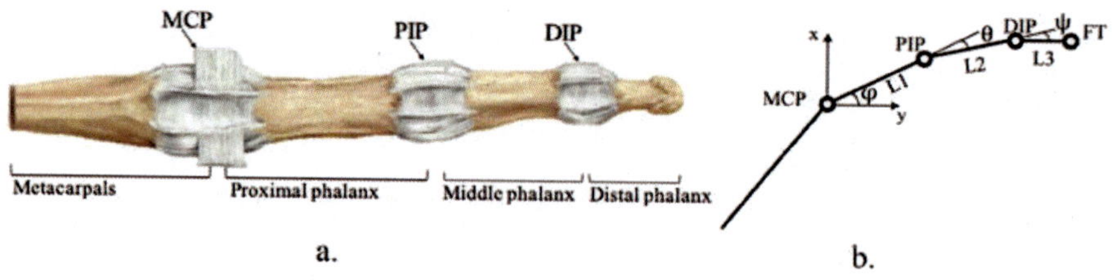

Fig. 24.1 Anatomy of the index finger: **a**. structure of the finger, **b**. kinematic scheme

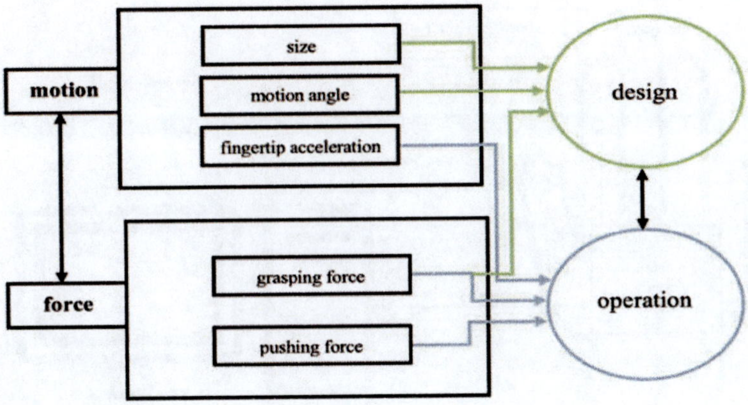

Fig. 24.2 A scheme for requirements

Table 24.1 Values of reference for the requirements and parameters

Parameter	Design	Operation
Size	5–8.8 cm	5–8.8 cm
Motion angle	0–30 deg	0–25 deg
Fingertip acceleration	1–10 m/s^2	1–5 m/s^2
Grasping force	1–50 N	1–25 N
Pushing force	1–100 N	1–30 N

fingertip acceleration, finger force with proper values in grasping and pushing modes. All those aspects, that can be evaluated with reference values in Table 24.1 from the expected average functioning [1], can be used both for design and operation solutions.

24.3 Lab Testing of Index Finger Application

Figure 24.3. presents the lab setup that has been used to conduct a series of tests for determining the functioning parameters of a human hand. The setup is based on two force sensors FSR400 [9], and one IMU sensor BMI160 [10], with the purpose of achieving high accuracy when measuring the parameters of a human hand operation.

The lab setup design in Fig. 24.3 shows also the mechanical design and the connections among the components, whereas the miniaturized size of the sensors permits their installation on a glove with electrical cables connecting to a Nano Arduino of 18 × 45 mm on the glove yet. The communication with a PC for data elaboration and visualization is worked by cable although, it can be by Wi-Fi or Bluetooth.

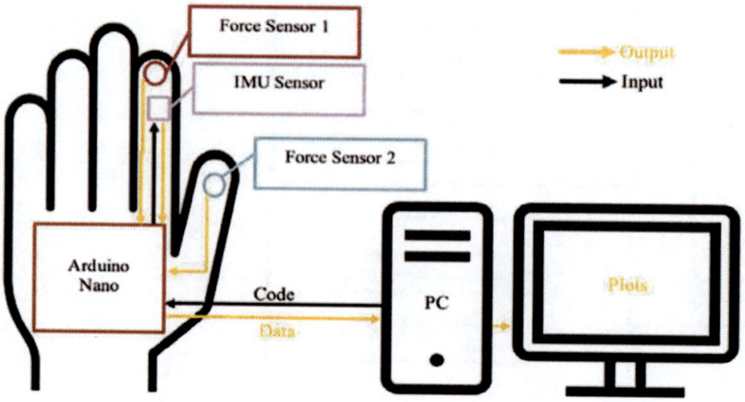

Fig. 24.3 A scheme of the lab setup with the designed sensored glove

Figure 24.4. shows an example of the testing with the sensored glove while grasping with finger and thumb reporting motion data and grasping forces with a sample acquisition rate of 50 Hz. The grasped object is a USB plug with 60 g weight. The test was repeated 3 times for statistical significance of acquired data with similar results.

Figures 24.5 and 24.6 give an example of satisfactory acquired results from the tests, as in Fig. 24.4. In Fig. 24.5 one can note the smooth motion of the index fingertip in terms of acceleration less than 4.2 m/s^2 with significant noise that is very likely due to movable fixture of the sensor on the finger, compliance of finger tissue, hand natural tremor. Figure 24.6 shows the grasping effects in terms of the forces exerted by the fingertips of index and thumb with a proper time evolution up to a firm grasp at about 2.6 N by index and at 1.1 N about by the thumb. The shown differences in the test repetitions can be due to the above reasons and not a very precise repetition conditions that nevertheless give the characteristics performance behavior and values of the human finger grasping.

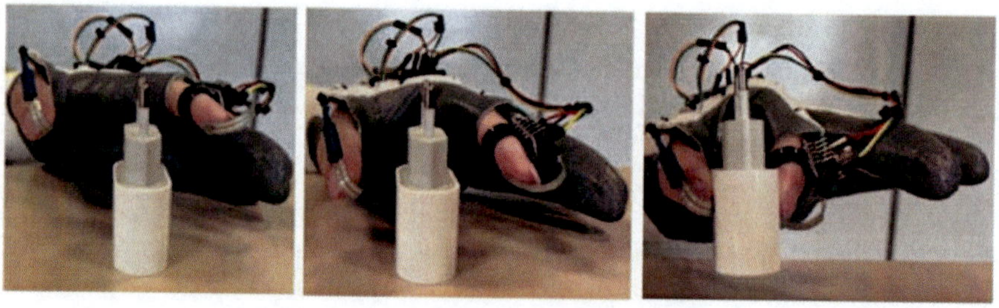

Fig. 24.4 A snapshot of a testing of grasping motion with the sensored glove in Fig. 24.3

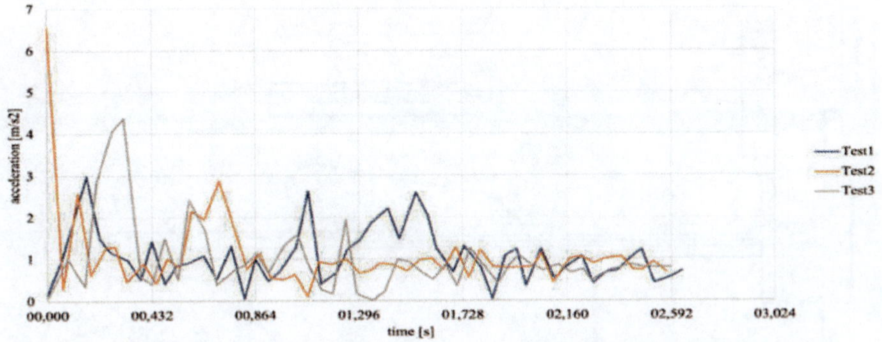

Fig. 24.5 Acquired acceleration of the index fingertip during the grasping test in Fig. 24.4

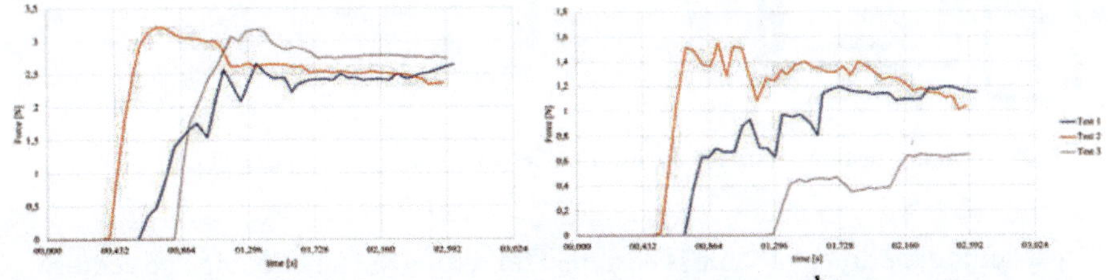

Fig. 24.6 Acquired force during the grasping test in Fig. 24.4 by: **a**. Index; **b**. Thumb

24.4 Lab Testing with ExoFinger Exoskeleton

The mechanical design of the ExoFinger exoskeleton in Fig. 24.7a is based on an Italian patent [8], and is formed by an eight-bar linkage mechanism. The used exoskeleton in Fig. 24.7b is designed with two interconnected four-bar mechanisms as a linkage of two DOFs to provide independent motion to the metacarpus phalange and distal phalange [6, 7].

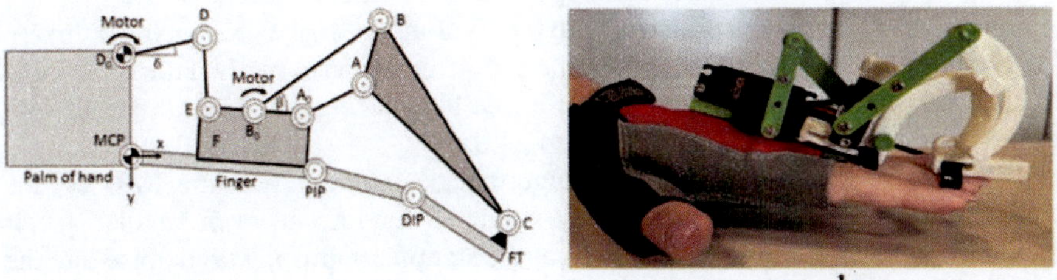

Fig. 24.7 The ExoFinger exoskeleton, [6–8]: **a**. a kinematic scheme; **b**. a prototype

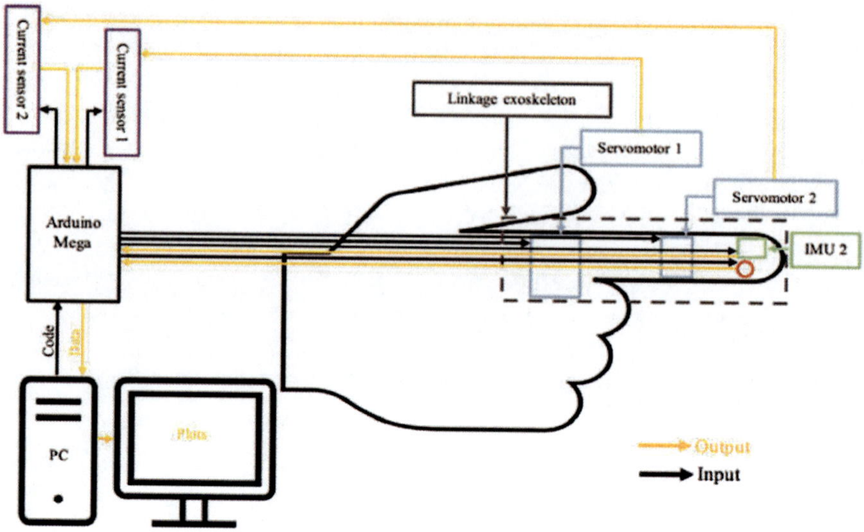

Fig. 24.8 Designed testing set up for the ExoFinger exoskeleton prototype in Fig. 24.7

A specific experimental setup is designed for testing the functioning and acquiring operation parameters as reported in Fig. 24.8 with a similar overall design structure for the sensored glove in Fig. 24.3.

The ExoFinger exoskeleton is driven by two servomotors, each used to drive one DOF. The first servomotor MG995 [12], moves the MCP with the whole finger while the second one MG90S [13], with smaller size and power being installed on a phalanx, moves the PIP together with the DIP. The two servomotors can be actuated either simultaneously in a coordinate way for a coordinated motion of the finger phalanxes or separately for individual motion of the phalanxes. The exoskeleton is provided with three different types of sensors to acquire information on the assisted motion operation. Two IMU sensors BMI160 [10], measure angle displacements, angular velocities, and linear acceleration of the metacarpus and distal phalanges; two current sensors ACS712 [11], measure the power consumption of each motor; and the finger forces are acquired through two force sensors FSR 400 [9].

Figure 24.9 shows a grasping test with the ExoFinger exoskeleton on index finger. The grasped object is a 60 g USB plug. The test are designed to verify if the ExoFinger exoskeleton gives a similar operation of a free human finger as from the tests with the sensored glove in Figs. 24.4, 24.5 and 24.6.

Figure 24.10 shows the acquired fingertip acceleration from the IMU on the extremity of the exoskeleton with very similar behavior and even smaller acceleration value from 2.5 m/s^2 to near zero at the stationary grasp, with a noise due the similar reasons as per the plots resulted from the experiment with the sensored glove.

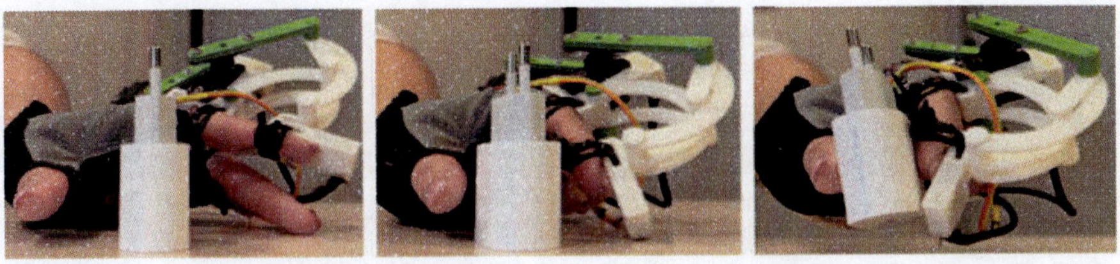

Fig. 24.9 A snapshot of a grasping test with ExoFinger prototype in Figs. 24.7 and 24.8

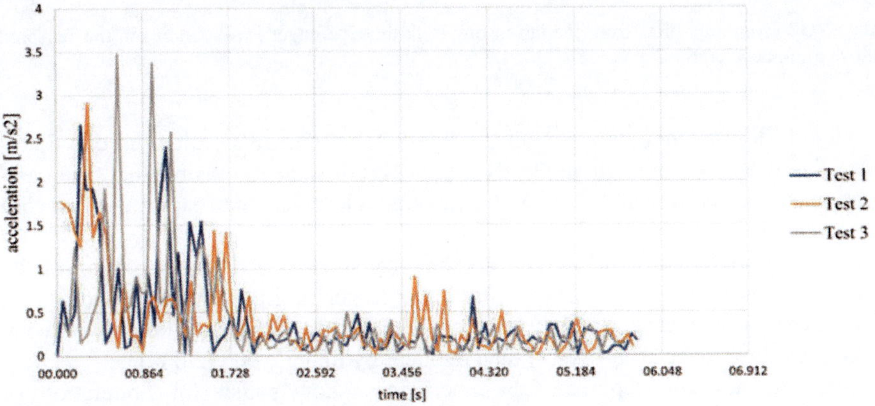

Fig. 24.10 Acquired fingertip acceleration during test with ExoFinger prototype in Fig. 24.9

Figure 24.11 a plots the acquired force data from the index fingertip during the grasping test, that led to a firm grasp of about 1.5 N. The total power consumption for the two servomotors was measured as in Fig. 24.11b with a maximum value of 8.2 W.

Comparing acquired results in Fig. 24.12a of the index fingertip's acceleration of both Experiments, we can determine that the peak acceleration of the fingertip

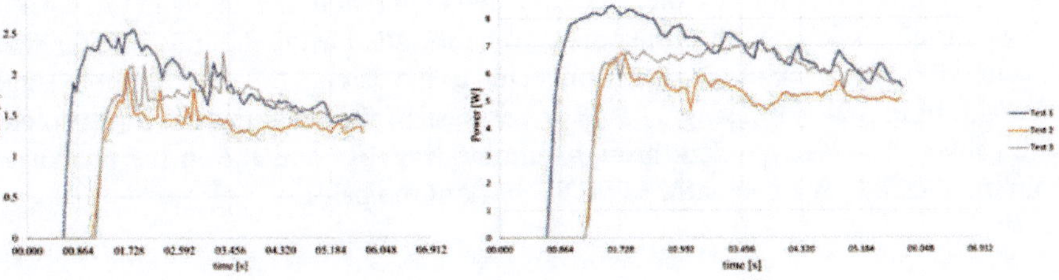

a. b.

Fig. 24.11 Acquired data during test with ExoFinger prototype in Fig. 24.9: a. index fingertip force; b. total power consumption

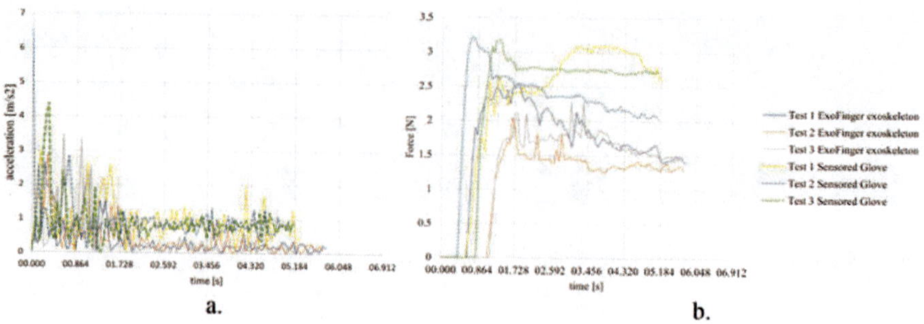

Fig. 24.12 Acquired plots from the lab testing with the ExoFinger exoskeleton and the sensored glove of: **a**. acceleration; **b**. Force

is slightly lower when driven with the exoskeleton (2.5 m/s^2) than the one determined with the sensored glove (3 m/s^2). However, it is clearly visible that the quasi-stationary acceleration at the end of the movement is lower when experimenting with the exoskeleton.

The reported results in Fig. 24.12b show satisfactory efficient operation of the ExoFinger exoskeleton in assisting the index finger in grasping. As expected, the authors observe that the grasping force of the index fingertip is slightly lower when driven by the exoskeleton (1.5 N) than the healthy finger's one (2.6 N).

In summary, the reported lab experiences show satisfactory operation of ExoFinger exoskeleton in assisting human finger motion with similar human characters, but its operation looks not fully reliable and quite complex for an autonomous operation with not good comfort features.

24.5 Design Consideration for New Solutions

The result analysis of the reported lab experiences may suggest improvements in the design and operation of the ExoFinger exoskeleton that could be conveniently implemented in new design solutions. Figure 24.13 shows conceptual design for future work in two designs for one DOF assisted finger motion, namely in Fig. 24.13a using a simple linkage with proper sizes to be installed aside the index finger with an actuator on the back of the palm connected to the driving link by a suitable small gear, and in Fig. 24.13b using a cable driven system for actuating the fingertip cup with cables c1 and c2 guided through suitable rings R1 and R2 on the phalanxes from the flange L of an actuator M on the back of the palm.

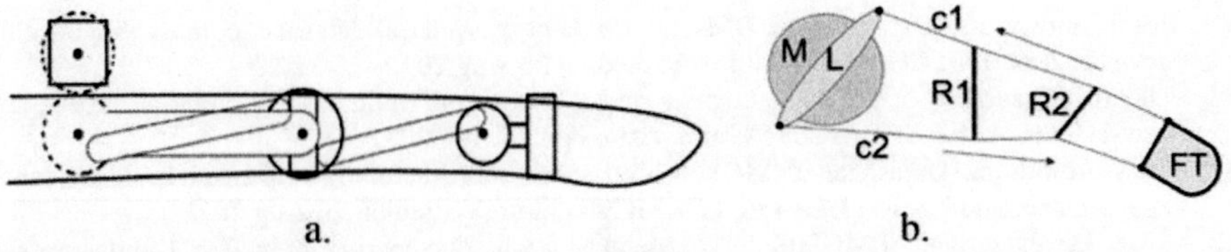

Fig. 24.13 Conceptual design for a new exoskeleton for index finger motion assistance: **a** with linkage; **b** with cable actuation

24.6 Conclusions

Results are discussed from lab testing with human fingers and available ExoFinger prototype to highlight characteristics and problems in exoskeleton assistance of index finger operation. It is found that although of ExoFinger exoskeleton operates with satisfactory operation in assistance of finger motion and grasping, improvements of ExoFinger exoskeleton in terms of conform and user's easy usage will require at the most a redesign of the linkage solution with one DOF. In addition, alternatives are outlined with conceptual design of new solutions, for future work.

Acknowledgements The first two authors gratefully acknowledge Polytechnic University Timisoara through Erasmus internship program permitting their period of study at LARM2 of Tor Vergata Rome University in the summer of 2022.

References

1. Schuenke, M., Schulte, E., Schumacher, U.: Atlas of anatomy. Thieme, New York (2020)
2. Ang, B.W.K., Yeow, C.H.: Print-it-yourself (PIY) Glove: A fully 3D printed soft robotic hand rehabilitative and assistive exoskeleton for stroke patients. In: 2017 IEEE/RSJ International conference on intelligent robots and systems (IROS). Vancouver, BC, Canada
3. Agarwal, P., Fox, J., Yun, Y., O'Malley, M.K., Deshpande, A.D.: An index finger exoskeleton with series elastic actuation for rehabilitation: Design, control and performance characterization. Int. J. Robot. Res. **34**, 1747–1772 (2015)
4. Cempini, M., Cortese, M., Vitiello, N.: A powered finger-thumb wearable hand exoskeleton with SelfAligning joint axes. IEEE/ASME Trans. Mechatron. **20**, 705–716 (2015)
5. Bijo,S., Ben-Tzvi, P., Refour, E.M., Xu, W., Pradhan, S., Guo, Y.: Robotic exoskeleton glove system, (2020). Patent US 2020/0375287 A1
6. Carbone, G., Gerding, E.C., Corves, B., Cafolla, D., Russo, M., Ceccarelli, M.: Design of a Two-DOFs driving mechanism for a motion-assisted finger exoskeleton. Appl. Sci. **10**(1–23), 2619 (2020)
7. Ceccarelli, M., Morales-Cruz, C.: A prototype characterization of ExoFinger, a finger exoskeleton. Int. J. Adv. Robot. Syst., 1–8
8. Gerding, E., Ceccarelli, M., Carbone, G., Cafolla, D., Russo, M.: Mechanism for finger exoskeleton. Patent no. IT. 102018000003847. (2020)
9. Interlink Electronics.: DataSheet of FSR 400 series force sensing resistor. https://pdf1.alldatasheet.com/datasheet-pdf/view/1246135/ETC1/FSR400.html. visited on 26 Aug 2022

10. Bosch Sensortec.: Datasheet of BMI160 IMU. https://pdf1.alldatasheet.com/datasheet-pdf/view/1132066/BOSCH/BMI160.html. Visited on 26 Aug 2022
11. Allegro.: Datasheet of ACS712 current sensor. https://pdf1.alldatasheet.com/datasheet-pdf/view/168326/ALLEGRO/ACS712.html Visited on 26 Aug 2022
12. Lofty Ambition.: Datasheet of MG995 high speed servo actuator. https://pdf1.alldatasheet.com/datasheet-pdf/view/1132435/ETC2/MG995.html. Visited on 26 Aug 2022
13. Tower Pro, Datasheet of MG90S Servo, Metal gear with one bearing, https://pdf1.alldatasheet.com/datasheet-pdf/view/1132104/ETC2/MG90S.html. Visited on 26 Aug 2022

Part V
Robot Control and Applications

Chapter 25
Development of a Force Feedback Control for Robotic Assisted Liver Cancer Treatment

B. Gherman, D. Horvath, E. Mois, C. Vaida, F. Graur, A. Burz, C. Popa, P. Tucan, N. Al Hajjar, A. Pusca, and D. Pisla

Abstract Liver cancer has become the third most common cause of cancer-related deaths, being ranked as the sixth most common cancer globally. The aim of this paper is to present the development of an innovative medical robotic system that can be used for the minimally invasive treatment of liver cancer through radiofrequency ablation. The robotic system was designed to be controlled by the user through teleoperation with the help of a commercially available haptic device and a graphical user interface developed in MATLAB, which also serves as a control point for the radiofrequency ablation instrument.

B. Gherman (✉) · D. Horvath · C. Vaida · A. Burz · P. Tucan · A. Pusca · D. Pisla
Technical University of Cluj-Napoca, Cluj-Napoca, Romania
e-mail: bogdan.gherman@mep.utcluj.ro

D. Horvath
e-mail: daniel.horvath@mep.utcluj.ro

C. Vaida
e-mail: calin.vaida@mep.utcluj.ro

A. Burz
e-mail: alin.burz@mep.utcluj.ro

P. Tucan
e-mail: paul.tucan@mep.utcluj.ro

A. Pusca
e-mail: alexandru.pusca@mep.utcluj.ro

D. Pisla
e-mail: doina.pisla@mep.utcluj.ro

E. Mois · F. Graur · C. Popa · N. Al Hajjar
Iuliu Hatieganu University of Medicine and Pharmacy, Cluj-Napoca, Romania
e-mail: emil.mois@umfcluj.ro

F. Graur
e-mail: florin.graur@umfcluj.ro

C. Popa
e-mail: calin.popa@umfcluj.ro

I. Doroftei et al. (eds.), *Proceedings of SYROM 2022 & ROBOTICS 2022*, Mechanisms and Machine Science 127, https://doi.org/10.1007/978-3-031-25655-4_25

Keywords Medical robot · Teleoperation · Force feedback · Cancer treatment · Needle insertion

25.1 Introduction

Hepatocellular carcinoma (HCC) is the most common types of liver cancer, accounting for 80–90% of liver cancer cases, being the third most common cause of cancer-related deaths [1]. For patients with metastatic cancer or HCC, less invasive treatments are required to eliminate any unnecessary strain on the patient [2], which most likely, is already undergoing chemotherapy [3].

Radiofrequency ablation (RFA) represents one such viable minimally invasive percutaneous treatment for cancer that involves the destruction through heat (between 50 and 100 °C) [1]. To eliminate undesired errors, such as natural hand tremors that can cause local hemorrhaging, robotic systems can be designed to hold, position, orient and insert the needle with high precision and accuracy at the target points, such as [4, 5], while being visually guided by CT, MRI, or US [1]. Such a system may be controlled using a variety of methods, such as the use of haptic technology which enables the user to teleoperate the robot and can provide haptic feedback from the work environment of the robot [6].

The aim of this paper is to present an innovative teleoperated medical robotic system consisting of the Kuka iiwa LBR 7 collaborative robot and a custom designed robotic instrument for the insertion and actuation of the radiofrequency ablation needle. The teleoperated robotic system uses a force feedback capable haptic device complemented by a reliable force and torque sensor with the scope of providing useful feedback interaction with the operating environment for the surgeon.

The paper is structured as follows: Sect. 25.2 presents the architecture of the robotic system, with details the teleoperation and force feedback aspects, user interface and the RFA instrument; Sect. 25.3 illustrates the experimental tests which were performed with this robotic system; Sect. 25.4 presents the conclusions and future works.

25.2 The Robotic System Architecture

The hardware architecture of the teleoperated robotic system is presented in Fig. 25.1 and consists of the following components: an Omega.7 haptic device [7], a master computer, a Kuka iiwa LBR 7 collaborative robotic arm, an RFA instrument and a FT 300 sensor [8]. The serial structure of the robot presents the advantage of a larger workspace while modularity provides the opportunity of interchanging different medical instruments based on the desired procedure.

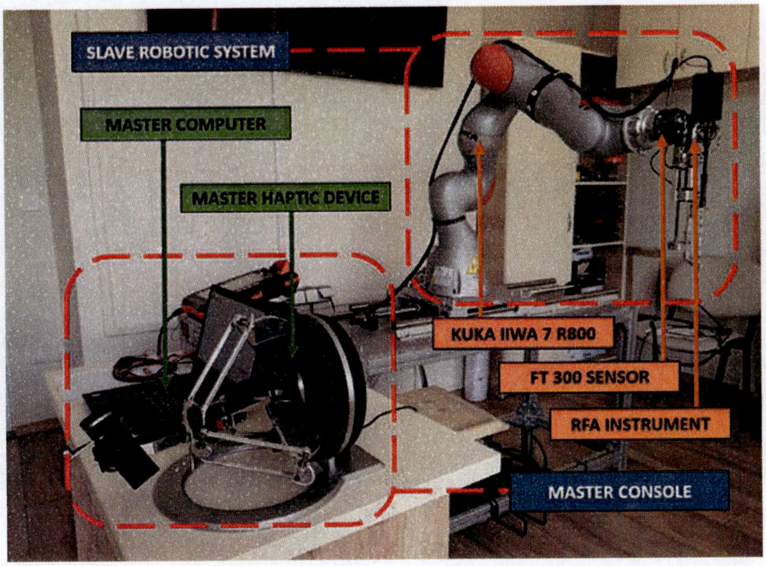

Fig. 25.1 Hardware components of the robotic system

25.2.1 The Control of the Robotic System

Generally, a haptic bilateral teleoperation system consists of the following elements: a human operator that handles a master haptic device, a slave teleoperated robotic system which interacts with the remote work environment and a communication channel between the two devices. An effective haptic teleoperation system is one in which the transparency between the robot and the operator is maximized, while stability is maintained [4]. Telepresence is another important characteristic which should be expanded upon in such a system, and this describes the extent to which an operator will feel as though they are conducting the task directly at the remote site [6]. The force feedback capability most often requires the use of force and torque sensors. Implementation of such a feature will add to the telepresence, which is felt by the operator, increasing the quality of control.

The haptic teleoperation control system is presented in Fig. 25.2. Based on a teleoperated architecture, where the master is the operator console, the system contains the haptic device controlled by the operator, a computer running a MATLAB interface that connects the data from the haptic device and the force sensor mounted on the robot flange, with the slave control system of the Kuka iiwa robotic arm that receives the input coordinates (x, y, z, α, β, γ) and guides the RFA instrument towards the desired insertion point. As a result, control is imposed from the master console onto the slave robotic system. Communication with the Omega.7 device is established

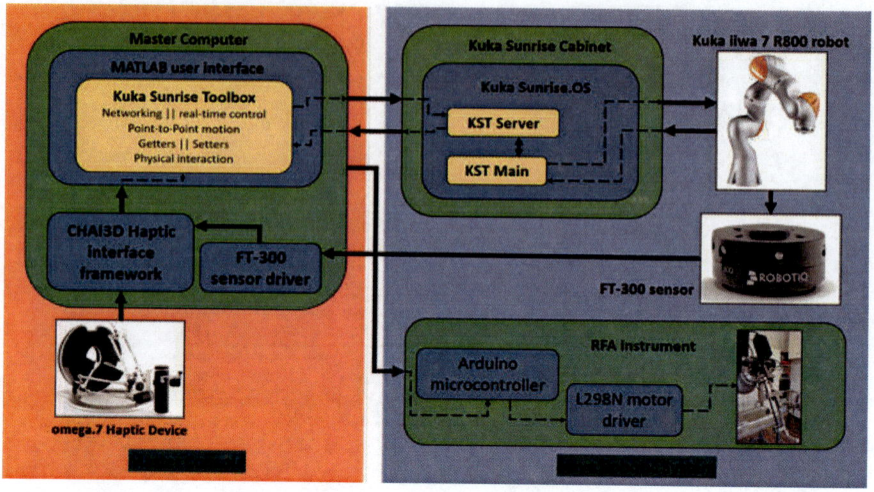

Fig. 25.2 The control system architecture

through a haptic interface based on the open-source CHAI3D cross-platform simulation framework which offers a variety of modules, functions and commands that can read data from the haptic device or control its static and dynamic behavior.

The interface consists of two threads: the main thread responsible for acquiring data from the haptic device and the force-torque sensor, processing that data, and sending the haptic data to the graphical user interface by means of a socket TCP/IP connection; and the haptics thread that deals mainly with applying the force-feedback of the device using the data acquired in the main thread and runs in parallel with the main thread. To effectively communicate the motion commands to the robot in real time, the KUKA Sunrise Toolbox [9] library was employed. It provides the required functions used to send the calculated target position in real time to the Kuka iiwa's controller. This toolbox also allows the user to read the position and orientation data from the robot which can be used for further functionalities and computation.

25.2.2 *The RFA Instrument and Control System*

To achieve efficient and good communication between the Arduino Nano microcontroller and the MATLAB interface (Fig. 25.3), a communication protocol was implemented (Fig. 25.4). Messages are transmitted by the user interface to the USB port to which the Arduino Nano is connected on the master computer, and this is read by the RFA instruments' microcontroller. The microcontroller is essentially programmed to "listen" and interpret the incoming commands sent through the serial port by the MATLAB interface.

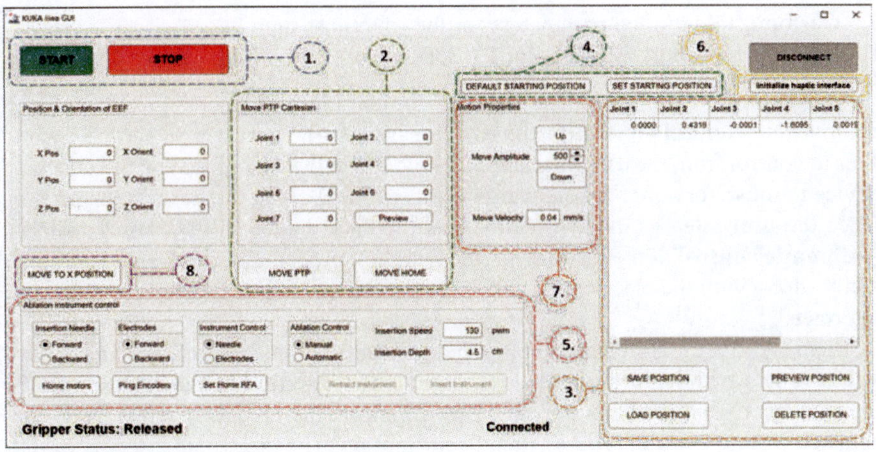

Fig. 25.3 The MATLAB graphical user interface

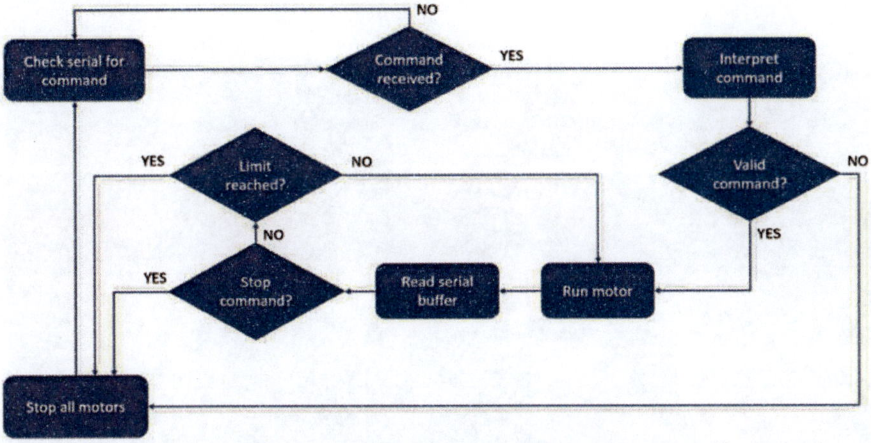

Fig. 25.4 RFA instrument communication protocol loop diagram

The interface was designed to offer the surgeon a variety of control options for both the robotic arm and the ablation instrument itself:

1. Real-time cartesian teleoperation of the Kuka iiwa robotic arm;
2. Point-to-point motion of the robotic system with visualization of the target positions;
3. Saving, loading and previewing of robotic joint coordinates;
4. Custom robot teleoperation starting position setting;
5. Manual/Automatic control of the ablation instrument;
6. Initialization of the haptic interface;

7. Customizable motion parameters for the robotic system;
8. Position control of the robot for FT 300 sensor calibration.

Two control strategies for the RFA instrument are implemented using encoded commands. **Manual control** of the two linear axis: the user can select the desired axes to control from the user interface and hold down the user switch on the Omega.7 device to move forwards or backwards the instrument along the linear transmission. Once the user releases the switch, the instrument is halted in its current position. **Automatic control** of the instrument: the insertion procedure is performed automatically, along with the electrode insertion, at the desired depth selected from the user interface.

Figure 25.5 presents the main components of the RFA instrument. The mechanical structure has two degrees of freedom, one for the insertion of the ablation needle into the liver, and the other for the deployment of the electrode array. These are actuated through two micro-DC metal gear motors. The electronic system consists of an Arduino Nano v3.0 microcontroller to control the motors for each axis, and an L298N driver. Two inductive sensors are used to determine the travel limit and home positions of each axis.

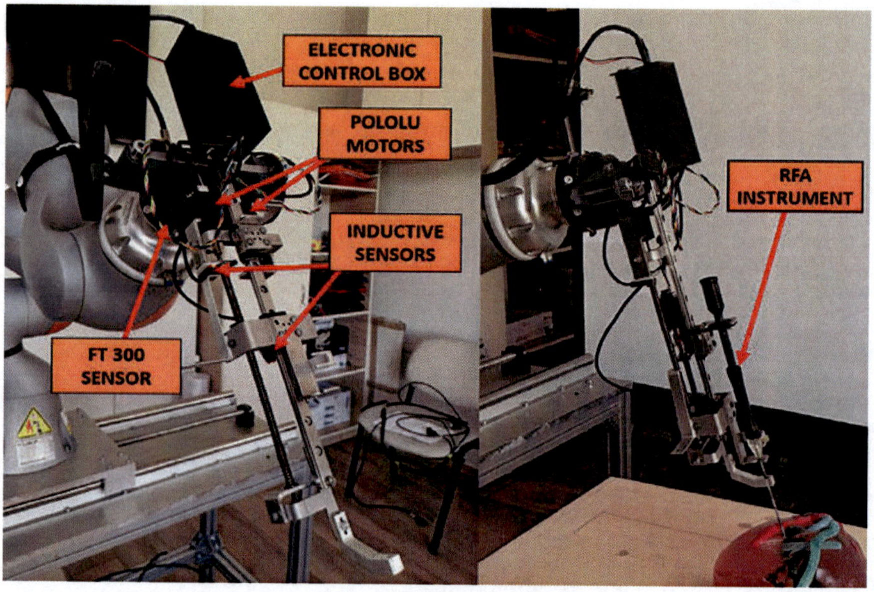

Fig. 25.5 The RFA instrument structure (left–without ablation needle, right–with ablation needle)

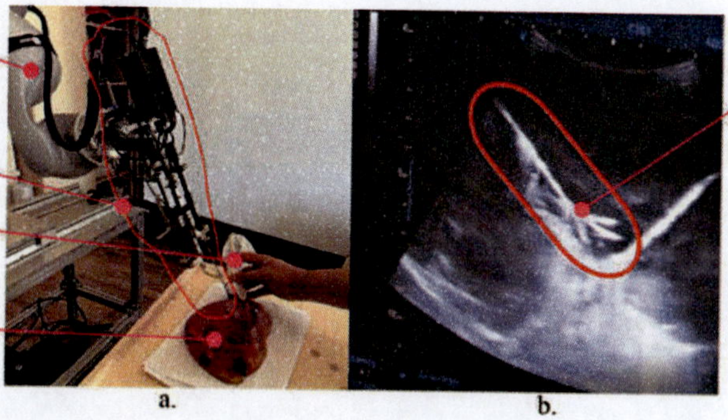

Fig. 25.6 Experimental testing of the system in laboratory conditions: **a** Experimental setup; **b** Ablation needle and electrode array as seen on the US screen

25.3 Experimental Tests

To assess the degree of usability of the robotic system, several ex-vivo tests were conducted using a phantom liver made from ballistics gel. The purpose of these tests was to show that the Kuka iiwa robotic arm could be teleoperated safely in the vicinity of the patient and that an RFA curative procedure could be safely and accurately conducted once the ablation instrument was positioned and oriented at the target points, while haptic feedback [10, 11] would also be provided to the operator via the force-torque sensor mounted on the robot.

During these tests, the robotic arm was guided using US imaging to an approximate target position, as seen in Fig. 25.6a, and then the RFA instrument was inserted into a phantom tumor in the ballistics gel liver, where the electrode array of the RFA instrument was actuated, as seen in Fig. 25.6b. The experiments show that the system can be successfully teleoperated, oriented along a desired trajectory and positioned at an insertion point using the implemented force feedback capabilities using the Omega.7 haptic device handle during operation.

Recorded torque measurements used as force feedback input for the Omega.7 haptic device are presented in Fig. 25.7, using the force and torque sensor mounted on the flange of the robot when the ablation needle touched the liver surface. During the experiment, the needle was maneuvered inside the patient (close to the liver surface) to correct the insertion trajectory based on US images, which generated the Z and Y axis torques.

25.4 Conclusions

An innovative modular teleoperated robotic system for radiofrequency ablation of HCC has been developed in this paper. The robotic system consists of the Kuka iiwa LBR 7 robot and a 2 DOF robotic instrument for accurate insertion and actuation of

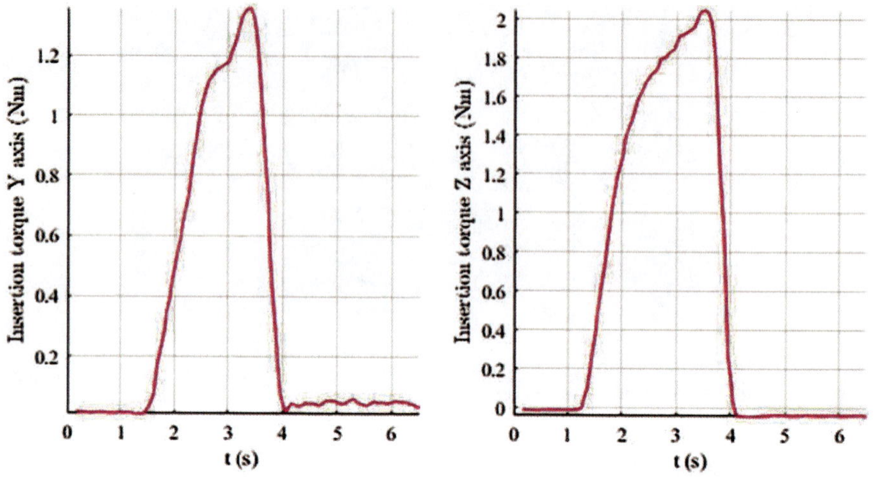

Fig. 25.7 Force and torque sensor measurements

the RFA needle, teleoperated by the Omega.7 haptic device and controlled via a user interface. Initial tests in laboratory conditions reveal promising results, the robotic system being able to place the RFA needle according to the input provided through the ultrasound device. Future work targets the development of a second module to guide the US probe and automatically provide the target coordinates of tumors.

Acknowledgements This work was supported by a grant from the Romanian Ministry of Research and Innovation, CCCDI—UEFISCDI, project number PN-III-P2-2.1-PED-2021 2790/694PED/2022 (ENHANCE) within PNCDI III.

References

1. Seinstra, B.A., van Delden, O.M., van Erpecum, K.J., van Hillegersberg, R., Mali, W.P., van den Bosch, M.A.: Minimally invasive image-guided therapy for inoperable hepatocellular carcinoma: What is the evidence today? Insights Imaging **1**(3), 167–181 (2010)
2. Yang, Y., Yu, H., Tan, X., You, Y., Liu, F., Zhao, T., Ji, Q., Li, J., Feng, Y., Zhu, Q.: Liver resection versus radiofrequency ablation for recurrent hepatocellular carcinoma: a systematic review and meta-analysis. Int J Hyperth **38**(1), 875–886 (2021). https://doi.org/10.1080/02656736.2021.1933218
3. Gherman, B., Hajjar, N.A., Tucan, P., Radu, C., Vaida, C., Mois, E., Burz, A., Pisla, D.: Risk assessment-oriented design of a needle insertion robotic system for non-resectable liver tumors. Healthcare **10**, 389 (2022). https://doi.org/10.3390/healthcare10020389
4. Abdullah BJ, Yeong CH, Goh KL, Yoong BK, Ho GF, Yim CC, Kulkarni A (2014) Robot-assisted radiofrequency ablation of primary and secondary liver tumours: early experience. Eur Radiol 24(1):79–85. https://doi.org/10.1007/s00330-013-2979-7. (Epub 2013 Aug 9. PMID: 23928933; PMCID: PMC3889281)

5. Vaida, C., Plitea, N., Hajjar, N.A., Burz, A., Graur, F., Gherman, B., Pisla, D.: A new robotic system for minimally invasive treatment of liver tumours. Proc Roman Acad Ser A **21**(3), 263–271 (2020)
6. Villard, C., Soler, L., Gangi, A.: Radiofrequency ablation of hepatic tumors: simulation, planning, and contribution of virtual reality and haptics. Comput Methods Biomech Biomed Eng **8**(4), 215–227 (2005). https://doi.org/10.1080/10255840500289988
7. https://www.forcedimension.com/products. Accessed March 2022
8. https://assets.omron.eu/downloads/manual/en/v2/robotiq_ft_300_force_torque_sensor_for_omron_tm_series_robots_instruction_manual_en.pdf. Accessed June 2022
9. Safeea, M., Neto, P.: KUKA sunrise toolbox: interfacing collaborative robots with MATLAB. IEEE Robot Autom Mag **26**(1), 91–96 (2019). https://doi.org/10.1109/MRA.2018.2877776
10. Birlescu, I. et al.: Experimental testing and implementation of a force–torque sensor in automated percutaneous needle insertion instruments. In: International Conference on E-health and Bioengineering (EHB), pp. 1–6 (2021). https://doi.org/10.1109/EHB52898.2021.9657712
11. Birlescu, I., Husty, M., Vaida, C., Plitea, N., Nayak, A., Pisla, D.: Complete geometric analysis using the study SE(3) parameters for a novel, minimally invasive robot used in liver cancer treatment. Symmetry **11**, 1491 (2019). https://doi.org/10.3390/sym11121491

Chapter 26
Perspectives on Originally Designed Eco-Friendly Robotic Cell for PCB Dismantling

I. Doroftei, S. Cazan, A. Burlacu, and D. Chirita

Abstract Printed circuit boards (PCBs) contain a lot of precious and important components. According to different statistics, in the future decades, more strategies for the conservation of expensive resources and recycling policies will be implemented. There exist a wide range of devices and machines that were designed to recycle electronic components, but the latest trends regarding their functionality involve a demand for a more diligent procedure of materials recovery. On one hand, it is important to recover precious metals like gold and silver but, on the other hand, different ceramic and plastic components still remain on the e-waste piece to be recycled, which does not seem like a fulfilled approach from an environmental perspective. This paper presents an originally designed eco-friendly robotic system for PCB dismantling whose particularity of parts removal is based on mechanical approaches. Unlike other parts removal procedures, in this case, the mechanical procedure removal is accomplished using linear motions of the rotary tool whose portable head allows a local contact exactly between the part and the portable head. In this case, adjacent zones of the board are not affected. Considering time and financial factors as key components of current industrial applications, this paper presents further possibilities of the originally developed robotic cell.

Keywords Robotic plant · Eco-friendly · Dismantling · e-waste

---

I. Doroftei (✉) · S. Cazan · A. Burlacu · D. Chirita
"Gheorghe Asachi" Technical University of Iasi, Iasi, Romania
e-mail: idorofte@mail.tuiasi.ro; ioan.doroftei@academic.tuiasi.ro

S. Cazan
e-mail: stelian.cazan@academic.tuiasi.ro

A. Burlacu
e-mail: adrian.burlacu@academic.tuiasi.ro

I. Doroftei et al. (eds.), *Proceedings of SYROM 2022 & ROBOTICS 2022*, Mechanisms and Machine Science 127, https://doi.org/10.1007/978-3-031-25655-4_26

26.1 Introduction

Recycling activities represent a basic ingredient in materials and resources preservation policies. This is the reason why different devices and mechanisms that are used in this activity have been enhanced with the main scope of better efficiency. An optimized version of this kind of machine or plant would follow not only the goal of precious metals recovery, but also the elimination of plastic and ceramic elements. According to [1], around 7% of the gold quantity in the world gold might be contained in electronic waste. Chemical or thermal recycling technologies [2, 3] can obtain metals of high purity, offering economic return and good efficiency. As a matter of fact, in these cases, there are needed high pressures and high temperatures, which can cause pollution through hazardous fumes. An environmental solution for dismantling is presented in [4]. The study reveals the appropriateness of a chemical-electrochemical process of dismantling. One of the conclusions states that about 90% of the copper would be extracted if a complementary step consisting of mechanical pre-treatment of the electronic circuit boards was added.

In [5], the approach is also mechanical, the main functioning being that of printed circuit board milling. An air classification test bench whose functionality is dedicated to examining the separation of the resulting plastic particles after milling was developed. It was ascertained that the developed stand is useful for the dissociation of particles whose dimensions do not exceed 2 mm. A thorough examination was carried out regarding the chemical composition of circuit board powder particles after milling. Another conclusion that reveals the necessity of mechanical approaches for circuit board dismantling is presented in [6]. The study represents a brief presentation of the main recovery techniques for materials from circuit boards. Different techniques for precious metals recovery are also presented in [7], with the aspect of robotic arms implementation in PCB dismantling being emphasized. The wide spread of industrial activities related to the usage of industrial robots is presented in [8], the main conclusion of this study being that of the replacement of human factor with the capabilities of industry 4.0.

It is emphasized in [9] that the mechanical-physical recycling techniques are not only economically feasible, but they are also suitable from an environmental perspective. The study points out the necessity of polymer composites and geopolymers to be considered when PCB components are fabricated. Similar studies reveal that the most environmental procedure of boards recycling is based on mechanical procedures.

Even though high temperatures are needed in the recycling method proposed in [10], environmental aspects are found in the small-molecule assisted method based on dynamic reaction. A wide range of PCBs can be recycled using this method. The eco-friendly characteristics come from the fact that the recycling solution could be used multiple times. The study makes a point of the importance of using polymers that contain ester groups as board substrates. This would be effective for efficient recycling, not only from an eco-friendly standpoint.

26.2 The Description of the Originally Developed Robotic Cell

The main utility of the robotic cell is to finish the step before the immersion of the circuit boards in the leaching solutions for precious metals recovery in the most eco-friendly way. The rotary tool that is attached to the robot has the main role of eliminating rough components off the board. A broad range of PCB types was used for the laboratory tests [11].

26.2.1 Robotic System Architecture

The robotic process of dismantling electrolytic capacitors from PCBs involves the use of a video camera to identify the capacitors, a manipulator-type robot, on which a device with a special tool for their extraction by twisting has been mounted, such as a vacuum cleaner, which will remove the capacitors extracted from the robot's work area (Fig. 26.1).

The process involves going through the following stages corresponding to a complete operating cycle of the robotic electrolytic capacitor dismantling system: manual mounting and centering of the board with electronic components on the conveyor belt; moving the plate within the range of the video camera; identification of electrolytic capacitors and their positions on the board; transmission of information resulting from image processing to the manipulator robot controller; moving the PCB in the action area of the robot; carrying out the program that includes the

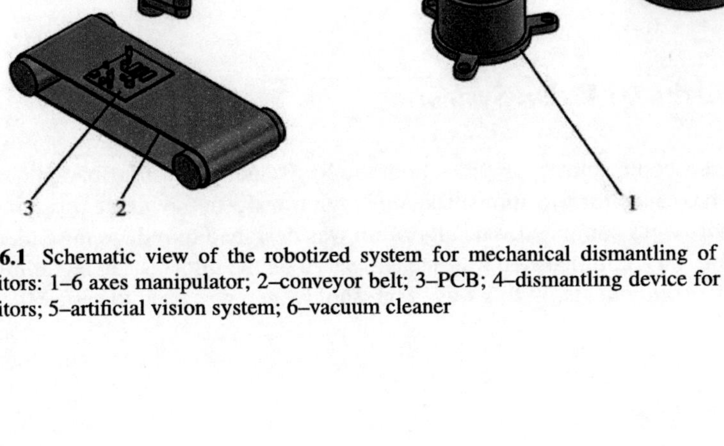

Fig. 26.1 Schematic view of the robotized system for mechanical dismantling of electrolytic capacitors: 1–6 axes manipulator; 2–conveyor belt; 3–PCB; 4–dismantling device for electrolytic capacitors; 5–artificial vision system; 6–vacuum cleaner

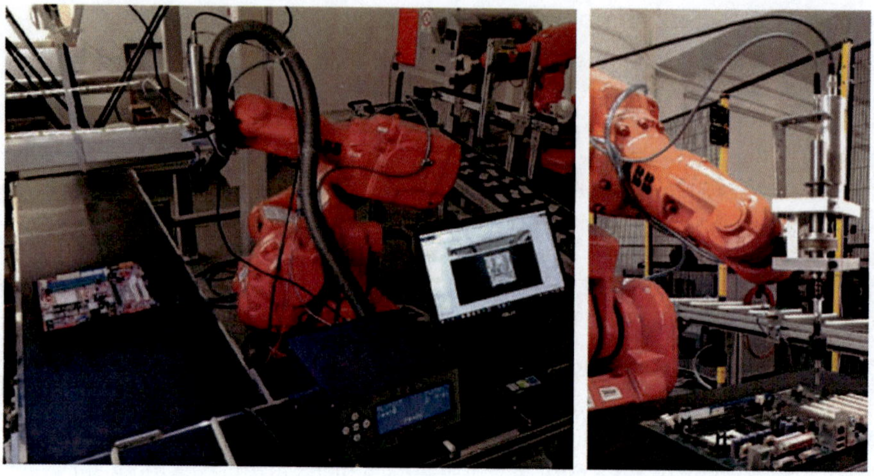

Fig. 26.2 Real robotized system for mechanical dismantling of electrolytic capacitors

movements of the robot and the dismantling tool by twisting, as well as the suction of the dismantled capacitors, as a result of the algorithm implemented in the robot controller, in correlation with the control algorithm of the special dismantling and suction device, until the dismantling of the last capacitor identified by the video system; moving the PCB (by the conveyor belt) outside the area of action of the robot; manual removal of the plate.

By applying this procedure, the following advantages are obtained:

- automating the process of extracting electrolytic capacitors and removing them from the PCBs;
- avoiding the destruction of all the electronic components on the board, a situation that occurs in the case of other mechanical disassembly methods;
- avoiding the mixing of toxic substances from electrolytic capacitors with other materials, which are to be recovered.

The real robotized system used for tests on mechanical dismantling of electrolytic capacitors is shown in Fig. 26.2.

26.2.2 Artificial Vision System

A ZED stereoscopic camera provides both the RGB and depth information needed to identify the capacitors positions. Using Hough transform the image positions for the capacitors were computed. The algorithm was designed to reduce false identifications. Thus, an additional step was needed, based on the information regarding the depths of the components. In this way, a threshold value was used to assure the real

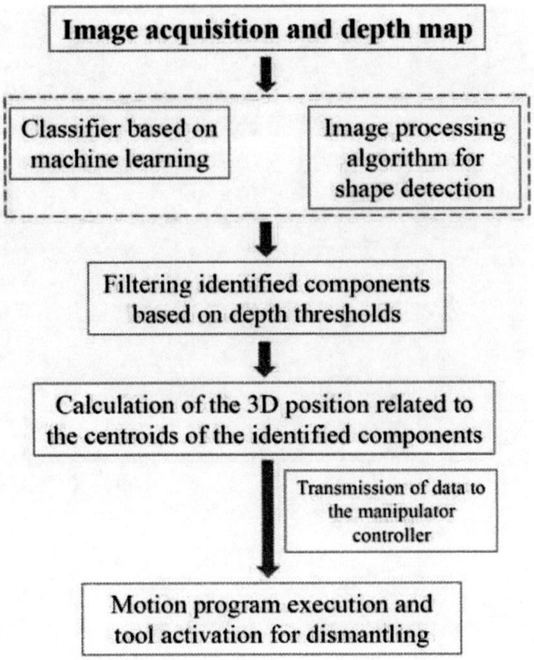

Fig. 26.3 Block diagram of the visual identification stage of electrolytic capacitors

location of a component. In the end, depth information was added to recover the 3D coordinates of the elements to be dismantled. The algorithm for capacitor position detection was developed using Matlab.

The block diagram of the visual identification stage of the electrolytic capacitors is presented in Fig. 26.3. A 86% efficiency of the vision system has been achieved during tests (Fig. 26.4).

26.2.3 Dismantling Device

An original dismantling devise has been design. Figure 26.5 illustrates both the 3D CAD model and the real prototype.

For capacitors dismantling, the portable head of the dismantling tool was chosen so that it has geometrically defined cutting edges (Fig. 26.6). Once the vision system identifies the capacitor, a rectilinear motion driven by the manipulator is executed and the head of the capacitor is perforated. When the capacitor is perforated, it means that the rotary moment of the tool is sufficient to dismantle the piece. In this way, a sensor triggers the tool and the head begins its rotations. After a certain number of rotations, the capacitor is dismantled and the vacuum collects the piece. A vertical rectilinear motion is performed upside by the manipulator so that the vision system can identify another capacitor to be dismantled.

Fig. 26.4 Capacitors identification using the vision system [11]

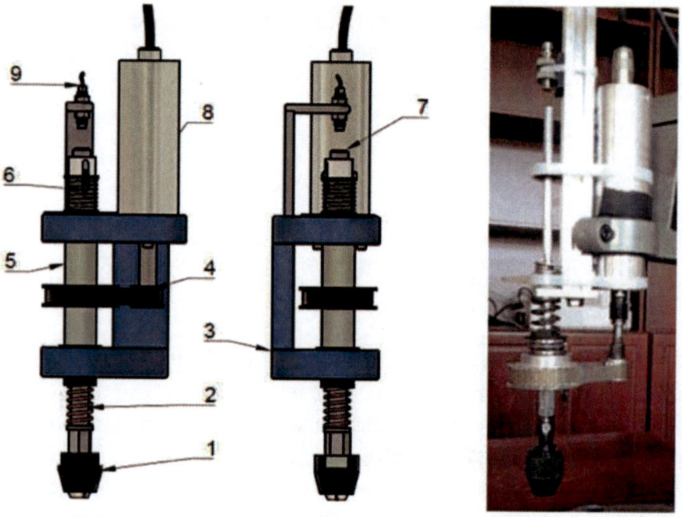

Fig. 26.5 Custom device for dismantling capacitors: 1–mandrel; 2–fastening spring; 3–device support; 4–toothed belt transmission; 5–rotating shaft; 6–spring to hold the spindle in the working area; 7–sliding shaft; 8–programmable screwdriver; 9–capacitive proximity sensor with possibility of adjustment in the working area

Fig. 26.6 Dismantling tool for electrolytic capacitors

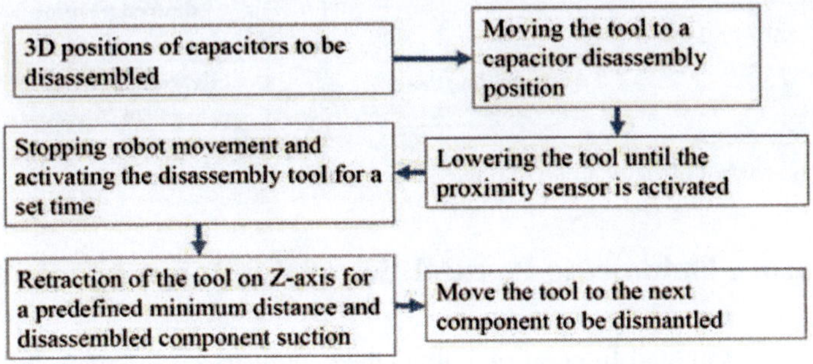

Fig. 26.7 Block diagram of the stage of mechanical disassembly by twisting the electrolytic capacitors and their suction

For screws removal, the tool head was replaced with another one whose geometry is very similar to that of a tubular key. The coordinates of the middle of the tool head and the coordinates of the middle of the screw head must be perfectly matching. Once the head of the screw is completely touched by the tubular key, the sensor triggers the tool and rotations begin. A pre-established number of rotations are executed so that all types of screws are removed accordingly. An electromagnet was built for screws removals after the pre-established number of rotations.

Block diagram of the stage of mechanical disassembly by twisting the electrolytic capacitors and their suction is shown in Fig. 26.7.

26.2.4 The Coordinating Process Device

This device has the main function of keeping some parameters in standards and to receive or send signals to the components of the plant. For example, the number of rotations of the rotary tool, for different types of portable heads, is pre-established using this device. The motions of the robot are synchronized with this device. The artificial vision system sends signals to this device so that the manipulator effectuates the necessary motions. The sensor that triggers the rotary tool is synchronized with this device. All the components of the robotic cell communicate with each other using this specific device.

The control block diagram of the robotic system is shown in Fig. 26.8.

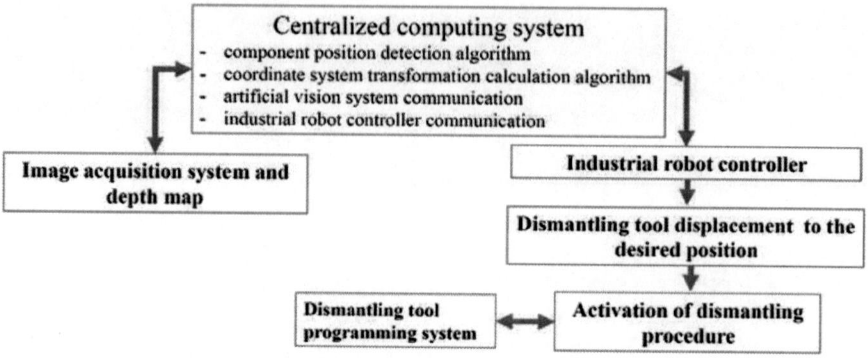

Fig. 26.8 The control block diagram of the robotic system

26.3 Future Enhancing Possibilities of the Robotic Cell

As cloud computing and Internet of Things have evolved a lot, most robotic plants in big factories base their functionality on this. For the present robotic cell, cloud computing principles can also be applied.

A consistent improvement regarding image processing data has been observed, [12]. More companies that develop SaaS and RaaS services focus their attention on image processing services, [13, 14]. Applications of cloud computing services for image processing are widely spread, a good appliance being observed in medical fields, [15]. Even though the current image processing system records 86% efficiency, different algorithms for image processing can be developed. Image processing in cloud computing proved to be a cheaper and a more effective way of using artificial vision.

If lower price is an important criterion, the robot manipulator IRB 140 can be substituted for a cheaper and an easier to program one. For example, Cartesian robots can move along linear axes, making them easy to be handled. If it were only vertical motions of the rotary tool, probably there would have been chosen a Cartesian robot. The decision of keeping IRB 140 manipulator was related to some other additional motions regarding screws removal.

26.4 Conclusions

There are numerous studies, like those presented in the introduction, that emphasize that the mechanical approaches of circuit boards dismantling are the most eco-friendly and feasible. As a consequence, an original eco-friendly robotic plant for capacitors and screws removal has been designed. The main aspect related to the environmental perspective is that of a non-destructive technique of the board using a rotary tool with a portable head so that another zone of the board is not affected. In this way, any component that is placed on a PCB can be removed using this type of tool. Ceramic and plastic components can be removed using this procedure. Unlike other devices that use distributed heat, high temperatures, high pressures and tools that can affect nearby areas, this procedure is based on the contact exactly between the portable head of the tool and the component to be replaced. In this case, the board is not affected and there is a higher possibility of better efficiency of plastic components removal and precious metals recovery.

The robotic cell consists of an IRB 140 robot, a coordinating device system, an originally developed image processing system, different fixing devices, a rotary tool with portable heads for capacitors and screws removal, a vacuum and a components storage box. It has been observed that the diameter of a capacitor can cause difficulties in the process of capacitor dismantling as well as of tilted capacitor cases. The image processing system identifies the capacitors and the screws placed on the board and then it triggers the robotic arm to perform vertical upside-down motions for the procedure to be accomplished. The image processing system provides 86% efficiency, while the mechanical procedure of capacitors and screws removal provides 75% efficiency. As future targets, more aspects related to tilted and capacitors with smaller diameters will be taken into account. The eco-friendly aspect, as a consequence, is fully achieved, as long as the local contact between the capacitor and the portable head can eliminate any piece placed on a PCB, while the board is not affected and future operations of recycling can be achieved accordingly. Future perspectives related to plants like this are strongly tied to the development of the 4.0 industry and the development of cloud computing applications.

References

1. https://ecelectronics.com/news/managing-and-reducing-e-waste-from-pcbs
2. Kopacek, B., Kopacek, P.: Extracting rare materials from electr(on)ic scrap. In: IFAC Conference Paper Archive, pp. 48–24, 157–161 (2015)
3. Korolev, I., Altınkaya, P., Halli, P., Hannula, P.M., Lundström, M.: Electrochemical recovery of minor concentrations of gold from cyanide-free cupric chloride leaching solutions. J. Clean. Prod. **186**, 840–850 (2018)
4. Fogarasi, S., Imre-Lucaci, A., Imre-Lucaci, F.: Dismantling of waste printed circuit boards with the simultaneous recovery of copper: experimental study and process modeling. Materials **14**(18), 5186 (2021)
5. Kers, J., Goljandin, D., Vilsaar, K., Tall, K., Mikli, V., Zuo, X., Zhang, L., Shuman, T.: Characterization of disintegrator milled electronic waste powders for materials recovery. In: The 6th International DAAAM Baltic Conference Industrial Engineering (2008)
6. Vermesan, H., Tiuc, A.E., Purcar, M.: Advanced recovery techniques for waste materials from IT and telecommunication equipment printed circuit boards. Sustainability **12**(1), 74 (2019)
7. Maurice, A., Din, K., Charpentier, N., Brambilla, A., Gabriel, J.-C.: Dismantling of printed circuit boards enabling electronic components sorting and their subsequent treatment open improved elemental sustainability opportunities. Sustainability **13**(18), 10357 (2021)
8. Javaid, M., Haleem, A., Singh, R.P., Suman, R.: Substantial capabilities of robotics in enhancing industry 4.0 implementation. Cogn. Robot. **1**, 58–75 (2021)

9. Wang, Q., Zhang, B., Yu, S., Xiong, J., Yao, Z., Hu, B., Yan, J.: Waste-printed circuit board recycling: focusing on preparing polymer composites and geopolymers. ACS Omega **5**, 17850–17856 (2020)
10. Chen, Z., Yang, M., Shi, Q., Kuang, X., Qi, H.J., Wang, T.: Recycling waste circuit board efficiently and environmentally friendly through small-molecule assisted dissolution. Sci. Rep. **9**, 17902 (2019)
11. Doroftei, I., Chirita, D., Stamate, C., Cazan, S., Pascal, C., Burlacu, A.: Robotic system design and development for automated dismantling of PCB waste. Ind. Robot. **48**(5), 720–725 (2021)
12. Pandey, N., Diwakar, M.: A review on cloud based image processing services. In: 2020 7th International Conference on Computing for Sustainable Global Development (INDIAcom) (2020)
13. Liu, J., Wu, J., Sun, L., Zhu, H.: Image data model optimization method based on cloud computing. J. Cloud Comput. Adv. Syst. Appl. **9**, 31 (2020)
14. Zou, Q., Li, G., Yu, W.: Cloud computing based on computational characteristics for disaster monitoring. Appl. Sci. **10**, 6676 (2020)
15. Mirarab, A., Fard, N.G., Shamsi, M.: A cloud solution for medical image processing. Int. J. Eng. Res. Appl. **4**, 74–82 (2014)

Chapter 27
Cloud-Based Digital Twin for Robot Health Monitoring and Integration in Cyber-Physical Production Systems

Th. Borangiu, S. Răileanu, F. Anton, I. Lenţoiu, and R. Negoiţă

Abstract The paper describes an aggregate cloud-based digital twin (DT) that models an industrial robot by virtualizing its physical structure, motion and action control, operating modes, parameters and interactions with shop floor devices. The DT functionalities are: tuning robot motion parameters, shop floor layout validation for collision avoidance, robot health monitoring and anomaly detection, concurrent job allocation to multiple robots, by implementing in cloud IAAS three DT-specific software configurations: simulation with software in the loop; robust deployment; embedded simulation faster than real-time. Experimental results are reported from data collecting ACE software for Omron-Adept robots.

Keywords Digital twin · Industrial robot · Production system · Cyber physical system

27.1 Introduction

The term Industry of the Future (IoF) or Industry 4.0 indicates the new industrial revolution initiated by a new generation of manufacturing systems designed to be adaptive, fully connected, highly efficient (optimized) and robust (fault tolerant, highly available). The global IoF model defines a new manufacturing stage fully automatized and robotized, using ever more advanced information, control and communication technologies (IC^2T). IoF works with intelligent factories, characterized by digitalization and interconnection of distributed manufacturing entities in a system

Th. Borangiu (✉) · S. Răileanu · F. Anton · I. Lenţoiu · R. Negoiţă
Department of Automation and Applied Informatics, University Politehnica of Bucharest, Bucharest, Romania
e-mail: theodor.borangiu@upb.ro

S. Răileanu
e-mail: silviu.raileanu@upb.ro

F. Anton
e-mail: florin.anton@upb.ro

I. Doroftei et al. (eds.), *Proceedings of SYROM 2022 & ROBOTICS 2022*, Mechanisms and Machine Science 127, https://doi.org/10.1007/978-3-031-25655-4_27

of systems approach: (i) new types of production resources (e.g., intelligent robots) strongly interconnected and self-organizing in the shop floor, products deciding upon their own route and resources; (ii) smart real-time decisions taken from production data collected from resources and products and processed with machine learning techniques. These issues are addressed in the scientific literature through three key enabling digital technologies related to Artificial Intelligence (AI): big data analytics, machine learning and digital twins [2, 4, 6].

The holistic view of the real capabilities, status and features of a shop floor entity (e.g., robot, product, process) including its digital representation, execution context, history of behaviour, time evolution and status can be encapsulated in a *digital twin* (DT) defined as an extended virtual model of that entity, that is persistent even if its physical counterpart is not always online/connected; this extended digital model can be shared in a cloud database with other shop floor entities [7].

The DT concept includes three main components: (i) a physical object in real world, (ii) a virtual object in the informational space and (iii) a set of connections between the real world and the informational system [8]. In terms of manufacturing, the next definition for a digital twin is generally accepted: A DT consists of a virtual representation of a production asset (e.g., a robot) or system (a robot team) that can run different simulations and is characterized by the synchronization between the virtual and real system thanks to sensed data and connected smart devices, mathematical models, and real-time data elaboration. Its typical role within Industry 4.0 manufacturing systems is to exploit these features to predict resource (robot) state, performances and use cost, and optimize the production system with reality-awareness at each product and factory life cycle phase in real time [3, 5].

From the point of view of modelling reality there are two types of DT: (a) DT with a physical counterpart or *data-driven* DT that collects data in IoT framework; this model is used to synchronize the virtual twin with the physical counterpart, and (b) DT without a physical counterpart or *model-driven* DT which is a digital simulation used for modelling, design, analysis, and forecasting. In the second case there is no IoT and no setup. Its use case comprises different processes and views outcomes prior to developing infrastructure (resources and technology).

The main benefits brought by the digital twin technology are: *visibility* in multi-resource manufacturing systems; *prediction* of resource states and unexpected events; *interaction with the model* by simulating conditions impossible put in evidence in real life; *documenting* to explain behaviours of interconnected resources; *integration* with backend business applications.

In this context the scope of the article is to describe a system representing an aggregate cloud-based DT for individual robot health monitoring and collaborative integration in manufacturing systems with multiple shared industrial robots.

27.2 Aggregate Robot Digital Twin Architecture

The proposed Digital Twin is an *aggregation* (DTA) of multiple Digital Twin Instances (DTI) that describe a single physical instance of an industrial robot and allows querying information from a group of robots included in manufacturing structures. This DTA is an *embedded digital twin* (EDT) involved in all activities that imply its physical twins–the robot manipulators and controllers and represents a single point of interaction and single source of information for the real-world counterparts. The connection between the physical asset and its related virtual robot model is established in dual mode: by generating real-time data from internal and external robot sensors and by issuing real-time (re) configuring commands generated by a decision-making entity executing in a private cloud platform. The embedded robot DT aggregates three DTIs that offer the functionalities:

f1. **Model-driven robot DT**: based on simulation and offline programming software; allows robot programming from a PC-type terminal to perform the tasks: (a) evaluate comparatively robot performances, (b) visualize robot trajectories in 3D, (c) check collision risk and optimize robot paths, (d) validate shop floor layouts.

f2. **Data-driven robot DT for health monitoring**: includes several DTIs that execute individually to track changes in robot behaviour and performance degradation, to detect anomalies and make decisions for predictive robot maintenance.

f3. **Aggregate cluster of model-driven robot DTs**: the DTs that virtualize the generic classes of factory entities: resources (robots), products and orders instantiate them at production run for collective workloads–concurrent activity (operation) scheduling and allocating to shared resources (robots) for products that add up to the factory's capacity of simultaneous product execution. This collaborative workload is typical for heterarchical production scheduling, which can be chosen either as scheduling mode for the entire batch or on short term at resource (robot) failure until reconfiguring of the resource team and global batch plan update.

The proposed robot DT technology and model synthetize the physical robot structure, state, behaviour and capabilities into a digital model; this model is updated with information collected from the environmental space extending individual robot control with reality awareness respectively support for optimal allocation of work tasks on robots in manufacturing systems. The DT stores in the cloud data about operating cost, performance and state to predict functioning anomalies, optimize operations allocation and take decisions about the need for predictive robot maintenance. The decisions are implemented in the physical robot space [1].

The proposed robot DT architecture is organized on **four layers** through which data flows in both directions as shown in Fig. 27.1: from the physical layer to the informational one where decisions are taken and backwards.

Layer 1 Edge Processing: Because the industrial robots used are powered by 220 V AC, this allows the use of smart power outlets with Wi-Fi connections to monitor power consumption. *Smart sensors* (in robot space: housing vibrations, contact

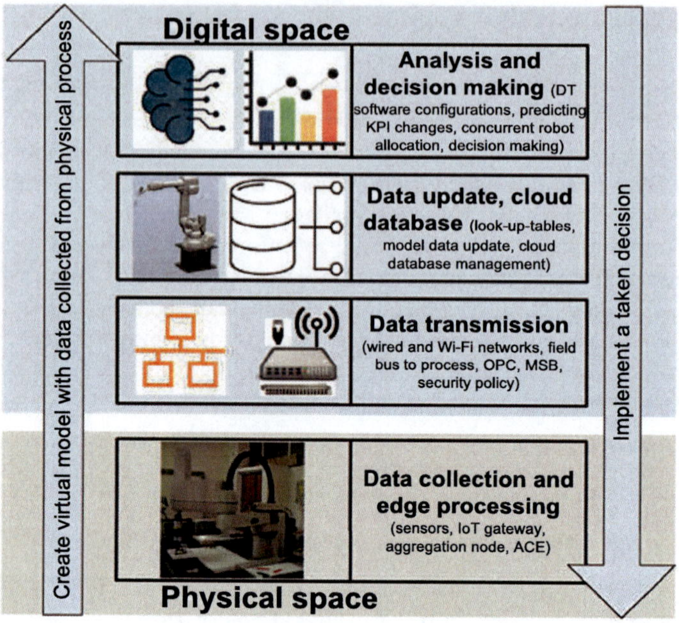

Fig. 27.1 Architecture of the cloud-based digital twin for robot health monitoring and optimal integration in manufacturing, and informational flow linking the physical and virtual spaces

effort in gripper, electrical parameters, and in work environment: temperature, scene lighting) and *IoT gateways* (energy consumption) are connected to *IoT aggregation nodes* integrated in edge computing for permanent robot instrumenting. Robot parameters such as supply voltages for various components, number of errors detected at engine level etc. can be collected using the robot proprietary software system, e.g., ACE for OMRON-Adept robots; this information is directly provided by the robot controller.

Layer 2 Data Transmission: Robot and MES (Manufacturing Execution System) virtualization involves migration of individual and collaborative workloads that were traditionally executed on physical machines to the private cloud platform as virtual workloads. These workloads are interconnected using the manufacturing service bus (MSB) software architecture that acts as service bus for robot and MES physical twins virtualized in the cloud, offering new features of robust queuing and consistent handling shop floor data bursts.

Layer 3 Model Update, Cloud Database Management: The data-driven DTI of an individual robot system and the cluster of model-driven DTIs integrated in the aggregate DTA executing on the MES layer concurrent operation allocation on available physical robots assume the creation of several look-up tables that are updated in a database managed in the cloud. Th tables' contents are updated in time dependent on DT configuration and functionality, operation type and robot work context. Covariance of various table parameters are used for health monitoring.

Layer 4 Analysis and Decision Making: Contains simulation and real-time emulation tools that execute the application programs for the functionalities f1-f3: robot trajectory validation, parameter tuning, health monitoring, documenting and support for concurrent allocation of operations on multiple robots. The robot and shop floor data streams aggregated in the models stored in the cloud database are used to extract knowledge and predict robot performances, parameter changes and usage costs to optimize job allocation at batch horizon on multiple robots.

27.3 Cloud Processing of the Robot DT Data

The two upper DT layers (Data update, Cloud database and Analysis and decision making) are implemented using container technology in an Infrastructure as a Service cloud system. The services implemented by the open source OpenStack platform are based on a series of pods that contain one or more containers hosting either databases or applications which communicate with the lower level of production equipment or with each other for data processing, store or decision-making). The automation of container and pod creation is done with the Ansible system that allows the automatic generation and configuration of DT solutions in a very short time (ms). Moreover, OpenShift allows using load balancing and high availability technologies for services offered through containers. For the complex DTA configurations with more services several pods with containers that communicate between them are used; this assures better application isolation and more effective control over communication channels, and overall, a better system security.

From the point of view of the computing architecture (Fig. 27.2) the system that is used for the implementation of the Aggregate Digital Twin solution is based on a private Cloud IaaS infrastructure IBM CloudBurst. The cloud infrastructure uses virtual machines hosted on IBM Blade HS 22 server systems installed with the hypervisor VMware ESXi server.

The OpenStack system has access to the cluster storage using several types of technology: Gluster, Ceph, etc. External connectivity is achieved through a routing level that allows connection to external applications or production infrastructure through several protocols that use VPN connections. The OpenShift infrastructure is managed using existing automation tools such as the Ansible system and integrated with versioning systems to allow DT system developers to retrieve container images directly from the versioning systems. The implementation of the virtual robot twin system becomes thus much easier for both production and testing environments because scalability tests do not affect production.

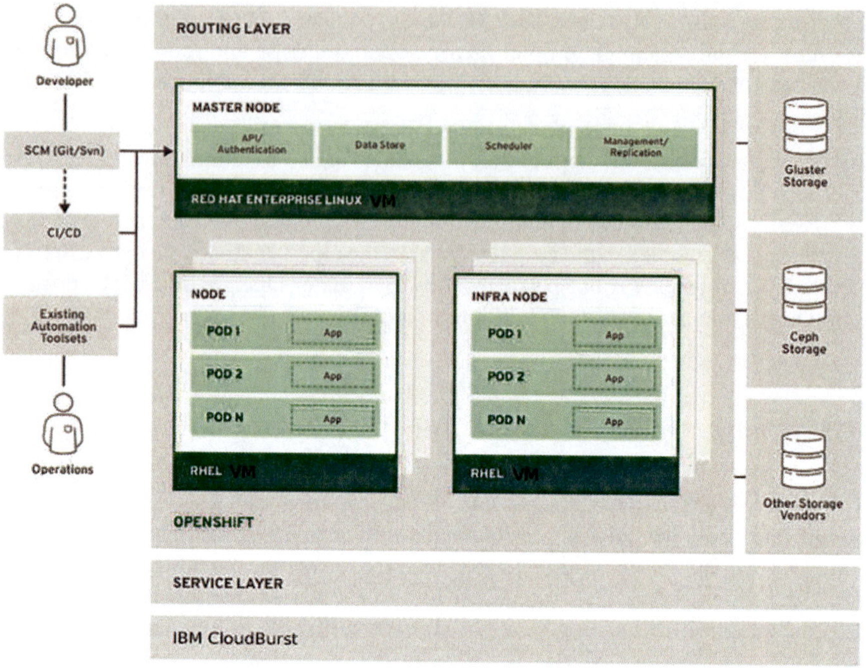

Fig. 27.2 Cloud IaaS architecture implementing the upper layers of the aggregate robot digital twin

27.4 Experiments with Cloud DT for Adept Robots

Experiments were performed to implement a 4-layer Digital Twin framework for Omron-Adept industrial robots in a private Cloud IaaS model. Robot data was acquired both from IoT gateways and from the Adept Automation Control Environment (ACE) that allows data storing into a log file. Adept ACE is based on a series of processes that run in the kernel space of the V+ robot controller operating system and have access to robot status data.

The data which was collected falls in three categories: (1) *Data characterizing the robot*: AC input [V]; amplifier bus voltage [V]; DC input [V]; amplifier temperature [°C]; encoder temperature [°C]; base board temperature [°C]; duty cycle [% limit]; harmonic drive usage [%]; peak velocity [RPM]; peak torque [% max torque]; peak position error [% soft envelope error]; (2) *Data for a robot-serviced belt*: belt velocity [mm/s]; instance count; instances per minute; active instances; latch faults; (3) *Data characterizing the process*: idle time [%]; processing time [%]; average total time [ms]; tasks processed/not processed per minute.

The Cygwin Linux tools package was used to exploit the data from the log file which can be processed in real time by a pipelined sequence of software tools. The pipeline is executed in a loop in a time interval set up by the **watch** command, and

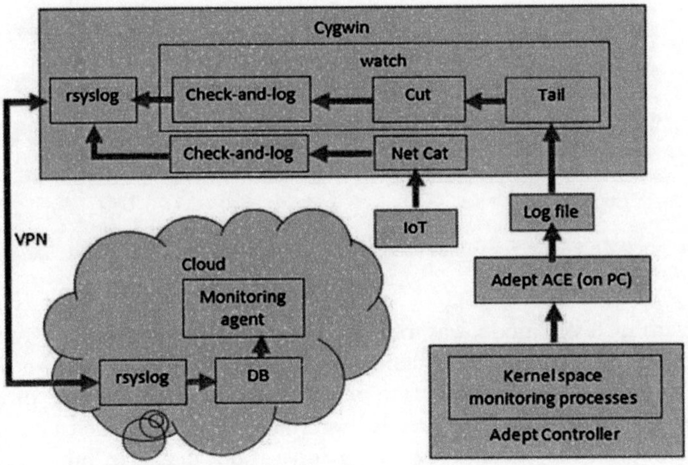

Fig. 27.3 Real time processing of robot data in cloud digital twin layers 3 and 4

consists of three commands: **tail**–extracts the last line from the log file; **cut**–extracts the individual logged values from the line taken by tail and sends the output to a custom script **check-and-log** which is based on a configuration file (Fig. 27.3).

The pipeline has two functions: (1) *compare the read values* with predefined thresholds; if the values exceed the thresholds the script sends a notification to the cloud DT in order to handle the situation and eventually stop the robot; (2) *generates log data* for the cloud system. In order to do this, the **logger** command is used to generate a standard log message which is sent to the **rsyslog** logging service.

The check-and-log script is executed in a pipeline to access robot ACE data and standalone in order to access other robot data (e.g., consumed energy) acquired by external IoT devices that use TCP-based Net Cat communication. When the messages generated by **logger** reach the local **rsyslog** server they get a timestamp (with ms resolution) and are sent then over VPN connection to the cloud where another **rsyslog** server gets the message and stores it into a MySQL database. This remote logging avoids local storage exhausting and if there are delays in receiving messages, their associated timestamp is not affected by the connection delay. The data from the physical robot stored in the cloud database is analyzed by a software agent in a heuristic parameter monitoring algorithm executed in two ways [9]:

(1) Learning a "normal" behaviour of the robot system (learning the variation of parameters during a robot cycle).
(2) Detecting anomalies in robot state based on collected data and trained model.

The first execution mode when the robot behaviour is learned should not generate false positive or false negative situations; model training is restricted to a single "static" robot cycle executed multiple times. A "static" robot cycle is composed by motions using fixed trajectories, constant load, speed and acceleration.

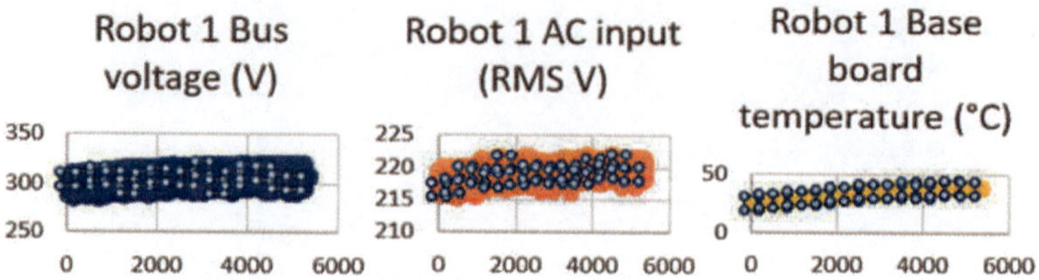

Fig. 27.4 Robot ACE parameter values used by the digital twin for behaviour training

For each robot task a model was trained and used to detect deviations from normal behaviour, based on three values defined for each parameter: *standard deviation, maximum* and *minimum value*. During experiments the digital model of the robot was created based on over 10 K reads. For some parameters such as voltage the model is based on the standard deviation for a robot cycle; in other cases (e.g., temperatures) the min. and max. values have been considered for the entire robot uptime. Figure 27.4. shows sets of 5600 data samples acquired for model training.

27.5 Conclusions and Future Work

The paper describes a layered architecture for aggregate digital twins of industrial robots integrated in multi-resource manufacturing systems. The instances of this 4-layer aggregated digital twin are embedded in individual health monitoring and predictive maintenance software programs and concurrent optimal allocation of product operations on robots at batch level. The robot DT layers span three implementing levels of Cyber-Physical Systems: the edge computing level for data acquisition, the fog computing level for data processing and analysis and the cloud computing level for machine learning and intelligent decision making.

The novelty of the research for the model of cloud-based Digital Twin for predictive robot integration in Intelligent Manufacturing Systems is represented by aggregating multiple virtual robot twins in semi-heterarchical scheduling of operations and allocation to resources through collective, global workload virtualization. This global control uses a software configuration with embedded simulation of batch production much faster than real time (performed by the cluster of virtual robot twins at computer time). The robot DTs are distributed at shop floor edge for data acquisition and alignment in time, and centralized in the cloud for anomaly detection and prediction of robot usage cost for optimal job allocation. The scientific contributions of the research are the design of the team robot DT embedded in holonic MES and its real time cloud implementation in a cyber-physical system.

Future research will focus on digital twin support for visual servoing of robots.

References

1. Borangiu, T., Răileanu, S., Anton, F., Anton, S., Iacob, I.: Embedded digital twin for ARTI-type control of semi-continuous production processes, service oriented holonic and multi-agent manufacturing systems for industry 4.0. Stud. Comput. Intell. **853**, 113–133 (2019). https://doi.org/10.1007/978-3-030-27477-1_9,Springer
2. Grieves, M., Vickers, J.: Digital twin: mitigating unpredictable, undesirable emergent behavior in complex systems. In: Book transdisciplinary perspectives on complex systems, pp 85–113. Springer (2017). https://doi.org/10.1007/978-3-319-38756-7_4
3. Kritzinger, W., Karner, M., Traar, G., Henjes, J., Sihn, W.: Digital Twin in manufacturing: a categorical literature review and classification. IFAC-PapersOnLine **51**(11), 1016–1022 (2018). https://doi.org/10.1016/j.ifacol.2018.08.474. Elsevier
4. Melesse, T.Y., Di Pasquale, V., Riemma, S.: Digital twin models in industrial operations: a systematic literature review. Proc. Manuf. **42**, 267–272 (2020)
5. Negri, E., Fumagalli, L., Macchi, M.: A review of the roles of digital twin in CPS-based production systems. Proc. Manuf. **11**, 939–948 (2017)
6. Peres, R.S., Rocha, A.D., Leitao, P., Barata, J.: IDARTS-towards intelligent data analysis and real-time supervision for Industry 4.0. Comput. Ind. **101**, 138–146 (2018). https://doi.org/10.1016/j.compind.2018.07.004
7. Redelinghuys, A.J.H., Kruger, K., Basson, A.H.: A six-layer architecture for digital twins with aggregation. In: Proceedings SOHOMA'18, Studies in Computational Intelligence, vol. 803, Chapter 32, pp. 412–423. Springer (2019) https://doi.org/10.1007/978-3-030-03003-2_32
8. Soldatos, J., Cavadini, F.A., Lazaro, O.: The Digital Shopfloor: Industrial Automation in the Industry 4.0 Era. River Publishers Series in Automation, Control and Robotics (2020). https://doi.org/10.13052/rp-9788770220408
9. Răileanu, S., Anton, F.A., Borangiu, T., Anton, S., Nicolae, M.: A cloud-based manufacturing control system with data integration from multiple autonomous agents. Comput. Ind. **102**, 50–61 (2018). https://doi.org/10.1016/j.compind.2018.08.004

Chapter 28
A Novel Design of a Robotic System for Biological Fluid Aliquoting

A. Voloshkin, L. Rybak, V. Cherkasov, and G. Carbone

Abstract This paper focuses its attention on the mechanical design of robotic aliquoting systems used to divide biological fluid into equal parts for the needs of the virology and biomedical industries. The article analyzes the existing mechanical structures, identifies problems in research and proposes a new layout of the robotic station and the design of the dosing device. The new design of the dosing device consists of a motor, a gearbox, a pneumatic cylinder connected to a rod on which the tip is fixed. The dosing device is mounted on the end effector of the delta manipulator. The attachment of the tip to the dosing device is carried out automatically, using a delta robot, due to the proposed design with a deformable sealing ring. The ratio of gear ratios of the reducer is formulated, which ensures the operability of the system. The tip attachment prototype is made of ABS plastic. The paper addresses the issues present in the current mechanical designs with a brief discussion on the system architecture.

Keywords Aliquoting system · Manipulator design · End-effector · Dosing system

28.1 Introduction

Over the past few years, demand for service robots capable of performing complex tasks without human intervention has sharply increased in almost all industries. The biomedical industry is no exception, which can be seen in a number of works [1–3]. Compliance of biosamples with high international standards for laboratory and clinical research can only be achieved by robot automation and reducing the duration of preparation and storage of initial samples at room temperature, minimizing the

A. Voloshkin · L. Rybak (✉) · V. Cherkasov · G. Carbone
Belgorod State Technological University (BSTU) named after V.G. Shukhov, Belgorod, Russia
e-mail: rlbgtu@gmail.com

G. Carbone
e-mail: giuseppe.carbone@unical.it

G. Carbone
University of Calabria, Arcavacata, Italy

I. Doroftei et al. (eds.), *Proceedings of SYROM 2022 & ROBOTICS 2022*, Mechanisms and Machine Science 127, https://doi.org/10.1007/978-3-031-25655-4_28

number of manual labor operations, and increasing the accuracy of the characteristics of the samples taken into account [4]. The need to reduce the number of production processes with the use of manual labor is also significant in order to prevent contamination of equipment and work surfaces with pathogens of infectious diseases and, consequently, to minimize the risk of infection of personnel. Several such stations have been developed to date, but most of them are directed to assembly line work. These automated stations have a typical architecture that is designed to aliquot a single body fluid in large quantities. The work of a blood laboratory in general medical institutions involves the aliquoting of biological fluids of various patients in small quantities. This results in occasional operator involvement to change body fluid tubes, as well as insertion of small well plates. To solve this problem, it is proposed to automate the work of the operator by installing a serial robot together with a delta robot.

28.2 Motivation

Aliquoting is the process of manipulating an accurately measured fraction of a sample (volume of solution) taken for analysis that retains the properties of the main sample. The process of aliquoting the blood of different patients involves dispensing biological material from a tube with a separated fraction (Fig. 28.1a) into small volume tubes (Fig. 28.1b). To do this, it is necessary to move the tube from the rack to the non-clear area (working area) where dosing into aliquots takes place (Fig. 28.1c). The quality of sample preparation is ensured at all stages, including aliquoting [5]. The work [6] shows the efficiency of automated aliquoting in comparison with the use of manual labor. Research in the field of application of robotic systems for dosing liquids and automation of laboratory processes for sample preparation is being carried out by many scientists. Active research is being carried out at the Institute of Automation, University of Rostock (Rostock, Germany), Center for Life Science Automation, University of Rostock (Rostock, Germany). Thus, in [7], a technology for determining the level of liquid in a given volume based on a pressure sensor is proposed, which makes it possible to ensure the stability of the platform for processing liquid and the accuracy of its dosing. The papers [8, 9] propose the use of a cognitive two-armed robot that works together with an operator and is remotely controlled.

There are already optimal preparation programs for fully automated pipetting and dilution of biological samples in favor of controlled bioassay [10]. Many generic solutions have been studied to better understand the problem. The main developers of robotic equipment for aliquoting biological samples are Dornier-LTF (Germany), TECAN (Switzerland), Hamilton. The laboratory equipment produced by them: PIRO robotic dosing station [11] manufactured by Dornier-LTF (Germany), robotic laboratory station [12] manufactured by TECAN, automatic dosing system (dosing robot) Microlab STAR manufactured by [13].

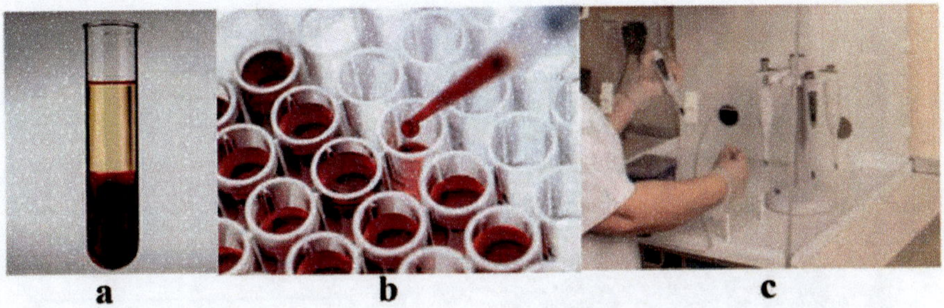

Fig. 28.1 Examples of aliquoting, **a** blood after the centrifuge, **b**, **c** liquid dosing

This equipment is designed to perform the same type of technological processes of aliquoting and includes the main elements: a robotic module with dosing multichannel pipettes and removable tips; robotic module with a gripping device for carrying tripods; auxiliary devices (pumps, barcode readers, thermo-shakers, UV emitters, detectors of the position of work surface objects, etc.), the number of which depends on the purpose of using the robotic station. The difference between the listed equipment lies only in the number of operations performed simultaneously, the systems have no structural differences. The structure of the monitored systems is designed to separate a single biological fluid into many aliquots, which is most often used in microbiology and virology laboratories. The structure of the aliquoting system must be more optimal to perform aliquoting of body fluids of various patients in general hospitals.

28.3 Design Considerations

After investigating the present designs, the following considerations were arrived upon for the robot's design:

1. It is necessary to reduce the participation of the operator in the aliquoting process in a non-clear area where biological fluid is present in open containers.
2. The station should be convenient for performing aliquoting operations, provided that the biological fluid of patients is changed frequently.
3. If the biological fluid is changed frequently, the manipulator should perform the movements more efficiently, since the optimal number of tips is no more than one.
4. The addition of technical vision will improve safety, as well as replace expensive contactless volume detection sensors.

As per the above considerations, it was decided that the robotic station should consist of two robots, a fast robot of a parallel delta structure and a collaborative serial robot with a gripper device. A dosing device will be installed on the delta manipulator. The dosing device must contain a system for creating a vacuum (vacuum) and

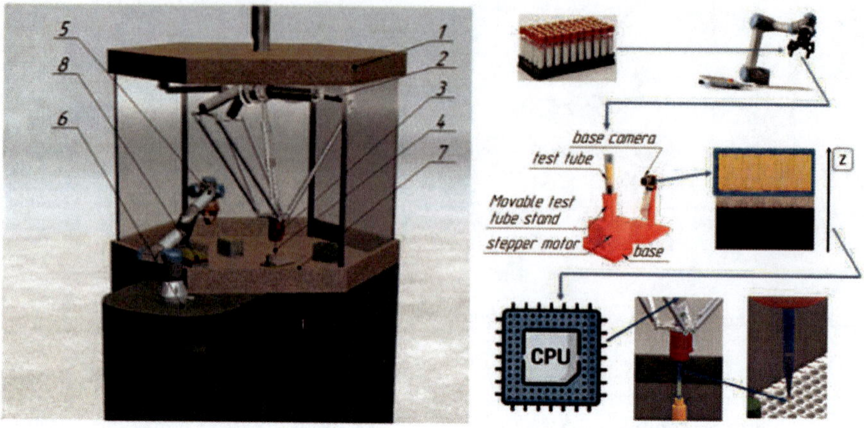

a b

Fig. 28.2 Robotic system for aliquoting biological material, **a** 3d model, **b** diagram of the operation of the robotic station

an automatic tip change system. The device must have a small mass, as this significantly affects the speed of movement of the delta robot. The relative position of the manipulators should ensure the intersection of workspaces for both manipulators to access the required objects [14]. A new design of a robotic system for aliquoting the biological fluid of various patients is shown in Fig. 28.2a. The RS (robotic system) includes: housing 1, in which a parallel delta manipulator 2 is located, which moves the dosing head 3, fixed in the center of the mobile platform of the manipulator and performing aliquoting of the biomaterial. The replaceable tip 4 on the dosing head is fixed with a rubber sealing ring. The serial 6-DOF UR3 manipulator 5 is mounted on a fixed base 6 and provides movement of the tablets 8 using a gripper device within the working area 7.

The device works as follows: UR3 moves the test tube from the tripod to the working area (non-clear area) where it is installed in a rotating stand for the test tube. The base camera loads data into the control device, where the volume of liquid is determined and the control signal of the delta manipulator is generated. When the manipulator moves the end effector to a given position, the biological fluid is taken and further divided into aliquots (Fig. 28.2b).

28.4 Dosing Head Design

The operations performed by the dosing head include fixing the pipette, taking the biological fluid, draining the liquid in equal parts and unloading the pipette. To perform these operations, the installation of several motors is automatically required: for changing the dispenser and for dosing. This would increase the mass of the

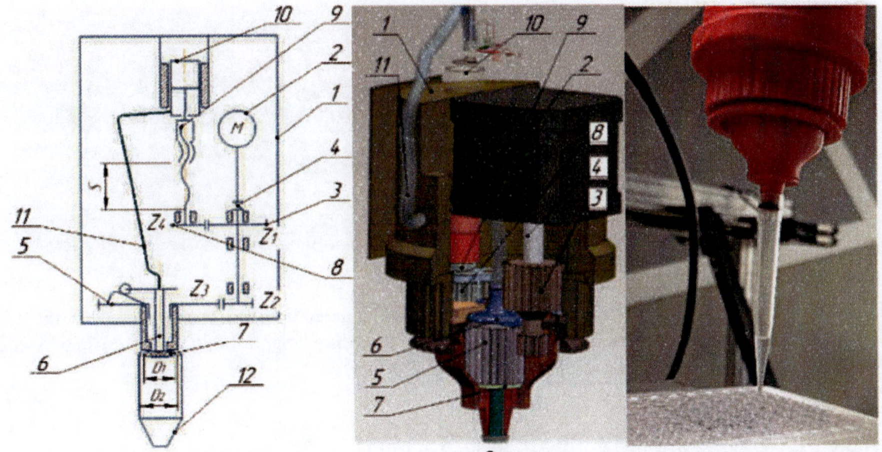

a b c

Fig. 28.3 Dosing device: **a** kinematic scheme, **b** 3D model, **c** liquid dosage of the plastic prototype

dosing head. To solve this problem, the following technical solution was proposed (Fig. 28.3).

The dosing head is shown in Fig. 28.3a and b and consists of a housing 1 in which an engine 2 is installed, connected to the drive gear 3, by means of a coupling 4. The drive gear provides rotation of the gear wheel with a cam 5, which in turn provides translational movement of the rod 6 and the sealing ring 7. The pneumatic part of the dosing head consists of a gear with a screw 8, a piston 9 and a pneumatic cylinder 10, which is connected to the rod by a sterile tube 11. Such a design of the dosing head allows to apply only one motor, greatly facilitating the design. The device works as follows: when the motor rotates, the gearbox rotates, as a result of which the cam of the gear wheel 5 lifts the rod 6. This leads to deformation of the O-ring 7, in which the diameter c of the original D_1 changes to the compressed D_2. At the same time, the inner diameter of the tip is larger than D_1 but smaller than D_2. The gear ratio is determined based on the number of teeth z_1, z_2, z_3, z_4 at the same time, there are n revolutions of z_4 per revolution of z_3, due to which the piston is displaced by the amount of S. The system performance is provided by the following dependencies

$$n_p = \frac{S}{t}, \tag{28.1}$$

where t is screw pitch, S is piston displacement. Let 's determine the number of revolutions n_p of the piston screw 9 for one revolution of the cam gear 4 according to the following dependence

$$n_p = \frac{z_1}{z_4} \cdot \frac{z_3}{z_2} \tag{28.2}$$

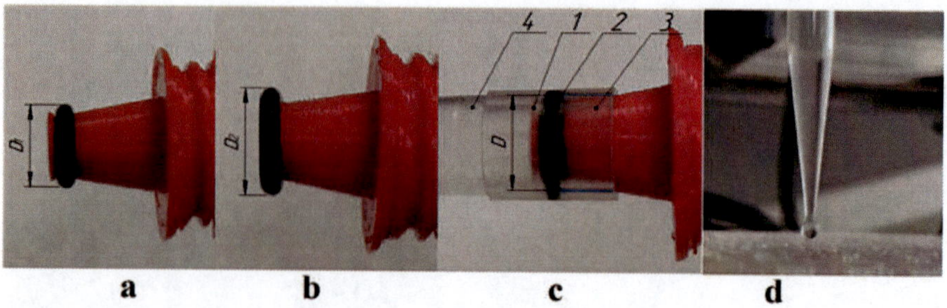

Fig. 28.4 Delta dosing head tip: **a** initial diameter, **b** diameter with the rod retracted, **c** fixing, **d** dosing operation and sealing of the pipette: 1–rod, 2–sealing ring, 3–body of the dosing head, 4–pipette

where z_1, z_2, z_3, z_4 are the numbers of gear teeth (Fig. 28.3a).

Using dependencies (28.1) and (28.2), we will select the optimal parameters of the gearbox, in which the full rotation of the gear wheel 5 will correspond to the full stroke of the pneumatic cylinder 10. As a result of the action of the transfer mechanism, a discharge is created in the cylinder 10, which allows the intake of liquid into the tip through the sterile tube 11. This makes it possible to achieve that the motor, rotating in one direction, ensures the fixation of the tip and, with further rotation, creates a vacuum in the cylinder for fluid intake. When the liquid is in the tip, the motor rotates in the opposite direction, thereby reducing the discharge and draining the liquid in the required volume. Rotation in the opposite direction will return the rod 6 to its original position, thereby changing the diameter of the O-ring 7 from D_2 to D_1. This causes the tip to reset.

Fixing the pipette at the end of the dosing head should be ensured so that the pipette is centered relative to the axis of the hole, and it is also necessary to ensure the tightness of this grip. With the help of 3D printing, a model of the tip was built and an analysis of fixation and tightness was carried out (Fig. 28.4). An experiment was carried out, as a result of which the tip was fixed and the liquid was taken. After that, two liquid dosages were transferred to small-volume test tubes. The dosing was carried out according to the specified number of engine steps.

28.5 Architecture

The general architecture comprises of the sensing and perception, motion planning, robot controller, and the actuation part shown in Fig. 28.5. The base camera is installed in a non-clear area where aliquoting takes place. The camera allows you to determine the boundary between the two phases of the biological fluid and give information about the volume of the liquid. The camera transmits the image to the processing algorithm in the system, which then sorts and produces the position of the target phase of the liquid. The robot controller optimizes the trajectory and speed of the

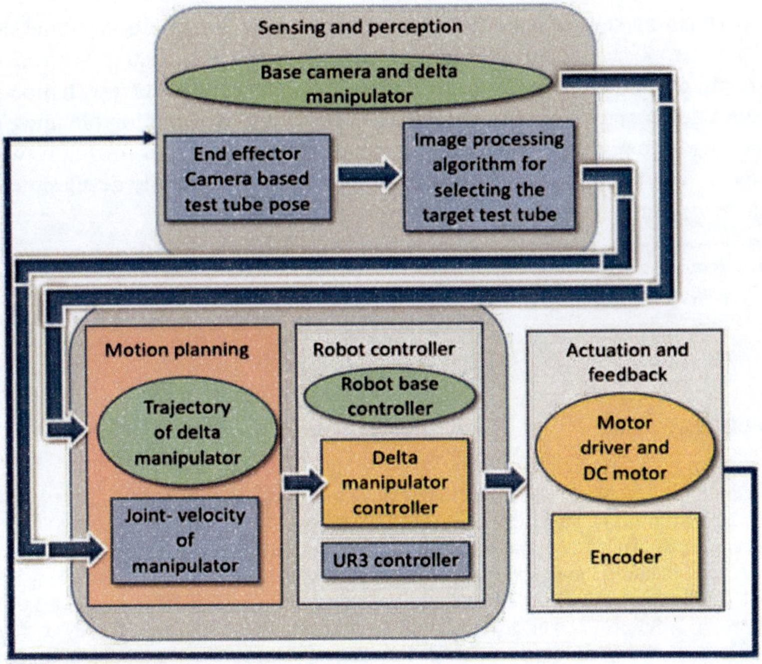

Fig. 28.5 General architecture of the system

biological fluid intake, thereby activating the motor drivers to achieve the desired goal.

The relative change in the pose of the robot and target test tube can be updated from the encoder feedback and camera feedback. The initial pose of the robot remains till the target is updated by the sensing and perception part.

28.6 Conclusions

This article proposes a new robotic system for aliquoting biological fluid, which will be applied to divide the phase of blood fluid into equal parts. The mechanical design with the system architecture has been discussed. The manipulators of the system are selected for mass production. The design of a dosing device for equipping the delta manipulator is proposed. The new design of the liquid dosing device made it possible to perform several operations using a single engine. The design of fixing the tip is proposed, for automatic alignment, fixing and sealing of the tip. A mock-up made of ABS plastic has been created to check the tightness of the joint during deformati on of the O-ring. The ring is made of polyurethane with good elasticity. The dependence of the gear ratios of the metering device gearbox is given, which allows performing several operations at a given number of revolutions of the stepper motor. The proposed technical solutions will automate the aliquoting process for various patients and protect the operator from the dangers of contact with blood. The paper does not account for the dynamic effects, control and motion planning along with the vision methods to be deployed, which remain reserved for future work in this project. The expected results of this project are the prototype development and laboratory testing.

Acknowledgements This work was supported by the state assignment of Ministry of Science and Higher Education of the Russian Federation under Grant FZWN-2020-0017.

References

1. Ceccarelli, M.: Problems and issues for service robots in new applications. Int. J. Soc. Robot. **3**(3), 299–312 (2011). https://doi.org/10.1007/s12369-011-0097-8
2. Suehara, S., Konishi, T., Fujimoto, K., Takeda, T., Fukuda, M., Koike, M., Inoue, S., Watanabe, M.: A combinatorial sample-preparation robot system using the volumetric-weighing method. Appl. Surface Sci. **252**(7), 2456–2460 (2006). https://doi.org/10.1016/j.apsusc.2005.05.082
3. Maccio, M., Bell, D., Davolos, D.: Modular automation platforms: a case study of a flexible NMR sample preparation robot. J. Assoc. Lab. Autom. **11**(6), 387–398 (2006). https://doi.org/10.1016/j.jala.2006.10.007
4. National Cancer Institute. NCI Best Practices for Biospecimen Resources. Biorepositories and Biospecimen Research Branch National Institutes of Health Table of Contents. National Cancer Institute (2016)
5. Malm, J., Fehniger, T.E., Danmyr, P., Végvári, A., Welinder, C., Lindberg, H., Appelqvist, R., Sjödin, K., Wieslander, E., Laurell, T., Hober, S., Berven, F.S., Fenyö, D., Wang, X., Andrén, P.E., Edula, G., Carlsohn, E., Fuentes, M., Nilsson, C.L., Dahlbäck, M., Rezeli, M., Erlinge, D., Marko-Varga, G.: Developments in biobanking workflow standardization providing sample integrity and stability. J. Proteomics **95**, 38–45 (2013)
6. Malm, J., Végvári, A., Rezeli, M., Upton, P., Danmyr, P., Nilsson, R., Steinfelder, E., Marko-Varga, G.: Large scale biobanking of blood-the importance of high density sample processing procedures. J. Proteomics **76**, 116–124 (2012)
7. Zhang, C., Huang, Y., Fang, Y., Liao, P., Wu, Y., Chen, H., Chen, Z., Deng, Y., Li, S., Liu, H., He, N.: The liquid level detection system based on pressure sensor. J. Nanosci. Nanotechnol. **19**(4), 2049–2053 (2019)
8. Fleischer, H., Baumann, D., Joshi, S., Chu, X., Roddelkopf, T., Klos, M., Thurow, K.: Analytical measurements and efficient process generation using a dual-arm robot equipped with electronic pipettes. Energies **11**, 2567 (2018)
9. Fleischer H, Drews RR, Janson J, Chinna Patlolla BR, Chu X, Klos M, Thurow K (2016) Application of a dual-arm robot in complex sample preparation and measurement processes. J lab Autom 21(5):671–681
10. Jiang, H., Ouyang, Z., Zeng, J., Yuan, L., Zheng, N., Mohammed, J., Mark, E.: A user-friendly robotic sample preparation program for fully automated biological sample pipetting and dilution to benefit the regulated bioanalysis. J. Lab. Autom. **17**(3), 211–221 (2012)
11. INTEGRA Biosciences. Freeling you from tedious multichannel pipetting tasks (2022). www.bionity.com/en/products/1128490/freeing-you-from-tedious-multichannel-pipetting-tasks.html
12. Freedom EVO platform (2022). https://lifesciences.tecan.com/freedom-evo-platform

13. Hamilton Broshure. Life science robotics. https://corefacilities.isbscience.org/wp-content/uploads/sites/5/2015/07/HamiltonBrochure.pdf
14. Malyshev, D., Rybak, L., Carbone, G., Semenenko, T., Nozdracheva, A.: Optimal design of a parallel manipulator for aliquoting of biomaterials considering workspace and singularity zones. Innov. Robot Des. Approaches **12**(4), 2070 (2022)

Part VI
Humanoid Robots

Chapter 29
Multimodal Perceptual Cues for Context-Aware Human-Robot Interaction

S. Paplu, H. Ahmed, A. Ashok, S. Akkus, and K. Berns

Abstract This paper highlights the relevance of context awareness and the understanding of intention in Human–Robot Interaction. Upon the spotting of conflicting/contradictory information in interaction, the humanoid robots ROBIN and Emah generate context-driven speech, gestures and facial expressions. A large number of situational gestures, postures and facial expressions have been implemented on the robots to cater for situational reactions. The addition of such a system in social robots aims to create a more intelligent interaction mechanism capable of adapting to the scenarios they are confronted with and responding in accordance to the context of the dialogue. Over the years, there has been much focus in the research of Human–Robot Interaction, particularly on the robot's autonomous behavior and surroundings. However, there has been little emphasis paid to the human's essential presence or engagement, which includes psychological, social, and deliberate components of human activity. Speech as well as vision are two modes of context taken into account while spotting anomalies in the interaction process. The system has been tested on several interaction scenarios. The evaluation results show a positive notion of cognitive skill of the robots.

Keywords Human–Robot interaction · Computational linguistics · Contextual anomaly detection · Cognitive robotics

---

S. Paplu (✉) · H. Ahmed · A. Ashok · K. Berns
Technische Universität Kaiserslautern, Computer Science, Kaiserslautern, Germany
e-mail: paplu@cs.uni-kl.de

H. Ahmed
e-mail: ahmedh@rhrk.uni-kl.de

A. Ashok
e-mail: aashok@cs.uni-kl.de

K. Berns
e-mail: berns@cs.uni-kl.de

S. Akkus
Technische Universität Kaiserslautern, Kaiserslautern, Germany
e-mail: sakkus@rhrk.uni-kl.de

I. Doroftei et al. (eds.), *Proceedings of SYROM 2022 & ROBOTICS 2022*, Mechanisms and Machine Science 127, https://doi.org/10.1007/978-3-031-25655-4_29

29.1 Introduction

Human communication has been characterized by Mehrabian's 7-38-55 model such that 7% of spoken text contributes towards personal communication [1]. Similar to in Human–Human Interaction (HHI), textual context-awareness enables a robot in Human–Robot Interaction (HRI) for optimal communication by adapting its behavior according to temporal, situational, spatial, social, and cultural contexts [2]. Contextual-anomaly detection is an important research in natural language processing and is considered to be a supervised learning task. During an interaction between a human and a social robot, detecting the interaction partners' abnormal statements and varying sentiment patterns enables the robot to exhibit context-awareness. The detected anomaly can be either used in direct context or as a new instance of knowledge. For example, the anomaly of greeting someone "good morning" after 12 pm can be interpreted as a mistake followed by correction of the person or maybe the new knowledge that it is indeed morning in this person's geo-location. This work attempts to spot a large number of contextual cues and contextual anomalies with the help of the perception system of the robot.

Attributing to a 55% contribution in the human communication model, our vision system detects the low-level percepts e.g., human skeleton, human joints, human identity, head pose, face, facial expression, facial action units, eye gaze, gender and ethnicity from the face, static hand gestures of both hands, dynamic hand gestures of both hands, and, human posture [3]. The high-level percepts include human feedback perception, context awareness, and human personality recognition [4]. Additionally, influenced by Paul Ekman's Facial Action Coding System (FACS), the robot has been programmed to express a large number of body gestures and facial expressions [5]. Attributing to a 7% contribution in the human communication model, our language processing module has focused on text manipulation algorithms [6]. Raw audio signals are analyzed to extract the high-level features extracted as output e.g., analysis of textual sentiment (happiness, surprise, fear, sadness, etc.), topic tracking, context-awareness etc. Finally, for conveying verbal dialogues or conversational responses, the robot utilizes a speech synthesizer called Acapela. Attributing to a 38% contribution in the human communication model, a paralinguistic module focuses on the tone of speech [7]. The detected speech data is converted into a descriptive feature vector. The high-level output of this module includes emotion labels of Neutral, Calm, Happy, Sad, Angry, Fearful, Disgust, and Surprise.

However, there has been very limited research exploration as to how the social robots should behave based on the contextual events or conflicting information provided by the interaction partners. The approach proposed in this work strives to utilize a diverse set of perceptual cues to find contradictions in the speech or the activities of human partners. Since the scope of contextual cues are vast, we focused on a limited number of anomalous events. The events are activated by our proposed system, with the robot generating appropriate behavior out of the event spotted. The rest of this paper is organized as follows. In Sect. 29.2, we discuss the state-of-the-art context awareness schemes for text data with the focus on the speech generation systems. The technical details of each component, and its corresponding implementation are discussed in Sect. 29.3. A description of the experimental validation and evaluation strategy coupled with results follows in Sect. 29.4, following which a short conclusion of our obtained results is discussed in Sect. 29.5.

29.2 Literature Survey

While HRI explores such interactions from a computational, design, and social standpoint, contextual information helps form the interpretation of personal intention during interactions. Context has been extensively researched in language usage, with "context" typically referring to the history of previous utterances (e.g. Ferstl [8]), but also encompassing other types of context. Holtgraves [9] discovered that the speaker's standing concerning the hearer influences whether or not the literal meaning of an indirect request is activated. According to Turne [10], we can define context and situation as follows. Any identifiable configuration of environmental and interlocutor-related features that has predictive power for an interlocutor's behavior. The term situation is used to refer to the entire set of circumstances surrounding an interlocutor. In this work, a multi-modal context verification scenario during Human–Robot Interaction is addressed. As discussed in [11–13] intelligent conduct necessitates context sensitivity. Context influences behavior by influencing the activities taken to attain goals, as well as their timing and style of execution. An intelligent agent may more swiftly pick suitable action to fulfill its goals by paying attention to its context, and it can more successfully concentrate its attention and respond to unforeseen occurrences by paying attention to its context. Hence, the motivation of our work is to understand this contextual information used as predictive and prescriptive information i.e. understand how to make predictions about the current context, including features that have yet to be seen, appropriately set behavioral parameters, assist the agent in focusing its attention on appropriate goals to achieve in the current situation, select appropriate actions to achieve its goals, and respond quickly and appropriately to unanticipated events.

In our work, we use a multi-modal approach for context awareness which makes it more robust and error-free. In a dialogue when an interlocutor specifies some arguments that are not correct the robot prompts the interlocutor regarding the argument and states that the interlocutor has stated wrongly and raises a flag specifying the particular type of anomaly found in the argument. We have tried to concentrate our work on rule-based scenarios using both speech and vision. This empowers the robot with context-awareness during the dialogue process.

29.3 Implementation

29.3.1 Perception System

Visual Perception

As mentioned earlier, we use OpenNI and NiTE libraries which enable the detection of humans by utilizing the depth and InfraRed (IR) sensor of ASUS Xtion. This camera sensor mounted on the chest of the robot utilizes algorithms not only to detect humans but also track them efficiently. The face tracker provides a count of the number of faces available in front of the robot. They extract human skeleton joints using NiTE library and convert them into angles. Feature vectors are generated using angles between each joint and classify them with Support Vector Machines (SVMs). The system uses low-level perception features to understand high-level perception behaviors, e.g., head gesture recognition [3], facial expression recognition [14], body posture [15] etc. Based on the depth information processed by the camera, proximity of the interaction partner is also calculated.

Auditory Perception

This social robot has been enabled to perceive acoustic features/audio signals from an interlocutor. The detected signals are then converted into written text (transcription) so as to form an understanding of the context and perform textual manipulation for obtaining knowledge for the robot [6]. Currently, English as well as German speech recognition capability is integrated with the dialog control system of the robot. In this paper, English speech recognizer has been utilized to obtain semantic and contextual information extracted from the user utterances with state-of-the-art information extraction library MITIE [16]. Some examples of auditory features include named entities (e.g., name of a person, place, organization), parts of speech tagging etc. Figure 29.1 illustrates the basic principle of NER.

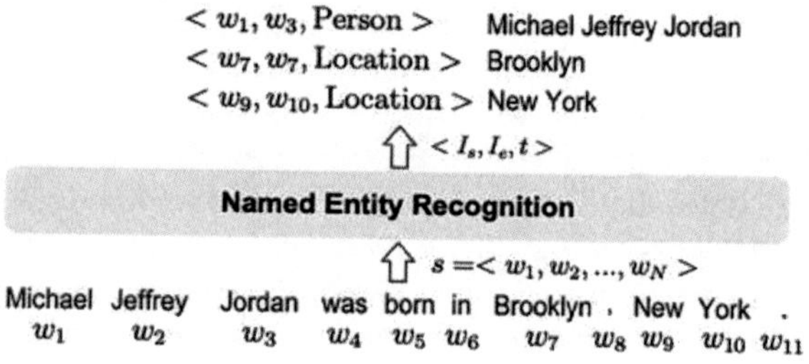

Fig. 29.1 Example sentence (user utterance) with NER applied on it

29.3.2 *Spotting Contradictory Information*

The proposed system as depicted in Fig. 29.2 takes audio-visual cues as inputs and generates context-aware robot behavior. The broader categories of conversational anomaly estimation are discussed below:

Temporal Anomaly

Current Clock Time: This event corresponds to auditory features. This event occurs when the interacting interlocutor states the time of day to the robot. The robot identifies whether the specified time is correct or incorrect. If the time is incorrect, a time anomaly event is raised. Contradicting responses are generated with the speech generation (chatbot) system. This keyword-based speech generation system can take an event as input and generate an appropriate response based on the input event. For example, "conflicting local time" events can trigger a response for the robot e.g., "I believe, the local time that you mentioned is incorrect. The local time is now @actual_time".

Current Date: This event relies on verbal cues of the interlocutor. This occurs when the interlocutor delivers a response corresponding to date. The robot is able to distinguish if the information is correct. If the date parameters are incorrect, a specific date anomaly is raised and a relevant response and gesture/facial expression is selected from the dialog file.

Month: This event occurs when the interacting interlocutor states the current month to the robot. Robot identifies if the specified month is correct or incorrect. If the current month is incorrect, a specific month anomaly event is raised and Incorporated with the chatbot system. This triggers an event-specific robot speech and its behavior in the form of gesture and facial expression.

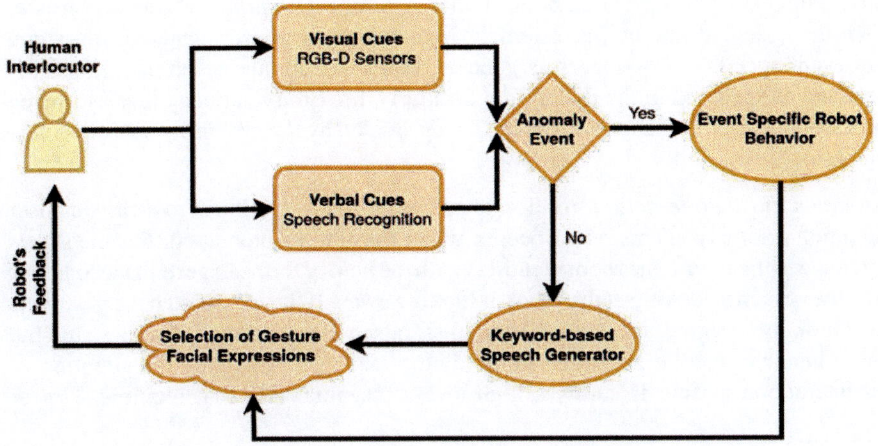

Fig. 29.2 Workflow of the context-aware HRI system

Year: System clock is used to extract the current year. Upon receiving contradictory responses from the human interaction partner, the robot verifies the given information against the genuine current year. Similarly, a flag is raised and robot's event-specific behavior generated.

Local Temperature: This event is independent of the visual system. Local temperature is extracted by our system from the web. The robot verifies the utterances that contain current temperature information. We maintained a threshold of $+5$ to -5 degrees on the exact temperature outsides. If the stated temperature is not in between this threshold boundary, a temperature anomaly flag is raised and the robot's behavior selected as specified in the dialog file.

Similarly, weather status, weekdays and seasons are verified against system data. All the weather information is extracted from a web API.[1] The validations with regards to weather information are validated against the data available on the web API.

Spatial and Situational Anomaly

Walking: This event corresponds to both visual and auditory cues. In this event, when the interlocutor states that he/she is walking, the robot identifies whether the recognized body movement is the same as walking or not. If the identified body activity is not recognized as walking, a walking anomaly is raised and incorporated on the chatbot system during the system run-time. The high-level visual proximity feature decides whether the uttered "walking" corresponds to the "walking" recognized by the vision system.

Number of people and proximity: This event also corresponds to not only visual but also auditory cues. This event occurs when the user specifies the number of people in the robot's known environment e.g., experiment room. The robot identifies this information through the vision system using the dynamic modeling of humans in the environment. It verifies whether the specified number of humans are in the environment or not. If the number of humans in the environment are less or greater than the specified number, an anomaly is raised in the speech generation system. An event specific robot speech is generated and appropriate gesture/expression is selected as specified in the dialog file. Similarly, proximity anomaly was estimated based on the user utterance and the the data provided by the proximity sensor (kinect camera).

Attention during interaction: Audio-visual cues have been utilized to estimate "user attention anomaly". This event occurs when the interlocutor utters that he/she is attentive to the robot. The robot identifies with the help of the head gestures recognizer that the person is looking at the robot or looking away. If the identified head gesture is not "looking straight", an attention anomaly is raised. Based on the event/flag, suitable robot behavior can be generated in the dialog manager. Assessing the attention of an interaction partner is surely a high-level perceptual task that requires a lot of

[1] https://openweathermap.org/.

contextual knowledge. In our case, the head pose of the human interaction partner played the key role in the assessment of attention. The real-time images captured with the Kinect camera have been utilized to estimate the head pose of the human.

29.3.3 Generation of Robot's Behavior

The behavior of the robot consists of three parts.

- Robot's verbal response to the user's utterances
- Gesture/body movements and
- Facial expressions

In order to control the robot's movements and facial-expressions based on the output of our sentiment analyzer, XML-based dialog files have been utilized. The generic dialogues are provided with the flexibility to change based on the user's speech and the anomaly events. As stated earlier, the basic actions that constitute the behavior of the robot include speech, body postures, and facial expressions. The synchronization of these actions is performed in the dialog files. The gestures/postures and facial expressions implemented in this work along with the speech of the robot have been used to make the interaction more realistic and engaging. In this work, we focused more on the interactions in which semantic and contextual information is flagged by the proposed system. The anomalies detected are the triggering events in most of the cases. The dialog system fetches the semantic or contextual information, making the interaction more dynamic. Based on the interaction partner's speech recognized by Julius, the dialog system triggers a response taking Named Entity Recognition (NER), Binary Relationship Detector BRD and specific contextual information into account [16]. The entities recognized by the system include name of a person, name of a place, name of an organization etc. The BRD establishes relations between two entities detected in the user utterances.

Speech: Speech is the spoken expression of textual information. For the robot's case, the internally implemented chatbot system [2] is the component in charge of generating textual information. The speech generation process is implemented based on a keyword-matching approach. It contains a knowledge base with all the possible outputs. Then, the input is matched with the corresponding output by using *fuzzy string matching*. This technique allows the identification of non-identical matches for a given target. Given the working principle from the chatbot, we can assume that it is mostly a reactive element. Text To Speech (TTS) system from Acapela Group[2] enables the robot to dictate the speech outcomes.

Gestures/Postures: More than 120 gestures/postures have been implemented on the robot. The selection of gestures are based on the emotion conveyed by them. For example, "come here" gesture can be executed by moving the entire arm or just

[2] https://www.acapela-group.com/.

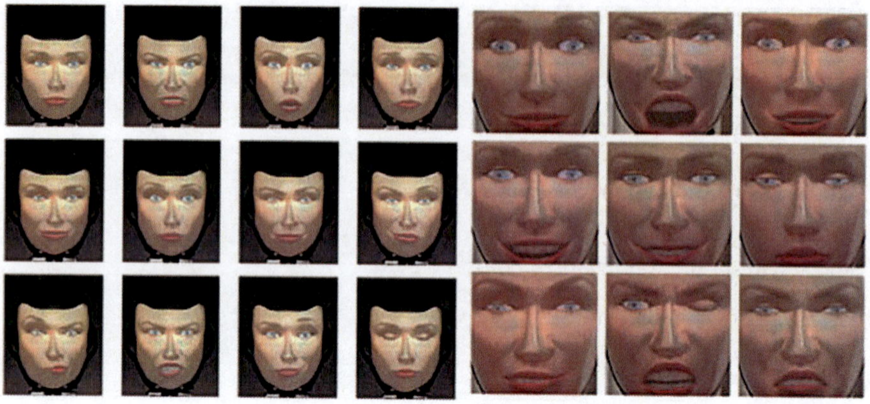

Fig. 29.3 The facial expressions implemented on the social robot ROBIN

using the fingers (depending on left/right handedness of the robot). There are four types of gestures developed so far on ROBIN: static, cyclic, trajectory, dynamic etc. The crucial parameters for gestures dealing with a certain trajectory are "time" and "repeat". A specific set of joint values are provided to make a specific gesture. Unlike the trajectory gestures, static gestures simply reach the desired set of joint values without following a specified pathway. However, each of the gestures have situational or contextual points of view and they keep running throughout the conversation process.

Facial Expressions: Facial expressions are particularly crucial in communicating emotional states and responsiveness during interactions [17]. They have been implemented in ROBIN using the Facial Action Coding System (FACS) developed by psychologist Paul Ekman [18]. This system supports that any facial expression can be described as a combination of several small anatomical factions called Action Units (AUs). ROBIN is able to display more than 140 different facial expressions. Figure 29.3 shows a set of basic and situational facial expressions implemented on ROBIN.

29.4 Experimentation and Evaluation

29.4.1 Experimental Setup

For testing the proposed system, the social robot ROBIN (Fig. 29.4a) has been extensively used. The robot Emah (Fig. 29.4b) has been used partially. Developed by Engineered Arts [19], the socially interactive robots are constructed in a way that they can mimic human movements and expressions. Overall, ROBIN is equipped with

34 Degrees of Freedom (DoF). Due to these, its movements and gestures are able to mimic those of the human body, interacting in a human-like way. The robot has a spine with 3 DoF provided by three DC motors in the artificial hip, which allows generic movement of the upper body. This spine supports ranges of 6° in tilting, 15° in bending, and 20° in turning.

A group of 16 participants, mostly university students, are invited to evaluate the system. The participants were informed of the contextual cues and the anomaly spotting mechanism. The subjects were briefly introduced with our anomaly estimation system. Each interaction partner was exposed to two scenarios: (i) The interlocutor invites the robot to his/her mother's birthday and then tries to provide incorrect information. The robot cross-checks the information with known environment parameters and responds accordingly. Temporal anomalies are tested in this scenario. (ii) This scenario was designed to verify all the multi-modal context verification events. The scenario tests all available spatial and situational anomalies. In this case, the interlocutor tries to provide incorrect information. The robot verifies the information with the auditory/visual sensors and responds accordingly. Below is an excerpt of the dialog for the first scenario.

Interlocutor: *Hello!*
Robot: *Hi, how are you doing?*

Interlocutor: I am doing good. I want to invite you. We will have a party soon.
Robot: That's Great to hear. What is the party for?

Interlocutor: It's my mother's birthday.
Robot: Nice, when is the party?

Interlocutor: The party date is today 18 Dec.
Robot: The information provided is wrong. Today is 17 Dec (current date).

Fig. 29.4 The social robots ROBIN and Emah used in the experimental scenarios

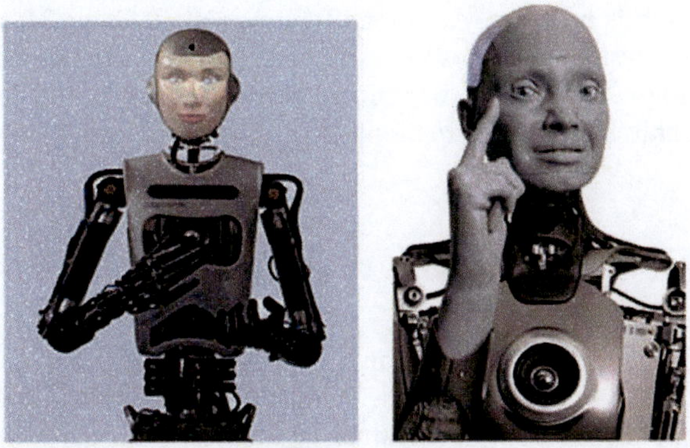

(a) Upper body of ROBIN (b) Upper Body of Emah

Interlocutor: Sorry, I mean tomorrow. I need to buy gifts.
Robot: Great! What do you plan to buy?

Interlocutor: As the season is Autumn, So I thought of buying her flowers.
Robot: I am sorry! This is not the current season. Currently, it's winter (current season).

Interlocutor: Thank you for correcting my mistake.
Robot: No worries! So, when will the party start tomorrow?

Interlocutor: It will start at the same time as now, at 19 O'clock.
Robot: But it's 16 O'clock. now. So, you mean that the party will start at 16 O'clock?

...

29.4.2 *Evaluation*

All participants are provided with a questionnaire comprising 5 questions. The questions include: (i) is the change in the robot's emotional state meaningful or realistic, (ii) is the period of time the robot takes to switch its behavior spontaneous? (iii) was the robot's speech, gesture and facial expressions synchronized properly? (iv) how was the display of facial expression on specific events? Each of these questions had 3 options to choose: realistic, unrealistic and unclear. Furthermore, there was an open-ended question, asking about the overall user experience with the interaction scenarios. User experience with the system implemented has been collected from the questionnaire. It can be observed that 67% of the participants considered the change in the robot's behavior during interaction to be realistic. However, the synchronization of speech, gesture and facial expressions was found to be a little bit unrealistic by a big chunk of users (33%). System latency could be one of the reasons for this problem. Additionally, it was unclear for 31% of the interaction partners if the system complies with a diverse set of conversational anomalies. The participants often expected the robot not to correct the users all the time, even if the users provide contradictory information.

29.5 Conclusion

The approach proposed in this paper presents a notion of context-awareness during human–robot interaction. The audio-visual perception system of the robot has been exploited to spot contextual anomalies, with triggering events generated. A large number of gestures and facial expressions for our social robots have been implemented to show reactions based on the contextual anomalies estimated. The experimental results showed that the robot, in most of the cases, managed to generate a suitable reaction based on the context-awareness mechanism proposed in this work. As a future work, additional situational and contextual cues can be incorporated. The inclusion of linguistic and paralinguistic cues can be another direction of future research that might strengthen the robustness of the system. Text summarization approach can be applied to estimate the content of conversational topics.

References

1. Mehrabian, A., Ferris, S.R.: Inference of attitudes from nonverbal communication in two channels. J. Consult. Psychol. **31**(3), 248 (1967)
2. Paplu, S.H., Arif, M.N., Berns, K.: Adapting behaviour of socially interactive robot based on text sentiment. In: 30th International Conference on Robotics in Alpe-Adria-Danube Region (RAAD 2021), pp. 221–228. Springer Nature (2021)
3. Saleh, S., Sahu, M., Zafar, Z., Berns, K.: A multimodal nonverbal human-robot communication system. In: Proceedings of the 6th International Conference on Computational Bioengineering (ICCB). Barcelona, Spain (14–16 Sept 2015)
4. Zafar, Z., Paplu, S.H., Berns, K.: Automatic assessment of human personality traits: a step towards intelligent human-robot interaction. In: 2018 IEEE-RAS 18th International Conference on Humanoid Robotics (Humanoids), pp. 670–675. Beijing, China (6–9 Nov 2018)
5. Paplu, S.H., Mishra, C., Berns, K.: Pseudo-randomization in automating robot behaviour during human-robot interaction. In: Proceedings of the 10th IEEE International Conference on Development and Learning and Epigenetic Robotics (ICDL-ER), pp. 120–125. IEEE, Valpara´ıso, Chile (26–30 Oct 2020). https://doi.org/10.1109/ICDL-EpiRob48136.2020.9278115
6. Paplu, S.H., Arif, M.N.I., Berns, K.: Utilizing semantic and contextual information during human-robot interaction. In: 2021 IEEE International Conference on Development and Learning (ICDL), pp. 1–6. IEEE (2021)
7. Ashok, A., Pawlak, J., Paplu, S., Zafar, Z., Berns, K.: Paralinguistic cues in speech to adapt robot behavior in human-robot interaction. In: Proceedings of the 9th IEEE RAS/EMBS International Conference on Biomedical Robotics & Biomechatronics (BioRob). Seoul, Korea (21–24 Aug 2022)
8. Ferstl, E.C.: Context effects in syntactic ambiguity resolution: the location of prepositional phrase attachment. Ph.D. thesis, University of Colorado at Boulder (1994)
9. Holtgraves, T.: Communication in context: effects of speaker status on the comprehension of indirect requests. J. Exp. Psychol. Learn. Mem. Cogn. **20**(5), 1205 (1994)
10. Turner, R.M.: Context-mediated behavior for intelligent agents. Int. J. Hum. Comput. Stud. **48**(3), 307–330 (1998)
11. Mellers, B.A., Cooke, A.D.: The role of task and context in preference measurement. Psychol. Sci. **7**(2), 76–82 (1996)
12. Wang, J., Ding, Z., Jiang, C.: Gaom: Genetic algorithm based ontology matching. In: 2006 IEEE Asia-Pacific Conference on Services Computing (APSCC'06), pp. 617–620. IEEE (2006)
13. Borrmann, D., Elseberg, J., Lingemann, K., N¨uchter, A.: The 3d though transform for plane detection in point clouds: a review and a new accumulator design. 3D Res **2**(2), 1–13 (2011)
14. Al-Darraji, S., Berns, K., Rodi´c, A.: Action unit based facial expression recognition using deep learning. In: International Conference on Robotics in Alpe-Adria Danube Region, pp. 413–420. Springer (2016)

15. Zafar, Z., Paplu, S.H., Berns, K.: Automatic assessment of human personality traits: a step towards intelligent human-robot interaction. In: 2018 IEEE-RAS 18th International Conference on Humanoid Robots (Humanoids), pp. 1–9. IEEE (2018)
16. King, D.E.: Dlib-ml: a machine learning toolkit. J. Mach. Learn. Res. **10**, 1755–1758 (2009)
17. Al-Darraji, S.: Perception of Nonverbal Cues for Human-Robot Interaction. Verlag Dr, Hut (2016)
18. Ekman, P., Friesen, W.: Manual of the facial action coding system (facs). Trans. ed. Vol. Consulting Psychologists Press, Palo Alto (1978)
19. Engineered Arts, U.: Robothespian-Humanoid Robot, the acting humanoid robot https://www.engineeredarts.co.uk/robothespian/. Accessed 14 09 2021

Chapter 30
An Experimental Characterization of LARMbot Torso V1

W. Gao and M. Ceccarelli

Abstract In this paper, a characterization of LARMbot torso V1 is presented by experiment result in one mode of functioning. Arduino nano is applied to control the whole system; PCA9685 board is used for connecting with the motors; sensors, including IMU sensor, force sensors, and current sensor, are applied for measuring the related parameters during the LARMbot torso's moving process. Results show that the LARMbot torso V1 can mimic successfully human motion with a low-consumption. The contact forces among the shoulders and cables are proper. Finally, operation flaws are identified so that improvements can be suggested for future work in new design solutions.

Keywords Robotics · Humanoid torso · Experimental characterization · Cable-driven mechanisms · LARMbot humanoid · Performance evaluation

30.1 Introduction

Humanoid robots play an important role in serving at home or factories. As a key part of a humanoid robot, a suitable design of humanoid torso can contribute to humanoid robots performing human-like motion. It is also fundamental for improving the working performance of a humanoid robot. Therefore, developing humanoid torso is important to give proper support to humanoid hands, head, and limbs.

Recently, some scholars obtained some achievements on developing humanoid torso with specific solutions. Huang et al. [1] proposed a humanoid torso with a 3-DOF (degree of freedom) PKM (parallel kinematic manipulator) structure by analyzing the characteristics of torso in humans and animals. A numerical simulations were conducted to prove the human-like movements of the proposed PKM torso. Reinecke et al. [2] introduced a parallel structure of actuation with tendons

W. Gao (✉) · M. Ceccarelli
Laboratory of Robot Mechatronics, University of Rome Tor Vergata, 00133 Rome, Italy
e-mail: wenshuo.gao@students.uniroma2.eu

M. Ceccarelli
e-mail: marco.ceccarelli@uniroma2.it

I. Doroftei et al. (eds.), *Proceedings of SYROM 2022 & ROBOTICS 2022*, Mechanisms and Machine Science 127, https://doi.org/10.1007/978-3-031-25655-4_30

and a weight compensation via a spring mechanism in a robotic torso system. The design is robust and can work efficiently, but the structure is complicated. Pencic et al. [3] proposed multi-joint flexible structure with stiff, low backlash and self-locking mechanism for robotic spine with some gears in the mechanism. Li et al. [4] proposed a humanoid torso with a six-degree-of-freedom serial mechanism with twist angles resulting a structure with complicated operation. Simulations validated that the proposed humanoid torso has high stiffness and easy-control. Meanwhile, several versions of humanoid torso are developed at LARM, such as CaPaMan-based trunk design in 2001, CALUMA design since 2003, PKM design in 2008, waist-trunk system by using two CaPaMan manipulators in 2010, waist-trunk system with CaPaMan 2bis structure, and trunk-waist design with two PKM modules, as reported in [5]. As a significant result, LARMbot torso V1 has good working performance [6, 7]. The current prototype is manufactured by 3D printer and assembled with market elements. However, an experimental characterization was planned also to look for improvements that can be conducted in the future.

In this paper, based on the current prototype of the humanoid torso of the LARMbot humanoid, an experiment in one bending situation is discussed with its results. Arduino nano is applied to control the whole prototype; an IMU sensor is used to measure the position parameters during the humanoid torso moving process; force sensors are set on the shoulders in order to test the contact force between the cables and shoulders; current sensor is applied to obtain the power consumption when the system is working. Results are discussed with acquired data on angular acceleration, roll and pitch angles, contact forces, and power consumption. Improvements are suggested for future work based on the testing results.

30.2 LARMbot Torso V1

LARMbot torso V1 consists of trunk, spine, and waist, as shown in Fig. 30.1. In Fig. 30.1a, it illustrates the scheme of LARMbot torso V1, which consists of vertebral disc, four cables, vertebral joint, actuators. The structure is based on a spine-like structure with three vertebra units, composed by market couplings as intervertebral discs and 3D printed plates as vertebras. The tension-muscle driving structure is made of cable actuated parallel manipulator.

Figure 30.1b shows a prototype that is manufactured by using 3D printer so that the prototype is light-weight and low-cost. The whole prototype is of 154 × 113.8 × 261.34 mm. It is a cable-driven mechanism which is actuated through four servo motors. Each motor drives a cable with it corresponding antagonism so that the structure works with 3 DoFs motion. As for spine part, it is made up three disk-coupling modules. Each vertebra unit is 53.14 mm height [6, 7].

It can work in several steps. At the beginning, the cables are pulled by the actuators. When the actuators move antagonistically, the humanoid torso can bend right, left, forward, and backward [7].

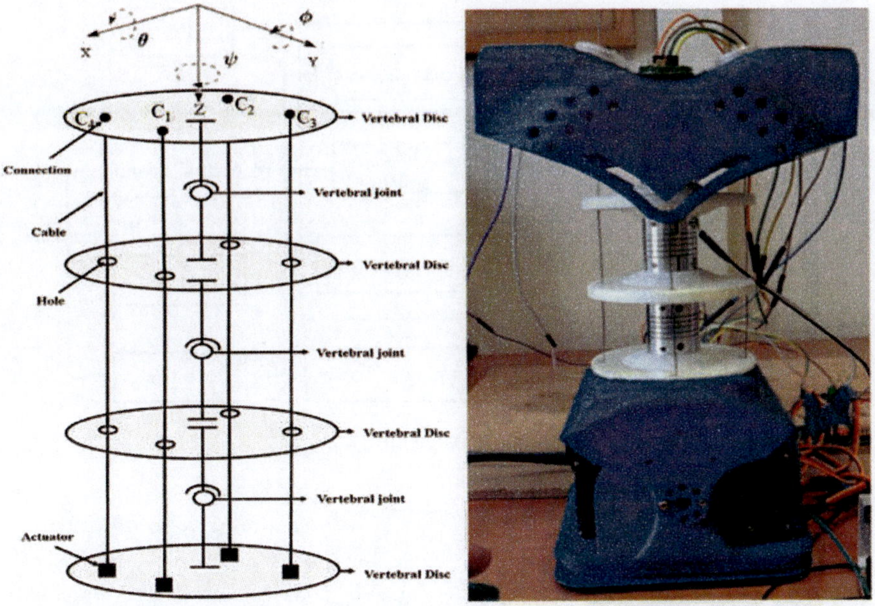

(a) (b)

Fig. 30.1 LARMbot torso V1 with cable-driven mechanism: **a** a scheme [7]; **b** lab prototype

30.3 Layout and Modes for Testing

To obtain a characterization of the LARMbot torso V1, an experiment is conducted in a mode with data summarized, as shown in Table 30.1. The mode 1 is planned for bending left and right. Significant parameters can be considered as the linear acceleration of the shoulder point H (shown in Fig. 30.3b and angles, the contact force between the cables and shoulder, and the power consumption. The testing layout is summarized in Fig. 30.2, indicating also data flows. The whole control system is designed as open-loop structure for easy-operation features also in testing with current prototype version.

The actuation system is selected referring to the human-like torso structure with the cable acting as actuating muscles of the spine-like structure, by designing and

Table 30.1 The testing mode of LARMbot torso V1

Mode	Action	Main parameters
1	Bending left and right	Acceleration: a_H (x, y, z axis) Angles: pitch and roll Contact force: F_{c3} and F_{c4} Power consumption: P

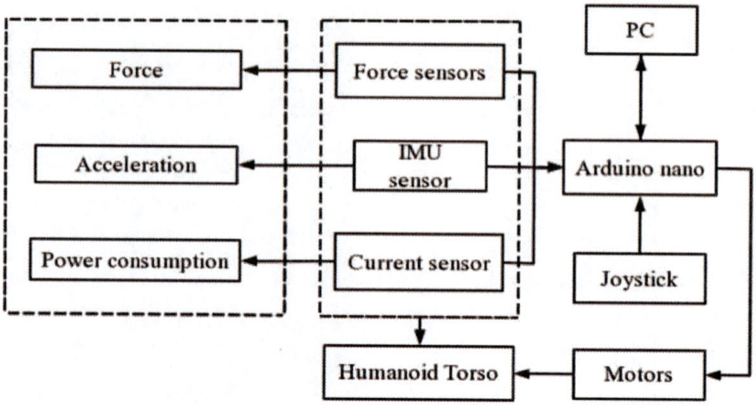

Fig. 30.2 A scheme of layout for testing LARMbot torso V1

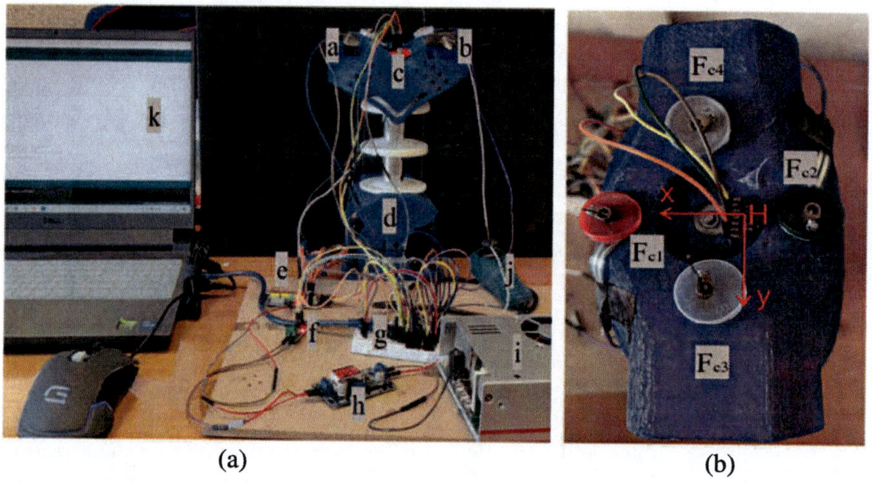

Fig. 30.3 Layout for testing LARMbot torso V1 in Fig. 30.2: **a** testing system; **b** distribution of sensors (a force sensor Fc4; b force sensor Fc3; c IMU sensor; d LARMbot torso V1; e PCA9685; f current sensor; g Arduino nano; h DC-DC convert; i power supply; j joystick; k laptop)

using a parallel architecture with the servo motors installed on the waist platform. The lab testing system is shown in Fig. 30.3a according to the design layout in Fig. 30.2. A laptop is used for transferring the input and output data. For measuring the designed parameters, an IMU sensor is applied for monitoring the motion of the LARMbot torso V1. The position of the IMU sensor is at the center of the shoulder, which is set as H point in Fig. 30.3b. Several force sensors are used to obtain the force between the cables and shoulder body with a distribution of force sensors shown in Fig. 30.3b. The PCA9685 board is connected with four servo motors. Arduino nano board is applied to control the whole system. A DC-DC convert is used to measure the voltage during testing. A joystick is used as an input element that a user can handle to move the LARMbot torso V1.

30.4 Results

Experiments are planned and executed successfully to replicate basic human motions with a direct motion planning in the servo motors driving the cables. Through handling the joystick, the LARMbot torso V1 can move as expected in mode 1, Table 30.1. The snapshots of a test are shown in Fig. 30.4. At first, the humanoid torso bends left, and then it moves back the home position. After that, it bends right and moves back. Finally, the LARMbot torso V1 completes the planned motion in mode 1. Significant parameters are acquired to characterize the test with illustrative results in Figs. 30.5, 30.6, 30.7 and 30.8.

In Fig. 30.5, the motion at point H is reported with required acceleration along x, y, and z axis. The curves of a_{Hx} and a_{Hz} show a similar timeline behavior since they do not change a lot during the bending motion process. However, the value of a_{Hy} increases until it reaches a maximum, about 2.8 m/s^2, at the beginning. Then, when the time is about 40 s, it starts to decrease and reaches to a minimum, about -1.6 m/s^2. After that, it increases again when the time is about 100 s. Finally, it remains the original level. During the whole process, the a_{Hy} changes from about 2.8 m/s^2 to about -1.6 m/s^2 with a timeline history that smoothly represent the smooth bending motion of the test.

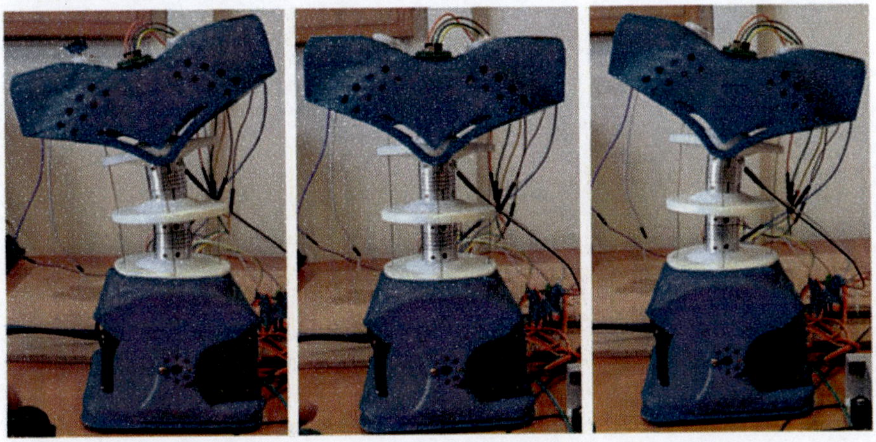

(a) (b) (c)

Fig. 30.4 Snapshots of the testing in mode 1: **a** bending left; **b** home; **c** bending right

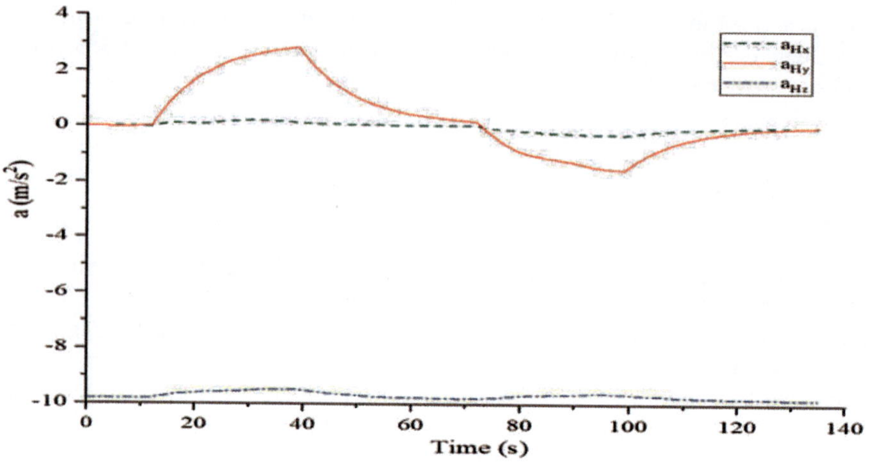

Fig. 30.5 Results of a lab test of testing mode 1, Table 30.1 and Fig. 30.4, in terms of linear acceleration

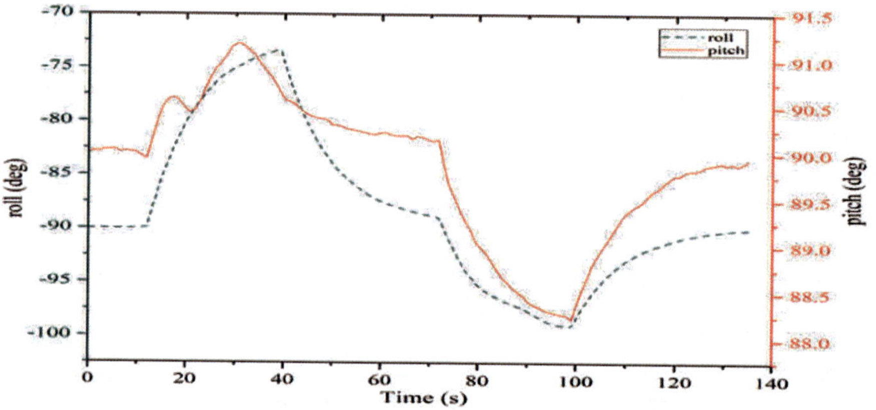

Fig. 30.6 Results of a lab test of testing mode 1, Table 30.1 and Fig. 30.4 in terms of roll and pitch angles

Figure 30.6 shows the angles changing of prototype at the point H when the experiment is conducted. The trends of curves are similar at roll and pitch direction. At the beginning, they increase to maximum, about 91.3° at pitch direction and about −73° at roll direction. When the time is about 40 s, they start to decrease. It is not until about 100 s that they decrease to minimums. After that, they increase again. Finally, the angle at pitch direction remains about 90°. The angle at roll direction remains about −90°. During the moving process in mode 1, the changing of angles in roll direction is about 26°. The changing of angles in pitch direction is about 3°.

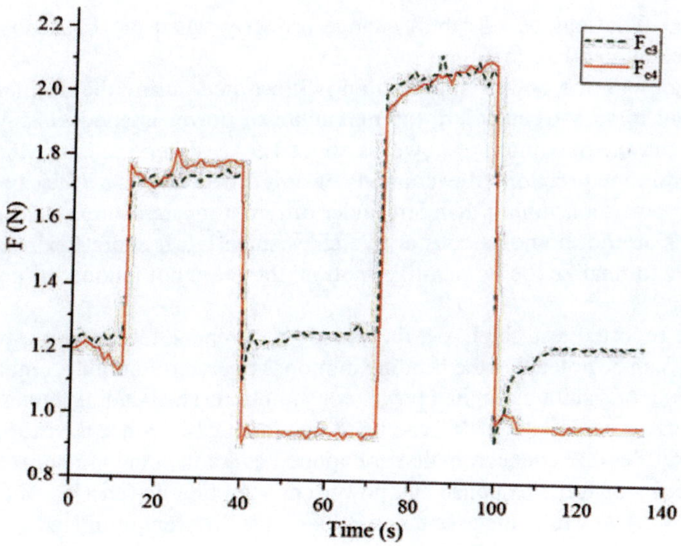

Fig. 30.7 Results of a lab test of testing mode 1, Table 30.1 and Fig. 30.4 in terms of contact force

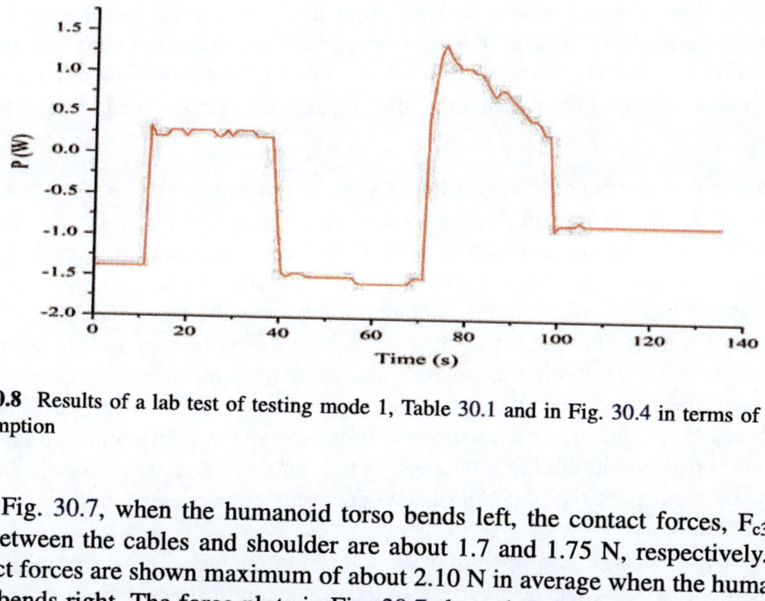

Fig. 30.8 Results of a lab test of testing mode 1, Table 30.1 and in Fig. 30.4 in terms of power consumption

In Fig. 30.7, when the humanoid torso bends left, the contact forces, F_{c3} and F_{c4}, between the cables and shoulder are about 1.7 and 1.75 N, respectively. The contact forces are shown maximum of about 2.10 N in average when the humanoid torso bends right. The force plots in Fig. 30.7 show that the tensions in the cables as measured at the connection on the shoulder body works properly with antagonist well-syncronized actions that are given by the syncronization of the corresponding actuators. The values of tensions are well below the stress limits of the used metal cables. Meanwhile, the force between the shoulder and cables can not damage the prototype. The lengths of cables change properly when the LARMbot torso V1 completes human-like motion.

In Fig. 30.8, the power consumption is illustrated during the testing. When the LARMbot torso V1 bends left, the maximum of power is about 0.25 W. When it bends right, the maximum of power is about 1.5 W. When the LARMbot torso V1 moves different directions, the required power is different. It is because bending right is more power-consuming than left under different tension forces. Meanwhile, the couplings are not homogeneous at two sides since left is more flexible than right. However, to finalize the human-like motion, there are not amount of power that is required.

In the reported test, the LARMbot torso V1 can bend left and right with similar angles. It can be noted that the bending motion is operated smoothly, requiring actuation torque for a suitable limited power consumption. The bending angles are convenient. It can bend left about 16° and bend right about 10°. Since the configuration of the market flexible connectors that are applied as joints, they are not fully symnet-icallin terms of force response and power consumption as function of the bending directions. Meanwhile, the assembled cables have different initial tensions so that it could not yet have the same bending performance as a human, 40° respectively, as reported in [5].

In summary, the LARMbot torso V1 shows a proper human-like motion that can contribute to mimic action of LARMbot humanoid in human-like operation. Sensors are applied to measure the performance of the LARMbot torso V1. Results are obtained to analyze the testing characterization of the prototype. However, the lab experiences highlighted problems that can suggest improvements in the future, such as.

A. Increase of the bending angles to get a larger workspace. It can make the torso work flexibly and benefit for the motions of humanoid arm, head, and limbs.
B. Revision of the torso structure both in the shoulder body and vertebra-like spine structure with more components of better stiff characteristics. A suitable structure can contribute to the cooperation between the torso and arms.
C. Improvements of the actuation both with better servomotors and a closed-loop control using the motion parameters and cable tension. Better servomotors can provide better torque and force and the cables can be driven with proper length. A closed-loop control system will make the prototype system stable and increase the robustness. In addition, decreasing the number of servo motors is also an important aspect for obtaining a low-cost and power-saving humanoid torso structure.
D. Revise the design and operation as to be in cooperation both with arms and locomotion body. The new version of the torso should support the other parts of the humanoid robot powerfully.

30.5 Conclusions

In this paper, experimental results are discussed to work out a characterization of LARMbot torso V1. The main contribution of the paper can be considered in the experimental characterization of the human-like capabilities of the humanoid torso for the LARMbot humanoid. In addition, sensors are used to measure the contact force between the shoulders and cables, power consumption, and angular motion. According to the results, the LARMbot torso V1 can operate with suitable performance in human-like motion. The range of bending angles is from about $-10°$ to about $16°$. The required power is less than 1.5 W. The contact forces are less than 2.1 N. A suitable control strategy is considered as future work for fully controlled motion so that efficient solutions can be obtained.

Acknowledgements The first author thankfully acknowledges to China Scholarship Council for Grant No. 202103250021 supporting his Ph.D. study at University of Rome Tor Vergata.

References

1. Huang, H., Ceccarelli, M., Zhang, W., Huang, Q., Yu, Z., Chen, X.: Biomimetic inspiration for PKM torso design in humanoid robots. In: 2015 IEEE International Conference on Mechatronics and Automation, ICMA 2015, pp. 1138–1143. Institute of Electrical and Electronics Engineers Inc. (2015). https://doi.org/10.1109/ICMA.2015.7237646
2. Reinecke, J., Deutschmann, B., Dietrich, A., Hutter, M.: An anthropomorphic robust robotic torso for ventral/dorsal and lateral motion with weight compensation. IEEE Robot. Autom. Lett. **5**(3), 3876–3883 (2022)
3. Penčić, M.M., Borovac, B.A., Kovačević, D.I., Čavić, M.V.: Development of the multi-segment lumbar spine for humanoid robots. Thermal Sci. **20**, S581–S590 (2016)
4. Li, T., Yao, P., Luo, M., Tan, Z., Wang, M., Guo, Z.: Design and kinematics analysis of a novel six-degree-of-freedom serial humanoid torso. Int. J. Adv. Rob. Syst. **15**, 1–10 (2018)
5. Ceccarelli, M.: LARM PKM solutions for torso design in humanoid robots. Front. Mech. Eng. **9**(4), 308–316 (2014)
6. Cafolla, D., Ceccarelli, M.: Characteristics and performance of CAUTO (CAssino humanoid torso) prototype. Inventions 2 (2017)
7. Cafolla, D., Ceccarelli, M.: An experimental validation of a novel humanoid torso. Robot. Auton. Syst. **91**, 299–313 (2017)

Chapter 31
Design and Performance of LARMbot Arm V3

R. Couprie, M. Ceccarelli, and M. A. Laribi

Abstract The LARMbot project intends to create a humanoid robot employing parallel mechanisms that are inspired by human anatomy. Solutions for humanoid arm are investigated combining the advantages of parallel and serial structures, as well as cable-driven architectures. A new solution has been designed for humanoid arm with a low-cost solution and user-oriented operation using commercial components and 3D printing manufacturing. Its feasibility is characterized by an analysis of operation performance via simulation whose reports are discussed also to suggest future improvements for construction and testing a prototype.

Keywords Robotics · Humanoids · Design · Performance analysis · Simulation

31.1 Introduction

Humanoid design has been developed at different levels mimicking the human anatomy [1], for several solutions around the world, with many prototypes available in laboratory frames for further research and design. Most of the current humanoid robots have been designed using serial kinematic structures, such as for example ASIMO robot [2], Wabot [3], HRP [4], HRB [5], and Nao [6]. Few other solutions have been designed as based on parallel kinematic architecture such as Waseda locomotor [7].

R. Couprie · M. A. Laribi
Department of GMSC, Pprime Institute CNRS, University of Poitiers, UPR 3346, Poitiers, France
e-mail: romain.couprie@etu.univ-poitiers.fr

M. A. Laribi
e-mail: med.amine.laribi@univ-poitiers.fr

M. Ceccarelli (✉)
Laboratory of Robot Mechatronics, Department of Industrial Engineering, University of Rome, Tor Vergata, 00133 Rome, Italy
e-mail: marco.ceccarelli@uniroma2.it

I. Doroftei et al. (eds.), *Proceedings of SYROM 2022 & ROBOTICS 2022*, Mechanisms and Machine Science 127, https://doi.org/10.1007/978-3-031-25655-4_31

LARMbot is a humanoid robot developed at the LARM2 laboratory of the University of Rome Tor Vergata based on parallel architectures [8]. Its legs have been built and validated using linear actuators to have great agility of movement while keeping a light and inexpensive structure. The trunk is made of a serial-parallel structure that is actuated by cables. The arms are designed with a first solution combining a tripod architecture with a serial arm connected to a revolute joint as elbow articulation [9, 10]. Payload and manipulation capability are constrained and indeed limited by the limb capabilities and particularly by the arm performance.

This paper presents a new solution for LARMbot arm V3 that has been designed to improve the upper arm using the already validated structure for the legs and adding a linear actuator for the elbow motion of the forearm. The designed solution is presented also with a performance analysis that was carried within an internship activity [11].

31.2 LARMbot Humanoid and Previous Arm

The LARMbot project intends to create a humanoid robot employing parallel mechanics that are inspired by human anatomy [1]. Using proper parallel mechanisms in torso, arms, and leg designs, a specific LARMbot design was designed since 2015 [8]. A prototype is built with low-cost user-oriented features utilizing commercial components and 3D-printed parts, Fig. 31.1a [8]. The legs design is inspired by human upper leg structure, which refers to a single parallel mechanism. Three actuators represent the main muscle group of the upper leg: hamstrings, quadriceps, and adductors. It is actuated by three linear actuators in the links, which converge to a single point of the ankle platform. The trunk design is inspired by the human torso bone-muscle complex. It is composed of parallel mechanisms with four cables and a central underactuated serial chain whose extremity joints are fixed in the center of the mobile shoulder plate of the parallel mechanism. The use of the linear concept is justified by high payload capacity and compact design for efficient operation and low-cost solution. Figure 31.1 shows an overview of the LARMbot robot and the last version of the arm with a tripod for the arm. In its first version, the LARMbot robot had anthropomorphic upper limbs as in Fig. 31.1a, but they could not perform complex movements with proper payload. In its last version, the LARMbot robot had upper limbs based on three linear actuators for the upper arm and a servomotor for the forearm [9, 10]. Figure 31.1b shows the last version of the arms with a tripod for the arm and fixing at the end of the platform a servomotor connected to a light bar acting as a forearm. This design performs human-like movements, but it does not allow a sufficient payload and the servo motor has a high-power consumption.

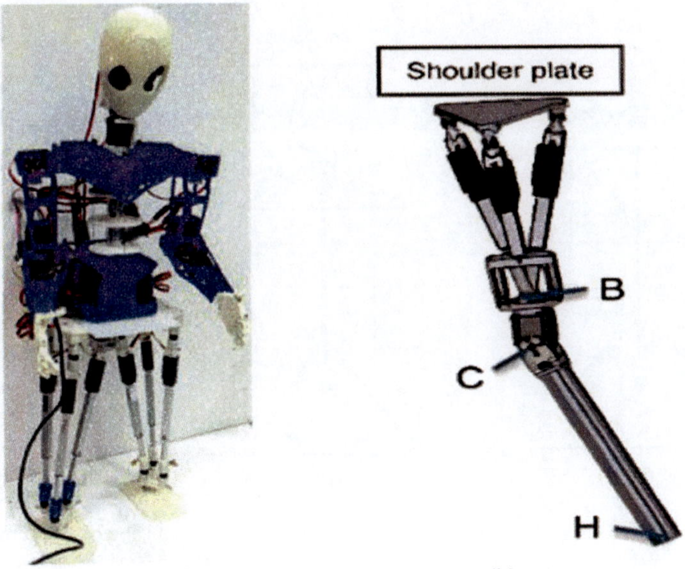

(a) (b)

Fig. 31.1 LARMbot humanoid [8–10]: **a** A prototype; **b** CAD design of arm V2

31.3 A New Design for LARMbot Arm

Improving the payload and reducing the power consumption of the LARMbot arms, are the objectives of the new solution. Different innovative solutions have been considered as summarized in Figs. 31.2 and 31.3. In Fig. 31.2a the LARMbot arm is designed with 3 cable-driven arm whose mobile plate carries the link body of forearm. In Fig. 31.2b the arm is designed with 3 cable-driven arm whose mobile plate is the base for 2-cable-driven forearm. In Fig. 31.2c the arm is designed with 3 linear actuators that join in one point and 2 other cables drive the forearm. A design with mechanical linear actuators is also considered as in Fig. 31.3, where each cable of the previous arm designs, is replaced by a single linear actuator.

In order to increase the payload capacity and to reduce the power consumption of the robot upper limbs, it was considered convenient a tripod system identical to the one used in LARMbot legs. The idea is to use the tripod for the arm, and to fix at the end of the platform a linear actuator connected to a light bar acting as a powered drive for the forearm. A kinematic diagram and a CAD design of the new LARMbot arm V3 are shown in Fig. 31.4a and b. Table 31.1 lists chosen dimensions to fit the upper limb with the built LARMbot humanoid, as characterized by the stroke of the linear actuators in the links (i = 1, 2, 3, 4).

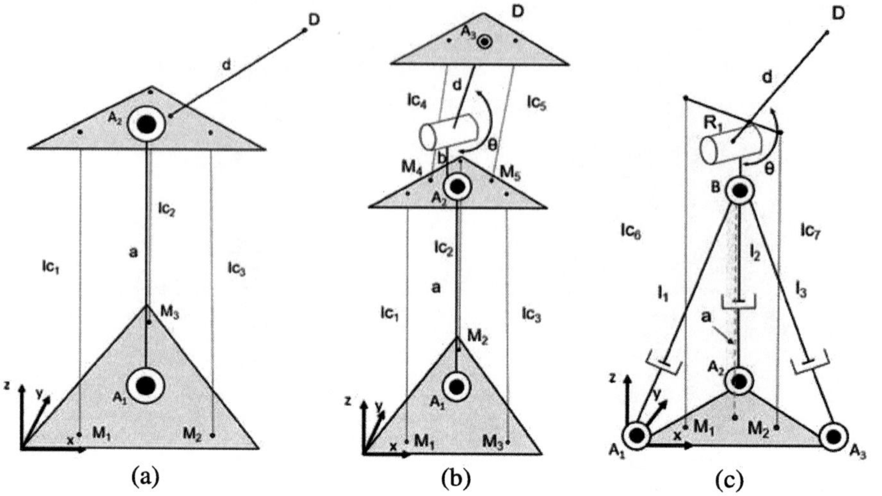

Fig. 31.2 Considered alternative kinematic cable-based solutions: **a** 3-cable-driven arm design; **b** Fully cabled-driven arm design; **c** Linear actuators and cables design

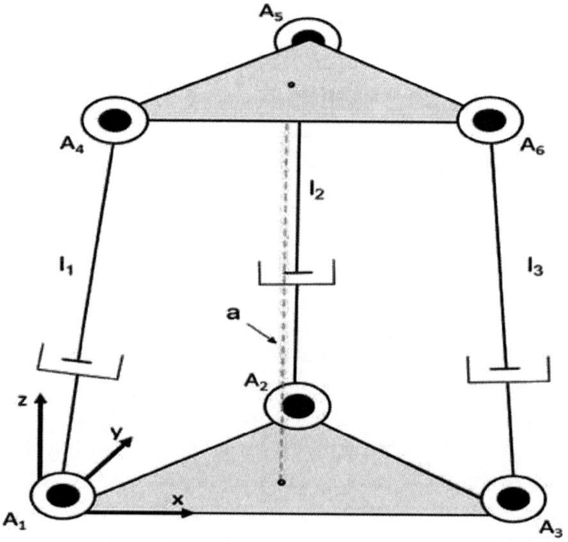

Fig. 31.3 Considered alternative solution with linear actuators

A prototype of LARMbot arm V3 is designed with the sizes in Table 31.1 using with proper low-cost, payload and mobility features four linear actuators from market after a performance analysis that has validate the design. This structure allows a light structure and a limited number of motors. On the platform corresponding to the elbow, one of the actuators is used to actuate the forearm in flexion/extension movements

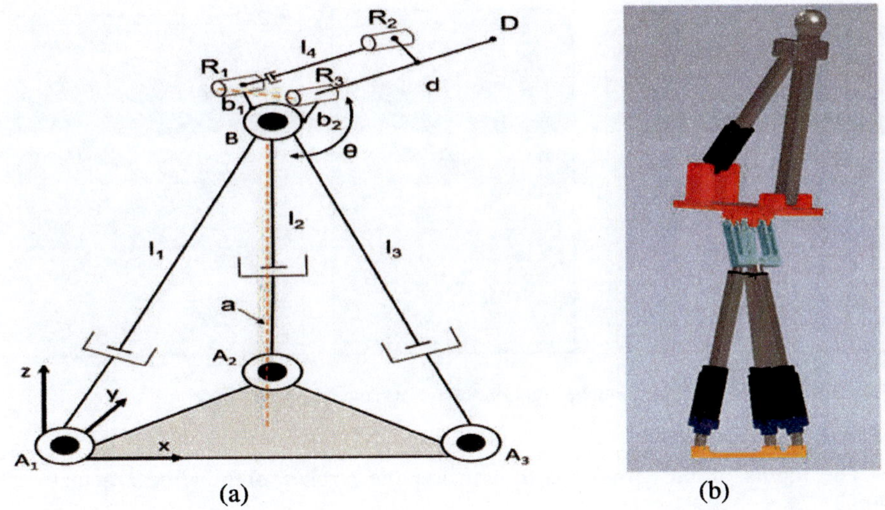

Fig. 31.4 The proposed solution for V3 design: **a** Kinematic scheme; **b** CAD design

Table 31.1 Dimension of parameters for the solution V3

I_i (i = 1,…4) (mm)	a (mm)	b_1 (mm)	b_2 (mm)	d (mm)	θ (deg)	ΔS (mm)	Δh (mm)	Δθ (mm)
168 + ΔS	250 + Δh	42	7.5	200	85 + Δθ	100	20	50

of the forearm. The tripod platform, which is an equilateral triangle, is designed to be fixed on the trunk of the LARMbot humanoid, and the three linear actuators meet on the elbow pad thanks to a special joint that ensure the point convergence of the 3 actuators.

The chosen linear actuators consist of a DC motor with a potentiometer sensor that gives feedback on the stroke, so that a control over the stroke can be planned for a smooth precise motion.

31.4 Performance Analysis

A dynamic simulation is carried on Autodesk inventor to verify the design feasibility and to characterize the operation performance. A movement is simulated on XY plane, when the tripod arm moves on the X axis meanwhile the forearm rotates around Z axis, as shown in Fig. 31.6.

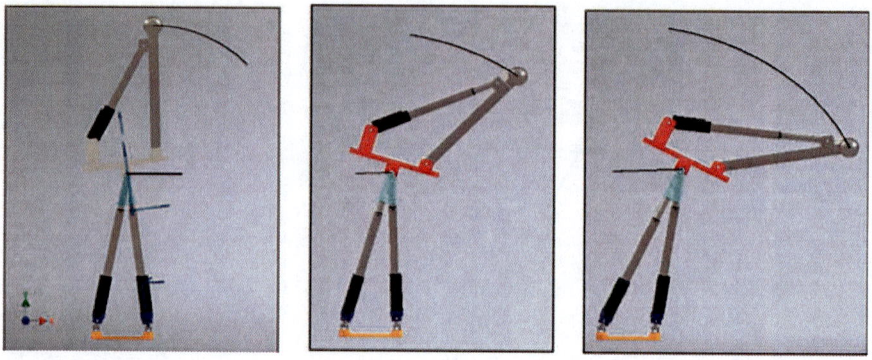

Fig. 31.6 Snapshots of the simulated motion for performance analysis

The inputs of the simulated motion are the strokes of the linear actuators, Fig. 31.7a.

The plot of computed results in terms of velocity Fig. 31.7b shows that the actuators 1 and 2 run with the same velocity. The actuator 4 is the fastest one while the actuator 3 is the slowest.

The trajectory in Fig. 31.8a shows that the forearm extremity point D traces a circular movement as expected. The plot of results in terms of acceleration of forearm point D in Fig. 31.9b shows that when the acceleration along Z is positive, the acceleration along X is negative and vice versa, as expected also with limited values.

The plot of results in terms of driving forces of the linear actuators in Fig. 31.9 reveals that the actuator force 2 and 3 have almost the same value during the simulation, reaching the maximum of about 12 N. The weakest force is in the actuator

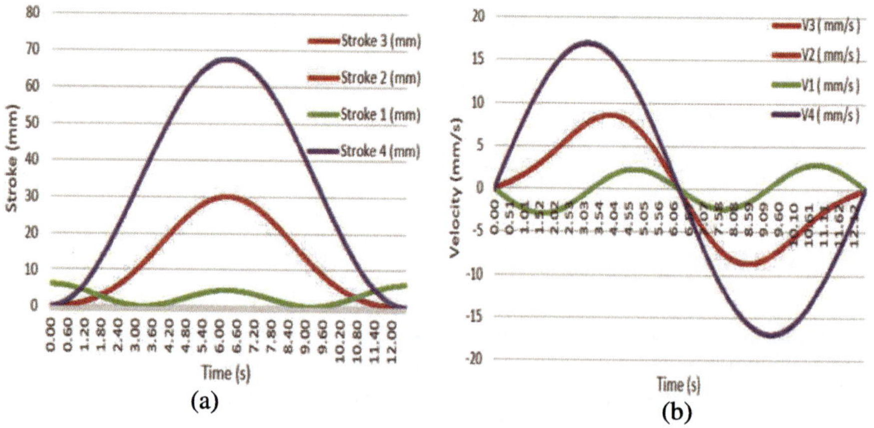

Fig. 31.7 Data for the actuators: **a** Input stroke; **b** Computed results of velocity from the simulation

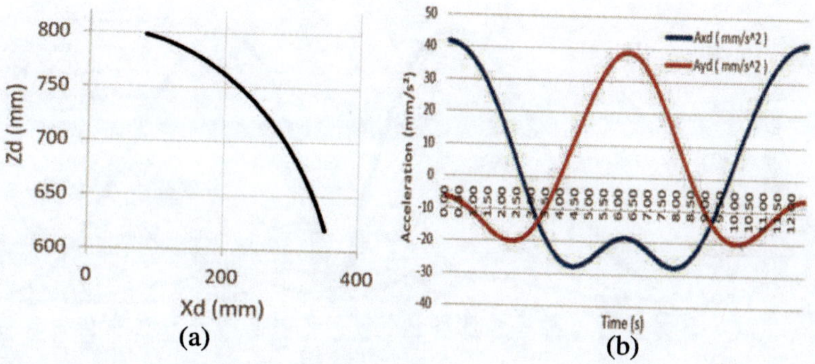

Fig. 31.8 Computed results of the simulation in Fig. 31.6 in terms of: **a** trajectory of D; **b** acceleration of D

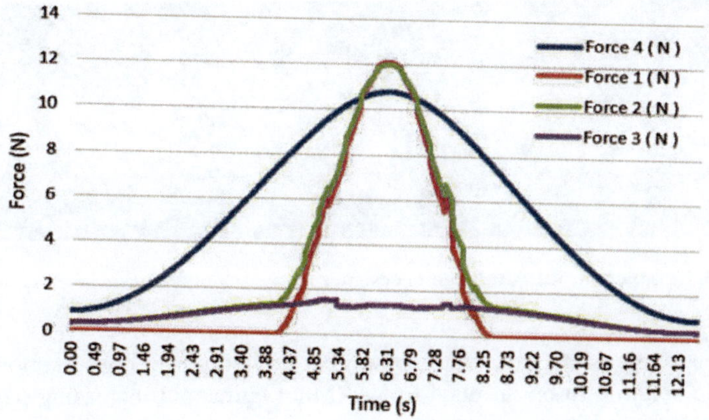

Fig. 31.9 Computed results of a simulation in Fig. 31.6 in terms forces of the actuators

3. Those values are much lower than the maximum payload capacity of actuators, confirming the feasibility of the actuator choice for the prototype construction.

The plot of computed power consumption of the four actuators is reported in Fig. 31.10 with maximum value of about 0.12 W. The total average power consumed by all the motors during the simulated motion 1.06 W that can be considered matching the LARMbot humanoid energy capability.

The prototype of the solution V2 was fully working and some of the components can be reused to build the prototype V3. The components than can be kept are the three linear actuators L16, the Arduino board mega, the motor shield L293D, the four current sensors and the two IMU sensors BMI160. The other parts are 3D printed or machined for the new solution V3 with a prototype assembly shown in Fig. 31.11 for next future testing. The overall cost of the prototype V3 was of 300 Euro.

The simulation results in Figs. 31.6, 31.7, 31.8, 31.9 and 31.10, as well as from other simulated motions, show that the LARMbot arm design V3 in Fig. 31.4 run

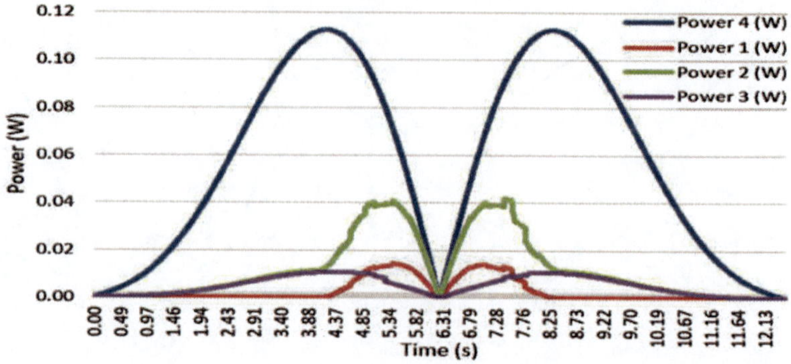

Fig. 31.10 Computed results in Fig. 31.6 in terms of power consumption of the actuators

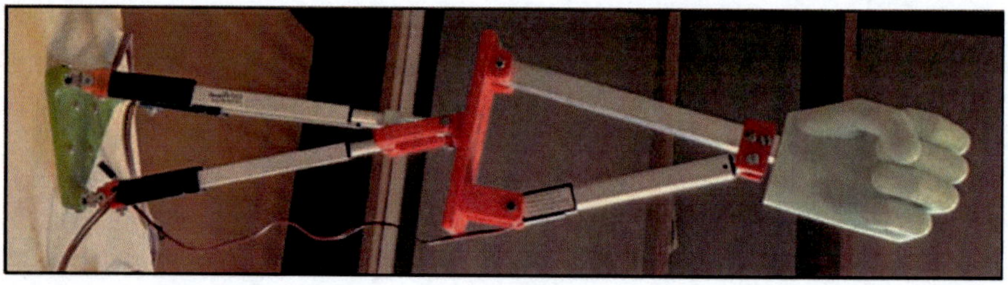

Fig. 31.11 A prototype of LARMbot arm design V3

with properly characteristics with convenient limited power consumption as being a feasible solution to be installed to LARMbot humanoid increasing payload and manipulation capability.

31.5 Conclusion

A new design for LARMbot arm is presented as version V3 with a tripod structure for the arm and a planar parallel mechanism for the forearm by using four linear commercial actuators. The design feasibility is evaluated by dynamic simulation whose results are discussed for a characterization of a proper low-power consumption operation improving payload and manipulation capabilities of LARMbot humanoid. Future work is planned for implementing and testing the proposed V3 arm design in a new version of LARMbot humanoid.

References

1. Ceccarelli, M., Matteo Russo, M., Morales-Cruz, C.: Parallel architectures for humanoid robots. Robotics **9**(75). https://doi.org/10.3390/robotics9040075
2. Sakagami, Y., Watanabe, R., Aoyama, C., Matsunaga, S., Higaki, N., Fujimura, K.: The intelligent ASIMO: system overview and integration. In: IEEE/RSJ International Conference on Intelligent Robots and Systems, vol. 3, pp. 2478–2483. https://doi.org/10.1109/IRDS.2002.1041641
3. Sugano, S., Kato, I.: WABOT-2: autonomous robot with dexterous finger-arm–finger-arm coordination control in keyboard performance. Proc. IEEE Int. Conf. Robot. Autom. **1987**, 90–97 (1987). https://doi.org/10.1109/ROBOT.1987.1088025
4. Kaneko, K., Kaminaga, H., Sakaguchi, T., Kajita, S., Morisawa, M., Kumagai, I., Kanehiro, F.: Humanoid robot HRP-5P: an electrically actuated humanoid robot with high power and wide range joints. IEEE Robot. Autom. Lett. 1. https://doi.org/10.1109/LRA.2019.2896465
5. Huang, Q., Yu, Z., Chen, X., Zhang, W., Yang, T., Liao, W., Ceccarelli, M.: Historical developments of BHR humanoid robots. Adv. Hist. Stud., Spec. Iss. Hist. Mach. Mech. **8**, 79–90 (2019). https://doi.org/10.4236/ahs.2019.81005
6. Gouaillier, D., Hugel, V., Blazevic, P., Kilner, C., Monceaux, J., Lafourcade, P., Marnier, B., Serre, J., Maisonnier, B.: The NAO humanoid: a combination of performance and affordability. In: Proceedings—IEEE International Conference on Robotics and Automation, Kobe, 2009, pp. 769–774 (2008). arXiv:0807.3223
7. Sugahara, Y., Hosobata, T., Mikuriya, Y., Sunazuka, H., Lim, H., Takanishi, A.: Realization of dynamic human-carrying walking by a biped locomotor. In: IEEE International Conference on Robotics and Automation, Proceedings. ICRA '04, vol. 3, pp. 3055–3060. https://doi.org/10.1109/ROBOT.2004.1307526
8. Ceccarelli, M., Cafolla, D., Russo, M., Carbone, G.: LARMbot humanoid design towards a prototype. MOJ Int. J. Appl. Bionics Biomech. **1**(2), 48–49 (2017). https://doi.org/10.15406/mojabb.2017.01.00008
9. Ceccarelli, M., Russo, M.: Arm mechanism for humanoid robot, patent, IT 102017000021920, Italy (2017)
10. Fort, A., Ceccarelli, M., Laribi M-A.: Prototype and testing of LARMbot PK arm. In: Advances in Mechanism Design III—Proceedings of TMM 2020, pp. 210–219. Springer, Cham, AG (2020). https://doi.org/10.1007/978-3-030-83594-1_22
11. Couprie, R., Ceccarelli, M., Laribi, A.: Design of new LARMbot arm based on a cable driven parallel robot. Internship report, University of Poitiers, Poitiers (2022)

Chapter 32
Mechanical Machines from Antiquity, the Middle Ages and the Renaissance at the Origin of Humanoid Robots

I. Staretu and S. Staretu

Abstract From ancient times, man has been concerned with creating an artificial structure similar to himself. This concern remained in the constant attention of the deepest and most creative thinkers from antiquity to the present day, when it is of the greatest interest. This paper aims to briefly present the most significant moments of evolution from the first legendary or real humanoid mechanical automata, from antiquity to today's humanoid robots. The paper also opens the perspective of defining the concept of an artificial humanoid useful both for human prosthesis, in case of certain disabilities, and for the development of a high-performance humanoid robot.

Keywords Mechanical automata · Antiquity · Renaissance · Humanoid robot

32.1 Introduction

Mechanical automata have preoccupied people since early antiquity if we consider, even if only legendary, *Talos*, built by *Hephaestus* [1–3]. Other mechanical automata of that period or late antiquity are generically attributed to educated people, who had knowledge or even then defined them, in several fields, which covered both theory: philosophy, oratory, physics, geometry, arithmetic, etc., and practice, which we can assimilate very well with engineering indeed with today's understanding. The ancient engineer was in fact a scientist who had various knowledge, or was concerned with finding new explanations for the phenomena noted or problems encountered to obtain the most appropriate solutions. Thus, it can be noted that in the year 350 BC the mathematician Arhitas built a mechanical bird (so mathematician-technician), called the *Pigeon*, which ran on steam. Then it is also recorded that in 322 BC,

I. Staretu (✉)
Transilvania University of Braşov, Brasov, Romania
e-mail: staretu@unitbv.ro

Romanian Academy of Technical Sciences, Bucharest, Romania

S. Staretu
Center for Studies, Theology and History of the Metropolitan of Transylvania, Cristian, Brasov, Romania

I. Doroftei et al. (eds.), *Proceedings of SYROM 2022 & ROBOTICS 2022*, Mechanisms and Machine Science 127, https://doi.org/10.1007/978-3-031-25655-4_32

Aristotle wrote, foreshadowing the development of mechanical automation, in the current sense of robotics: "*If each tool could do, on order or on its own initiative, the appropriate work ... then there would be no need for disciples for the workers of the master, nor for slaves for the masters*". At the same time, Ctesibios of Alexandria and Philo of Byzantium described several human-like and animal-like automatons. Heron of Alexandria also described a group of puppets that were used in miniature theatrical performances operated with a complex system of counterweights, and mechanical parts such as levers and even gears. Thus Archimedes, Thales of Miletus, even Pythagoras and many others were both theorists and practitioners, in the modern sense engineers [4].

32.2 University of Constantinople

Obviously, the foundation of the first schools-centers for the transmission and deepening of knowledge about nature and society was essential. Among these we can enumerate *the School of Pythagoras, the School of Archimedes, the School of Aristotle* and then of *Plato*, from ancient Greece [4], then from imperial Rome, as well as others about which we have information left (which we do not have for example from ancient Egypt, Crete or other areas close to Europe), which were the centers where knowledge of geometry, arithmetic, physics, geography, astronomy, medicine, architecture, oratory, theology, painting and engineering were synthesized, diversified and developed.

Although there are no significant data on ancient engineering, defined in the modern sense, it is enough to say that the achievements of the great temples of Egypt, the pyramids, simple machines (inclined plane, lever, pulley, screw, etc.), the first war machines used by the Romans, the building of complex constructions such as aqueducts, viaducts, bridges (for example bridges over the Danube, the first being that of Turnu Severin of Apollodorus of Damascus), water storage tanks, etc. they clearly needed complex engineering knowledge in architecture, construction, metal processing, design of devices and equipment so mechanical engineering, materialized especially by various mechanisms and more complex structures such as war machines, between these fields being an almost total symbiosis.

Suffice it to mention the Antikythera mechanism [5], dated around 150 BC, which proved to be a complex astronomical instrument [6] and the complex mechanisms of Nero's Palace (around 60 AD), where there was a rotating gazebo.

One can also refer to other devices or even equipment, although some seem legendary, such as *the Ark of the Law*, which according to tradition was a generator that produced electric sparks, an aspect still to be elucidated.

At the same time, this knowledge allowed the setting up of the great cities of late antiquity best exemplified by Alexandria, including the famous Lighthouse, considered a wonder of the world, but also the outstanding library, which was practically a space for acquiring and developing knowledge, which worked like a real *Scientific forum,* even in the current sense of the term, several hundred years.

Fig. 32.1 Image associated with the University of Constantinople [7]

The knowledge largely existing in ancient Egypt, in ancient Gecia, widely used by the Roman Empire (later called the Byzantine), led to the emergence of the first scientific centers, which were the germs of the first universities, attested in Athens, Alexandria, Antioch and then, in Constantinople.

The University of Constantinople founded in 425 AD, by Emperor Theodosius II, can be considered the first university-type institution in the true sense of the word, Fig. 32.1.

This, according to the founding act, had the role of transmitting the knowledge already accumulated until then but also to develop it, so enrich it, to use it for the general welfare of society, for the good of the people, those, but not only, of the Roman Empire. The university got at one point to have 31 departments, in fact faculties, which transmitted and developed various knowledge from the humanities to the practical-engineering ones [8]. In the run-up to the conquest of Constantinople (1453), much of the scientific treasure accumulated over a thousand years was transferred with many scholars to central and western Europe (Italy, Germany-Bohemia, even France), where they generated the phenomenon of the Renaissance, then continued by the rapid development of society, through the so-called *Industrial Revolution*, until today.

32.3 Mechanical Automata and Other Engineering Achievements in Constantinople Before 1000 AD

There are several accounts of mechanical automata or other technical achievements in Constantinople before 1000 AD, some of which have recently been noted.

We must first mention the automata, functional around the year 900, described by *Liutprando da Cremona*, a diplomat at the court of Emperor Constantine VII, which he reached in 949. He described a room called *Magnaura* (translated *the Great Zephyr*) in which he saw two cinematically active mechanical lions, made on a 1: 1 scale, which flanked the throne of the emperor of Constantinople, a throne that could be raised and lowered, and mechanical birds chirping. Obviously, these achievements must be linked to the *University of Constantinople*, which had become a large-scale scientific institution at the time. It is also worth noting that part of this description was taken and presented under the title *The Byzantine Robot*, by the well-known writer Umberto Eco, in the volume entitled The Book of Legendary Lands, published in 2013 and translated into Romanian [9].

This year analyzing another medieval text, namely *The Song of Roland and Other Poems of Charlemange* [10], which describes situations placed around the year 800 (when Charlemagne lived), or even a little later, taking into account the evocation of the king Hugo (attested in Constantinople during the Latin conquest), we personally identified in the poem *Charlmange's journey to Jerusalem and Constantinople*, the following text:

> *There was a statue of two children in copper and metal,*
> *Each of them holding in their mouth a white ivory horn.*
> *If a gust blows from the sea, or breeze, or other wind*
> *Buffeting the palace from a westerly direction,*
> *It makes it revolve rapidly and often*
> *Like a chariot's wheel rolling along the ground.*
> *Thus these horns sound and bugle and play*
> *Like drums or thunder or a huge hanging bell.*
> *The figures look at each other as if they are smiling,*
> *Which made you think that they were actually alive,*

This text, therefore, describes two mechanical automatons in humanoid form, of two young people, cinematically active, driven by air currents, which sing and move, getting even to seem alive. The achievement is amazing and is the first time it is presented from a technical-engineering point of view. Also in this poem reference is made to *axles* and *wheels* and to an anti-earthquake solution used in the construction of the imperial palace. It is obvious that the level of engineering shortly before the year 1000 AD at least in Constantinople was far above what we usually estimate.

32.4 Renaissance Automata

As already mentioned, much of the knowledge gained at the University of Constantinople until 1453 (when the university was not abolished but became the first Ottoman university, today the University of Istanbul) was transferred to Western Europe as the basis of the Renaissance.

In this context, it is easier to explain *Leonardo da Vinci's* rather successful attempt to create a mechanical lion but also a medieval knight in mechanical armor. This latter embodiment can be considered somewhat the forerunner of the current humanoid robot. In the following period, after 1500, there were multiple attempts to make various automata that mimic the activities, especially artistic, of man, of remarkable complexity and functionality. We highlight Vaucanson's flute singer (made and exhibited at the Louvre in 1738), the humanoid automaton capable of writing a certain text—The Writer of Pierre Jacquet Droz (from 1768); The harpsichord singer of the same master [11], Kintzing's timpani singer (1784), etc.

From the next period we can mention the set up of mechanical orchestras, the first is attributed to Maelzel, for whom Beethoven himself composed the first part of the composition *Opus 91, Victoria of Wellington.*

These orchestras included flutes, clarinets, trumpets, triangles, and strings, which were operated by air ducts and hammers. A mechanical violinist, probably inspired by Maelzel's automatons, was presented in 1838 at the Paris Conservatory, and was considered a real success.

Another desire expressed during the medieval automata was the creation of an automaton capable of playing chess, which materialized in a sufficiently efficient form, in the early twentieth century, by *Leonardo Torres y Quevedo*, a Spanish engineer and mathematician.

32.5 Humanoid Robots

After some calm in this direction, the concerns will be resumed relatively intensely in the middle of the twentieth century, after the launch of the robot concept, by Karel Capek in the play R.U.R., in 1920, by making several humanoid mechanical automatons, but quite rudimentary. The real start of producing, this time, humanoid robots, activities that continue today, with great perspective in the very near future, took place in the 70s in Japan, which culminated with the manufacturing of the well-known robot ASIMO, in continuous improvement at the moment too. In this context it can be mentioned that the humanoid robot can be approached from a modular perspective consisting of: head module, body module, arm module and movement module (lower limbs, in the case of bipedal robots). Each module can be developed relatively independently provided that in the end a unitary and at the same time very functional assembly is obtained. Thus ASIMO, the version from 2018, improved in terms of details to date, is a humanoid robot similar to a cosmonaut, with 57° of mobility, 1.30 m high, weighing about 48 kg, with a speed of 9 km/h (can walk on stairs and slopes and even dance), who speaks and understands Japanese and English, can make a presentation, can react to certain human gestures, can manipulate various objects and is able to shake hands [12].

Below are some significant aspects to create a suggestive image of today's *humanoid robotics*. First of all, we point out the intention, already partially materialized, of a Japanese company to hire 1000 robots as sellers in its stores, in order to sell coffee machines. This robot called Pepper [13], is 120 cm tall, moves on rollers, has a tablet-console interaction on the chest, understands 70–80% of spontaneous conversations, and was sold at a price of about $ 2000, one piece, in the first part of 2015. Robots of this type already work in more than 70 mobile phone stores owned by SoftBank in Japan, currently their job is to meet customers and from 2020 in Romania too.

Another robot is Robroy, considered one of the most advanced humanoid robots, made at the University of Zurich. This is a robot shaped by human structure, with tendons similar to human ones. It is a robot designed to perform services independently for humans, sharing the work space with them. It is equipped with artificial leather, which ensures a better integration in the environment in which it will work with human operators. Generally bipedal humanoid robots, or partially humanoid, on rollers, are expected to be used as nannies, nurses, emergency medical staff, waiters, drivers, nurses, accountants, shopkeepers, hotel staff, marketing staff, etc.

In this context, the statement of the well-known Bill Gates is suggestive, who claimed that: *Replacement by software, whether it is drivers, waiters or nurses... is happening at this time. Technology, over time, will reduce the demand for jobs, especially for people who will not be highly trained. In 20 years from now, the demand for work for a lot of jobs will be substantially reduced. I don't think people understand that.* Also, futurist Ray Kurzweil, who has written several books on technological singularity (when artificial intelligence will dethrone the human mind), estimates that machines will be smarter than humans by 2029 and by 2045 will be billions of times more powerful compared to human intelligence.

These aspects lead us to scenarios in which the labor market will be completely transformed by computer programs and robots, even if people overcome their physical limits with the help of highly evaluated mechatronic devices such as exoskeletons, but also sensory and intellectual to increase visual and auditory abilities, and even to increase cognitive abilities. In these conditions it can be mentioned that more and more people take advantage of the latest technological discoveries to improve their body and senses with various gadgets-devices, parts and mechatronic mechanisms, and it can be said they lay the foundations of a real new species: *Homo Roboticus*.

The humanoid robot will become an ordinary companion of man, becoming indispensable in its immediate vicinity both in the home environment [14], as a pet and support robot, and in the service-job, as a partner in solving tasks specific to the activities of most companies which we know today.

It is not without interest to mention here the intention of Tesla to launch a humanoid robot on the market in 2022 for boring, repetitive tasks and hard work.

Already today, utility robots are used for a wide range of services including surveillance and cleaning both in households and in hospitals and asylums. At the same time, given that the world's population is aging, this type of robot will also be able to help ensure the autonomy of the elderly. It is very important to use them in difficult times such as epidemics. Other areas of real interest for humanoid robots

Fig. 32.2 The humanoid robot as a human partner [15]

are entertainment and sports. There are in progress, or already made, several types of robots that can practice various artistic activities, such as robots that can play various instruments: flute, violin, etc.—modern variants of medieval automatons. Also, the robots destined to practice various sports, especially football, are of some relevance, being already organized football championships for robot teams, the robots are even expected to have their own Olympic Games, China being wiling to organize the first edition in Harbin.

It is also expected that pet human robots will be the next major step in consumer appliances. The category of humanoid robots also includes robots: Poppy, BioRob 2008, NAO, Icub, PETMAN, Aiko, Kobian, Albert Hubo, Roboray, Romeo, etc., each with various constructive and functional features that tend to bring them as close as possible to the human model (Fig. 32.2).

A particularly challenging field is the creation of humanoid robots that are very difficult to distinguish from humans. It is an area as challenging as it is real and with great perspective, which will be able to fundamentally change human society as we know it today. As an example, the latest achievement of the Japanese professor Hiroshi Ishiguro from Osaka University is presented, namely the robot called Geminoid F [16]. He can smile, move his mouth as he speaks and sings, and simulate 65 facial expressions, such as those that express amusement as well as enigmatic states, he can raise his eyebrows with a number of 12 motor structures under artificial plastic "skin". The price of this robot, exhibited in a shopping center in Hong Kong, is $ 110,000, it is still very high, but it will certainly drop sharply in the coming years. Obviously, such a robot can also be an actor, a profession that adds to those already specified that could be fulfilled by robots.

32.6 Conclusions

The following conclusions can be drawn based on those presented in this paper:

- service robotics, which includes humanoid robots, has an upward dynamic that is evident now and will be maintained in the future;
- humanoid robots derive directly from humanoid mechanical automata, some made practically others legendary, known since antiquity but also later in the Byzantine Empire and especially during the Renaissance;
- *robots sapiens*—humanoid robots tend to be very similar to humans, with advanced functionality and aesthetics;
- several models of humanoid robots are already commercially available and will soon become a common human companion, especially in critical situations;
- the humanoid robot is a complex mechatronic product that is a concrete application of transdisciplinarity and interdisciplinarity in engineering;
- concerns for the development of humanoid robot models are also of interest in Romania, which may have important results in this direction.

Acknowledgements Our gratitude to the Company Cloos Germany and its Romanian representative Robcon which partly sustained these works.

References

1. Whelen, E.: Automaton in the ancient world: the robots of Greece and Rome by Whelan on January 31, 2020. https://classicalwisdom.com/culture/history/automation-in-the-ancient-world/
2. Mayor, A.: Gods and robots: myths, machines, and ancient dreams of technology. Princeton University Press, Amazons (2018)
3. Shashkevich, A.: Stanford researcher examines earliest concepts of artificial intelligence, robots in ancient myths (2019). https://news.stanford.edu/2019/02/28/ancient-myths-reveal-early-fantasies-artificial-life/
4. Betts, G., Henry, A.: Complete Ancient Greek: A Comprehensive Guide to Reading and Understanding Ancient Greek, with Original Texts. John Murray Press (2017)
5. Information on: https://www.wikiwand.com/ro/Mecanismul_de_la_Antikythera. Accessed May 2022
6. Information on: https://www.descopera.org/ultimele-descoperiri-despre-masinaria-zeilor/. Accessed May 2022
7. Information on: https://www.foliamagazine.it/en/university-of-constantinople/. Accessed May 2022
8. Information on: https://en.wikipedia.org/wiki/University_of_Constantinople. Accessed May 2022
9. Eco, U. (ed.): History of realms and legendary places. RAO, Bucharest (2014)
10. Gaunt, S., Pratt, K.: The song of Roland and Other Poems of Charlemange. Oxford University Press (2016)
11. Information on: http://www.snipview.com/q/Jaquet-Droz...automata. Accessed April 2022
12. Information on: https://global.honda/innovation/robotics/ASIMO.html. Accessed April 2022

13. Information on: http://www.hngn.com/....-2015.htm. Accessed April 2022
14. Information on: http://www.nspyre.ro/news/3649/.... Accessed April 2022
15. Information on: http://www.activistpost.com/2013/07/...robot.html. Accessed April 2022
16. Information on: http://www.geminoid.jp/en/robots.html and https://education.nationalgeographic.org/resource/robot-geminoid-f. Accessed April 2022

Part VII
Mobile Robots

Chapter 33
Obstacle Crossing Strategies for High-Speed 4WD Small-Scale Vehicles

D. N'chot, R. Lenain, J.-C. Fauroux, L. Al Bassit, and P. Vaslin

Abstract Unmanned ground vehicle obstacle crossing generally relies on two strategies: (i) applying a wheel torque for climbing and (ii) modifying the vehicle shape by using a wheel-leg or wheel-paddle to lift the wheel on top of the obstacle. However, most of those strategies sacrifice speed in order to have a longer contact duration between the wheels and the obstacle. This paper investigates the behaviour of a 4WD high-speed vehicle while crossing a step obstacle using a design of experiment (DoE). A 3D multibody vehicle model is equipped with a novel 2-DoF suspension system, which horizontal damping coefficient is modified to dampen wheel motion in longitudinal and vertical directions with respect to the chassis, for a given speed and obstacle height. The DoE results allow proposing a novel high-speed obstacle crossing strategy based on three metrics: (i) the kinetic energy variation of the vehicle, (ii) the contact duration between the wheel and the obstacle, and (iii) the pitch rate at the beginning of the ballistic phase. Experimental functions are proposed to modify these metrics in real time.

D. N'chot (✉) · J.-C. Fauroux
Université Clermont Auvergne, Clermont Auvergne INP, CNRS, Institut Pascal, 63000 Clermont-Ferrand, France
e-mail: denis.n-chot@sigma-clermont.fr

J.-C. Fauroux
e-mail: jc.fauroux@mecabotix.com

R. Lenain · L. Al Bassit
Université Clermont Auvergne, INRAE, UR TSCF, 63000 Clermont–Ferrand, France
e-mail: roland.lenain@inrae.fr

L. Al Bassit
e-mail: lama.al-bassit@inrae.fr

J.-C. Fauroux
MecaBotiX, At Home/La Cité, 55 Avenue Louis Bréguet, 31400 Toulouse, France

P. Vaslin
Université Clermont Auvergne, Clermont Auvergne INP, EMSE, CNRS, LIMOS, 63000 Clermont-Ferrand, France
e-mail: philippe.vaslin@limos.fr

I. Doroftei et al. (eds.), *Proceedings of SYROM 2022 & ROBOTICS 2022*, Mechanisms and Machine Science 127, https://doi.org/10.1007/978-3-031-25655-4_33

Keywords Obstacle crossing · Design of experiment · Fitting function · Optimization problem

33.1 Introduction

Obstacle crossing of wheeled off-road vehicles starts from the contact between the wheel and the obstacle. When obstacle slope is high, the horizontal component of the ground reaction force reaches the same order of magnitude as the vertical component [1], which creates dynamic pitch, roll and yaw momentum, resulting in a high reduction of the vehicle speed and an increasing risk of vehicle tipping over. Obstacle crossing generally relies on two strategies: (i) applying a sufficient wheel torque to climb the obstacle and (ii) modifying the shape of the vehicle by using wheel + leg or wheel + blade system, for instance [2–4], with or without a passive or active suspension mechanism, in order to lift the wheels on top of the obstacle. As both these strategies require maintaining contact between the wheels and the obstacle, they sacrifice the vehicle speed [2]. Indeed, the wheel continues to climb over the obstacle when torque and friction conditions are met [3]. In the literature, the maximum height of an obstacle that can be crossed by a vehicle is equal to the wheel radius [4, 5]. Previous works [1, 6] have shown that the Susp-4D suspension system with 2-DoF: (i) reduces the deceleration of a vehicle when crossing an obstacle, (ii) allows crossing obstacle of greater height than the wheel radius and (iii) reduces the risk of tipping over. This paper presents a study on a step obstacle crossing at high speed by a 4WD vehicle equipped with Susp-4D and proposes novel strategies of obstacle crossing based on the control of the horizontal suspensions in order to optimize three criteria:

- Saving the vehicle kinetic energy;
- Lengthening the contact duration enough for allowing the vehicle to climb;
- Reducing the vehicle pitch rate at the beginning of the ballistic phase.

Section 33.2 presents the multi-body model of the vehicle and the design of experiment. Sections 33.3 and 33.4 present the results of the experimental design, discuss the control strategy and Sect. 33.5 concludes and proposes some future work.

33.2 Material and Methods

33.2.1 Multi-body Model Description of the Full Vehicle

The Adams multi-body modelling software is used to build a 3D model of a 4WD vehicle (Fig. 33.1). The vehicle consists of a rectangular chassis and four driven wheels. The chassis is constrained by a plane linkage (Fig. 33.2b: P2) in order the vehicle performs a 2D crossing. Both front wheels are connected to the chassis with Susp-4D, whereas both rear wheels are connected to the chassis with Susp-3D suspensions, which are Susp-4D with locked horizontal suspensions. The two first possible motions of front wheels are longitudinal and vertical with respect to the chassis, while rear wheels can only move vertically, according to the kinematics (Fig. 33.2a) of Susp-4D and Susp-3D [6], respectively. The other two degrees of freedom of the wheel are rotations around Yo and around Zo (Fig. 33.1a). The obstacle is a rectangular bar of height h_O fixed to the ground. On horizontal ground, the altitude Z_{CoG} of the vehicle centre of mass is at 0.13 m and the wheel radius wr is 0.0745 m. The vehicle tyres are made with an outer rubber envelope filled with elastic foam and the ground is asphalt. The wheel-obstacle contact model uses a spring damper formalism (Fig. 33.3), and corresponds to the Adams IMPACT function [7], with wr_1 the distance between the wheel centre and the contact point (Fig. 33.3a):

$$F_{IMPACT(t)} = \begin{cases} 0 \ \ if \ \ wr_1 > wr \\ k(wr - wr_1)^e - c_{max}\dot{wr}_1 * STEP(wr_1, wr - d, 1, wr, 0) \ \ if \ \ wr_1 \le wr \end{cases} \quad (33.1)$$

The stiffness value k of the contact force takes into account both the material properties and the contact geometry, and ensures the conformity of the simulation

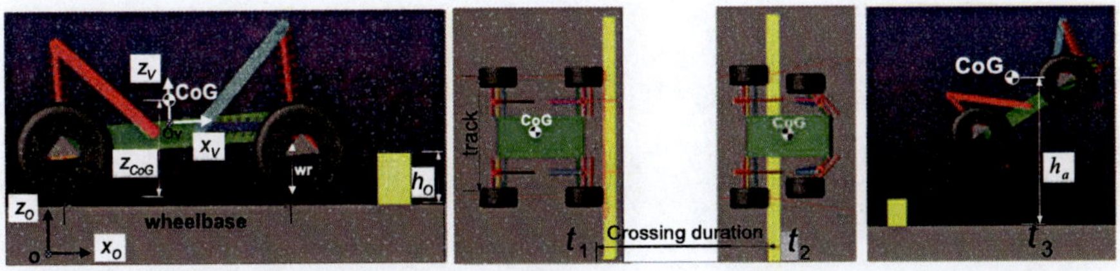

Fig. 33.1 3D model of the vehicle **a** before, **b** and **c** during and **d** after the obstacle crossing

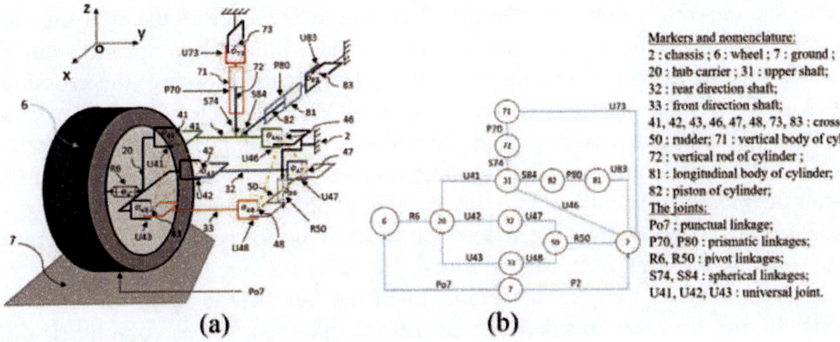

Fig. 33.2 **a** Kinematic diagram and **b** linkage graph of Susp-4D

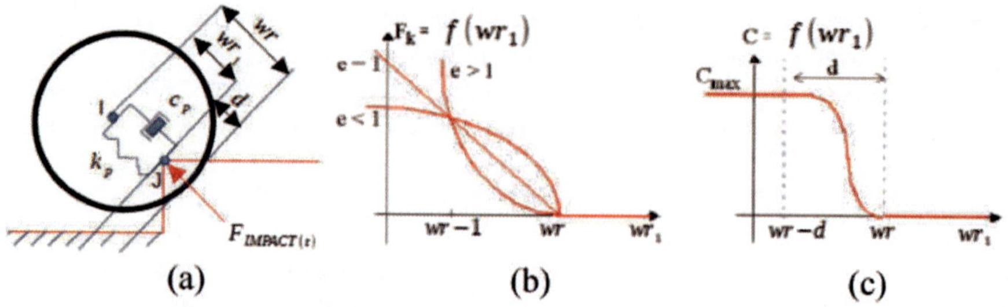

Fig. 33.3 Wheel-ground contact model in Adams **a** impact force **b** exponent selection charts **c** evolution of the damping coefficient

Table 33.1 Wheel-ground contact parameters used in Adams simulations

Stiffness (N/mm)	Force exponent	Damping (Ns/mm)	Penetration deph (mm)	Static friction	Dynamic friction
$k = 1000$	$e = 1.1$	$Cmax = 10$	$d = 0.01$	$\mu_s = 1$	$\mu_d = 0.95$

with the physical phenomenon [7]. The literature proposes several stiffness values for rubber [8, 9].

The exponent e of the stiffness force depends on the material. The value given in Table 33.1 corresponds to rubber [7]. The recommended damping value Cmax [7] in Adams is around 1% of the stiffness value; It is chosen to guarantee the penetration depth d. Static μ_s and dynamic μ_d friction coefficients are chosen among existing values in the literature [10, 11].

33.2.2 *Design of Experiment for Obstacle Crossing*

The obstacle crossing can be divided into three phases: (1) Before the crossing, all four wheels of the vehicle are rolling on the ground without slipping, the vehicle is moving toward the obstacle (Fig. 33.1a); (2) During the crossing, the crossing starts at time t_1 when the front wheels collide with the obstacle (Fig. 33.1b) and are in the pre-constrained position C_p; the collision phase is the time taken by the wheels to maximally compress the longitudinal suspension, the crossing ends at time (t_2) after the rear wheels collide with the obstacle or pass over it without touching it (Fig. 33.1c); (3) After the crossing, the vehicle reaches the apex of its aerial trajectory at time (t_3) before falling back to the ground (Fig. 33.1d).

This paper aims at studying the influence of the Susp-4D to optimize obstacle crossing. In this aim, we considered three metrics that qualify the crossing: (i) the kinetic energy variation ΔE_c between t_1 and t_2, which is an expression of the speed loss; (ii) the pitch rate (δ') at the end of the crossing t_2, which should be minimal at the beginning of the ballistic phase and (iii) the wheel-obstacle contact duration

Table 33.2 Numerical values of the relevant experiment parameters (wr = 0.0745 m)

Parameters	Value 1	Value 2	Value 3	Value 4	Value 5
h_O (m)	25%wr	50%wr	80%wr	90%wr	100%wr
vc (m/s)	3	6	9	12	15
c_{AV} (Ns/mm)	0.4	0.8	1.6	3.2	6.4

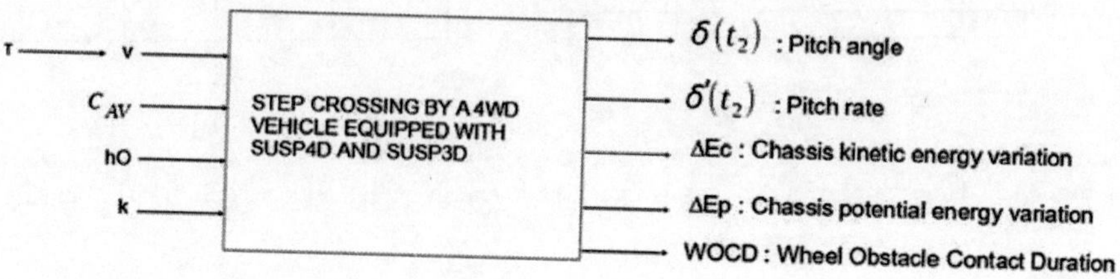

Fig. 33.4 Synoptic of the design of experiment

WOCD, which is the time given to the vehicle to climb over the obstacle. We chose five obstacle heights h_O, five vehicle speeds vc and five damping coefficients of the front longitudinal suspensions c_{AV}, which represented $5^3 = 125$ trials (Table 33.2). The torque applied to the wheels τ was monitored to maintain a constant speed before crossing throughout the tests and the stiffness k of the suspension was set to 2.26 N/mm (Fig. 33.4).

33.3 Results of the Experiment

33.3.1 Influence of the Damping Coefficient When Crossing

Let us only consider the example of a vehicle rolling at a constant high-speed of 12 m/s and crossing an obstacle of $h_O = 25\%$ wr ($h_O = 2$ cm) with three different damping coefficients, the minimum, the middle and the maximum value (Table 33.2). Results of these three simulations are presented on Figs. 33.5 and 33.6. During crossing, the relative kinetic energy loss (Fig. 33.5a) between t_1 and t_2, and the pitch rate at t_2 (Fig. 33.5b) decrease along with the damping coefficient. Moreover, the lower the damping coefficient, the lower the maximum applied torque between t_1 and t_2, after t_2 the vehicle is in ballistic phase (Fig. 33.6a) and the longer the horizontal displacement of the wheel with respect to the chassis (Fig. 33.6b).

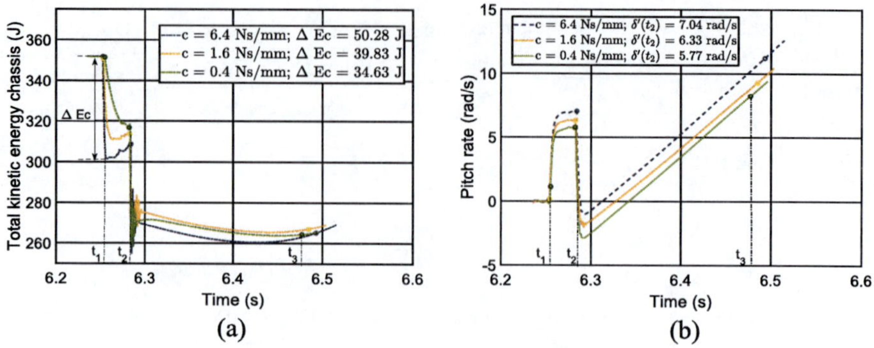

Fig. 33.5 Time series of **a** the chassis total kinetic energy Ec **b** the chassis pitch rate δ' during the crossing

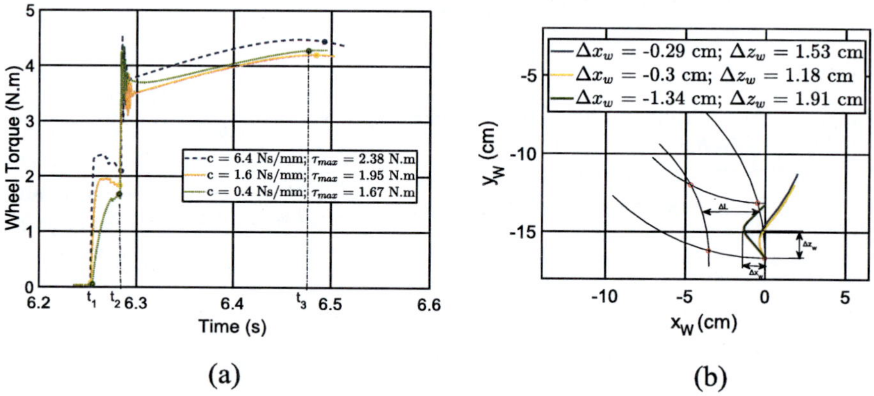

Fig. 33.6 Time series of **a** maximum torque applied to the wheels during the crossing between t_1 and t_2 **b** the trajectory of the wheel on the workspace of the mechanism during obstacle crossing with $\Delta L_{max} = 3.5$ cm

33.3.2 Influence of Simulation Parameters on Crossing Metrics

Relationships between simulation parameters and crossing metrics (ΔE_c, δ', WOCD) have been represented (Fig. 33.7).

Obstacle height, vehicle speed and damping coefficient influence the kinetic energy variation ΔE_c in the same direction. ΔE_c grows with respect to each of the three variables h_O, vc and c_{AV}, when the two others are kept constant (Fig. 33.7a).

Similarly to ΔE_c, δ' grows with the growth of each to the three variables h_O, vc and c_{AV}, when the two other are kept constant. However, the shape of the response surface looks paraboloid for ΔE_c and with a flatter top that could tend towards an

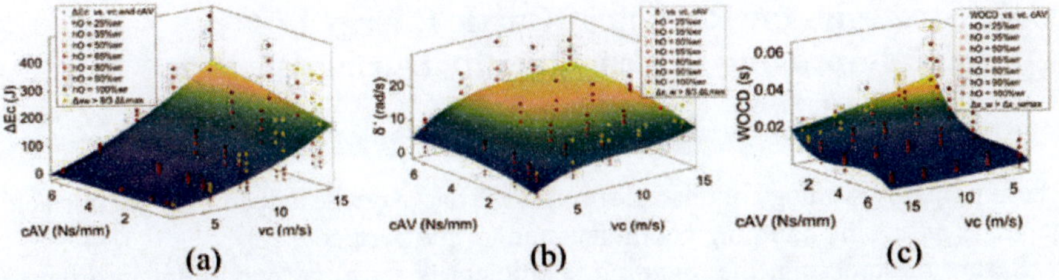

Fig. 33.7 Polynomial fitting function of the chosen metrics: **a** Kinetic energy variation. **b** Pitch rate. **c** Wheel-Obstacle Contact Duration (yellow dots for $x_w(v_c, c_{AV}) > x_{wmax}$)

asymptotic plane for δ'. The damping coefficient influences δ' in a parabolic way for given vc and h_O: the minimum pitch rate is reached within the range of 0.4 and 3.2 Ns/mm (Fig. 33.7b).

The obstacle height influences the wheel-obstacle contact duration (WOCD) in the same direction: the higher h_O (for given vc and c_{AV}), the higher WOCD (Fig. 33.7c). At a given h_O, vehicle speed influences WOCD in the opposite direction, for c_{AV} lower than 3.2 Ns/mm, and along a parabolic way, for c_{AV} higher than 3.2 Ns/mm. In this case, WOCD is higher for extreme speeds (3 and 15 m/s) and is minimal for intermediate speeds. Finally, as expected, for given vc and h_O, the damping coefficient influences WOCD in the opposite direction: WOCD increases as c_{AV} decreases.

Results of all the 125 simulated trials have then been used to compute fitting functions linking the three chosen metrics to vc and c_{AV} for a given h_O:

$$\Delta E_c(v_c, c_{AV}) = e_{00} + e_{10}v_c + e_{01}c_{AV} + e_{20}v_c^2 + e_{11}v_c c_{AV} \tag{33.2}$$

$$\begin{aligned}\delta'(v_c, c_{AV}) = {} & p_{00} + p_{10}v_c + p_{01}c_{AV} + p_{20}v_c^2 + p_{11}v_c c_{AV} + p_{02}c_{AV}^2 + p_{30}v_c^3 \\ & + p_{21}v_c^2 c_{AV} + p_{12}v_c c_{AV}^2\end{aligned} \tag{33.3}$$

$$\begin{aligned}WOCD(v_c, c_{AV}) = {} & d_{00} + d_{10}v_c + d_{01}c_{AV} + d_{20}v_c^2 + d_{11}v_c c_{AV} + d_{02}c_{AV}^2 \\ & + d_{21}v_c^2 c_{AV} + d_{12}v_c c_{AV}^2 + d_{03}c_{AV}^3\end{aligned} \tag{33.4}$$

where $d_{i,j}$, $e_{i,j}$ and $p_{i,j}$ coefficients are different for each obstacle height with i, j in {0,1,2,3}.

33.4 Strategies for Reducing Kinetic Energy Loss and Improving Vehicle Stability During Obstacle Crossing

The anticipatory strategy of obstacle crossing is based on the detection of the obstacle height to adjust the damping coefficient and torque parameters.

Before the crossing, the damping coefficient is set to its maximum in order to limit the chassis load transfer during acceleration. Assuming a minimum obstacle detection distance of 1 m, at a maximum speed of 15 m/s, the delay to reduce the damping coefficient and to apply a convenient torque (τ) is only 67 ms.

At this speed, during the crossing, the vehicle takes off the ground regardless of the applied torque and the damping coefficient. In this case, it is necessary to apply a torque at t_1 and t_2, for reducing the pitch rate of the chassis.

Thus, obstacle crossing by a high-speed vehicle can be seen as an optimisation problem whose objectives are to minimise ΔE_c during crossing while maximising WOCD in order to apply a convenient torque during the crossing that will minimise the vehicle pitch rate at t_2. Mathematically, this problem can be expressed as follows for a given h_O:

$$min\ _{cAV}\Delta E_c(v_c, c_{AV});\ min\ _{cAV}\delta'(v_c, c_{AV});\ max\ _{cAV}WOCD(v_c, c_{AV});$$

with the following constraints:

- The damping coefficient must be chosen in order to avoid reaching the stops and to keep the wheel centre horizontal displacement admissible (Fig. 33.6b):

$$x_{wmin} < x_w(v_c, c_{AV}) < x_{wmax} \tag{33.5}$$

  with x_{wmin} and x_{wmax} depending on the suspension stops.
- The torque should be sufficient to climb, but not so strong as to cause the tire to rip on the obstacle:

$$\tau \leq \tau_{max} \tag{33.6}$$

  with τ_{max} correspond to the maximum torque before slipping.

33.5 Conclusion

In this study, we simulated the behaviour of a multi-body 4WD vehicle equipped with Susp-4D on the front while crossing a step obstacle of five different heights h_O at five different speeds vc with five different damping coefficients c_{AV} of the longitudinal suspensions. The Design of this Experiment (DoE) produced 125 trials. Obstacle crossing has been evaluated by three metrics: vehicle kinetic energy variation (ΔE_c) and pitch rate (δ') at the beginning of the ballistic phase, and wheel-obstacle contact duration (WOCD). In all trials, a lower damping coefficient resulted in a less kinetic energy variation, a lower pitch rate and a longer WOCD.

From the DoE are extracted three polynomial relationships for the vehicle kinetic energy variation $\Delta E_c(v_c, c_{AV})$, the pitch rate $\delta'(v_c, c_{AV})$, and the wheel-obstacle contact duration $WOCD(v_c, c_{AV})$ in function of vehicle speed vc and damping coefficient c_{AV}.

Future work will be dedicated to find a cost function gathering these objective functions in order to compute the optimal c_{AV} that will minimize ΔE_c and δ', and maximize WOCD under two constraints of suspension stroke and non-sliding wheel-obstacle contact.

This optimal c_{AV} will then be implemented in a robust control law of the longitudinal suspensions to enhance/improve obstacle crossing. The applied torque (τ) will be integrated later in the control law.

Acknowledgements This work has been sponsored by the French government research program "Investissements d'Avenir" through the IDEX-ISITE initiative 16-IDEX-0001 (CAP 20-25) and the IMobS3 Laboratory of Excellence (ANR-10-LABX-16-01).

References

1. Fauroux, J.C., Bouzgarrou, B.C.: Dynamic obstacle-crossing of a wheeled rover with double wishbone suspension. In: Field Robotics. pp. 642–649 (2011). http://jc.fauroux.free.fr/PUB/ARTICLES/2011_CLAWAR_Fauroux_Bouzgarrou_DRAFT_Dynamic_obstacle-crossing_of_a_wheeled_rover_with_double_wishbone_suspension.pdf. Accessed 30 Aug 2022
2. Lauria, M.: Nouveaux concepts de locomotion pour véhicules tout-terrain robotisés. Thesis EPFL (2003)
3. Abad-Manterola, P., Burdick, J.W., Nesnas, I.A.D., Chinchali, S., Fuller, C., Zhou, X.: Axel rover paddle wheel design, efficiency, and sinkage on deformable terrain. In: 2010 IEEE International Conference on Robotics and Automation, pp. 2821–2827 (2010)
4. Kumar, S.: All-terrain mobile robot for extra-terrestrial applications. Bachelor dissertation, 50 p. (2012). https://doi.org/10.13140/RG.2.1.2331.1449. Accessed 30 Aug 2022
5. Wada, M.: Studies on 4WD mobile robots climbing up a step. In: 2006 IEEE International Conference on Robotics and Biomimetics. pp. 1529–1534 (2006)
6. N'chot, D., Fauroux, J.-C., Al Bassit, L., Lenain, R., Vaslin, P.: Performance evaluation of an innovative suspension system for obstacle crossing. In: Cascalho, J.M., Tokhi, M.O., Silva, M.F., Mendes, A., Goher, K., and Funk, M. (éds.) Robotics in Natural Settings, pp. 95–106 (2023)
7. Verheul, C., SayField International: Benelux ADAMS User Meeting, 19 p. (2012)
8. Gregory, M.J.: Measurement of rubber properties for design. Polym. Test. **4**, 211–223 (1984). https://doi.org/10.1016/0142-9418(84)90013-8. Accessed 30 Aug 2022
9. Yu, W., Zhijun, L., Baolin, L., Mingshuai, G.: 1180. Stiffness characteristic comparison between metal-rubber and rubber isolator under sonic vibration, **16**(11) (2014)

10. Wong, J.Y.: Theory of ground vehicles, 4th ed. Wiley (2008).
11. Engineers Edge. Coefficient of Friction Equation and Table Chart, Accessed 02 feb 2023 (2000).

Chapter 34
A Review on Mobile Robots with Multimodal Locomotion

F. Kadar and M. O. Tătar

Abstract This paper aims to study and research the state of the art of the mobile robots with multimodal locomotion. The first part of the paper presents introductive aspects and classification of the mobile robots with multimodal locomotion. The main multimodal locomotion methods used are presented according to the operating environment: aerial-terrestrial, aerial-aquatic, terrestrial-aquatic, terrestrial, aquatic-aerial-terrestrial. In the second part of the paper, hybrid terrestrial robot variants developed in the Mechatronics and Machine Dynamics Department at the Technical University of Cluj-Napoca are briefly described. The authors propose a conceptual design for a mobile robot capable to operate in multiple environments using multimodal locomotion methods.

Keywords Mobile robots · Multimodal locomotion · Aerial-terrestrial robots · Terrestrial-aquatic robots

34.1 Introduction

Mobile robots were created out of the necessity to safely perform dangerous tasks, transportation of heavy objects in harmful environments which present a high risk to the health and integrity of the human operators.

Mobile robots with multimodal locomotion are a particular category of mobile robots which possess at least two distinct locomotion methods.

The multiple locomotion methods allow the usage of a single mobile robot to perform wide range of applications in two or more operation environments. Also, the usage of the multiple locomotion methods assure a better adaptability to the locomotion surface particularities and eliminates the limitations of the conventional mobile robots.

F. Kadar · M. O. Tătar (✉)
Technical University of Cluj-Napoca, Cluj-Napoca, Romania
e-mail: Olimpiu.Tatar@mdm.utcluj.ro

F. Kadar
e-mail: Francisc.Kadar@mdm.utcluj.ro

I. Doroftei et al. (eds.), *Proceedings of SYROM 2022 & ROBOTICS 2022*, Mechanisms and Machine Science 127, https://doi.org/10.1007/978-3-031-25655-4_34

Since this particular category of mobile robots uses a single mobile robot in multiple operating environments it's an economically advantageous solution for performing multiple and different tasks in mobile robotics.

34.1.1 Structural-Functional Classification of the Mobile Robots with Multimodal Locomotion

Mobile robots with multimodal locomotion have multiple mechanical structure solutions to perform the multimodal locomotion process. Thus Fig. 34.1 presents a classification of them according to their mechanical structure.

Based on the study of the specialized literature we identify multimodal locomotion capable mobile robots with standard/conventional structure which have a single mechanical configuration that allow both locomotion methods to be performed simultaneously or one after another. A different category is represented by the multimodal mobile robots with reconfigurable structure which own two or more mechanical configurations to maximize the advantages provided by each different locomotion method used. Reconfigurability is achieved through human intervention for the manually reconfigurable robots or independently using a transition process for the auto-reconfigurable robots. We also identify mobile robots with modular structure capable of multimodal locomotion, which allow the functions and size of the robot to be modified.

An auto-reconfigurable mobile robot with transformable locomotion unit mechanism is presented in Fig. 34.2a. The robot can perform multiple locomotion modalities based on the locomotion units configurations: wheel, leg, tracks. The transformable wheel mechanism of the mobile robot change the shape of the rims to achieve an excellent terrain adaptability and locomotion efficiency in unpredicted work environments [1, 2].

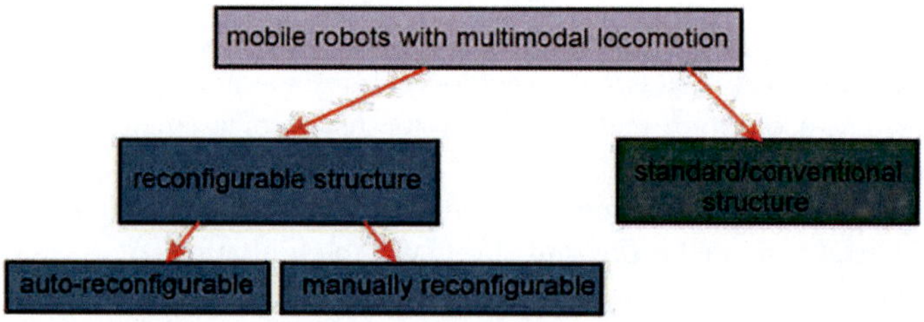

Fig. 34.1 Classification based on the mechanical structure of the multimodal mobile robots

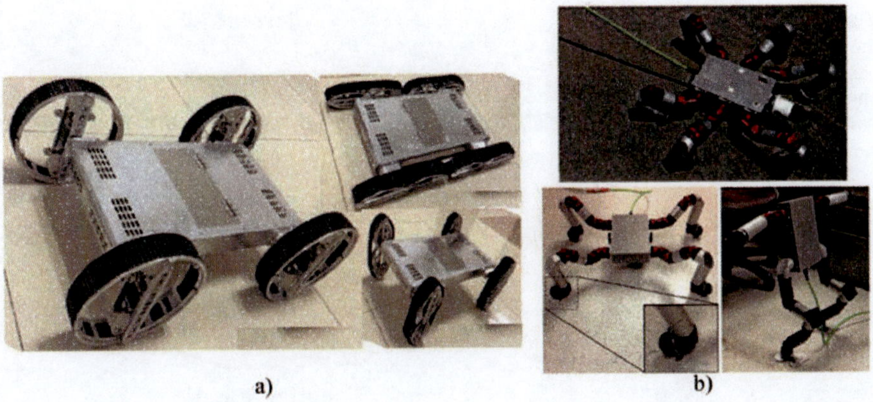

Fig. 34.2 Mobile robots with reconfigurable structure capable of multimodal locomotion [1–3]

The manually reconfigurable modular mobile robot, Snake Monster, is shown in Fig. 34.2b. Different mechanical configurations of the modular robot for the realization of different walking locomotion modes are achieved through the human intervention. The locomotion modalities of the robot are: hexapod, quadruped and bipedal walking [3].

34.1.2 *Classification Based on the Operation Environment*

As mentioned, multimodal locomotion allows the usage of a single robot in multiple operation environments, adapting to the particularities of each environment.

By studying the operation environments and the specifics of the locomotion processes, they can be classified in the following 5 main categories of mobile robots with multimodal locomotion: *aerial-terrestrial, aerial-aquatic, terrestrial-aquatic, terrestrial, aerial-aquatic-terrestrial* (Fig. 34.3).

Relevant examples for each category of the multimodal locomotion robots are presented in the following chapters. This special category of mobile robots can be used for a longer operation time period by optimizing energy consumption.

34.2 Aerial-Terrestrial Robots with Multimodal Locomotion

The aerial-terrestrial multimodal mobile robots move efficiently on the soil, or in the air, using several locomotion methods simultaneously or one after another. Thus, taking advantage of the provided benefits, specific to the distinct locomotion methods,

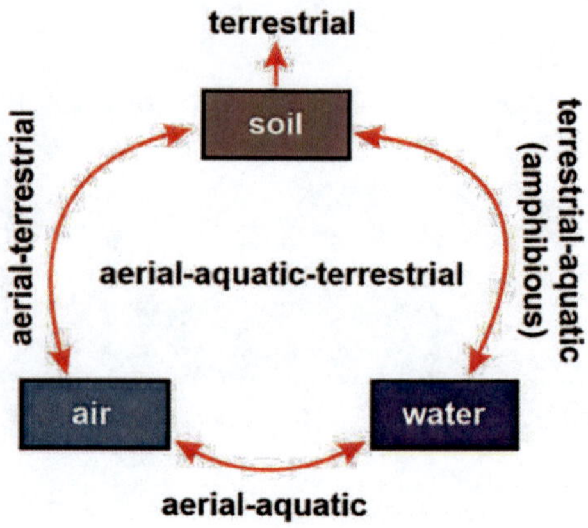

Fig. 34.3 Classification of the multimodal locomotion based on the operation environment

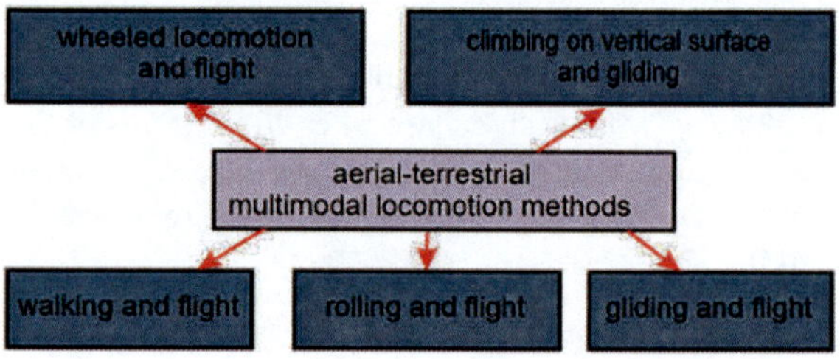

Fig. 34.4 Aerial-terrestrial multimodal locomotion methods used in mobile robotics

mobile robots move efficiently and quickly on the ground surface and easily overcome the obstacles by flight.

The main aerial-terrestrial locomotion modalities used by mobile robots are presented in Fig. 34.4.

In Fig. 34.5 two walking mobile robot variants capable of multi modal locomotion are presented. The left side of the figure shows the minirobot named BOLT, which moves using flapping-wing flight and bipedal walking [4]. On the right side of the figure is presented a robot that use the quadrupedal walking and flying locomotion methods. The flight is achieved by four propellers, similar to drones [5].

The quadrupedal walking remains challenging because the gravity center of the robot is continuously shifted during the locomotion process to maintain the robot in a stable position [6]. The quadrupedal walking modes are classified in changeover walking and galloping [6]. The changeover walking can be divided based on the leg pairs used in, trotting gait (diagonal pairs) and pacing gait (lateral pairs). The galloping gait, known also as bounding, used the front and rear leg pairs and has a higher energy consumption [7].

Fig. 34.5 Aerial-terrestrial multimodal mobile robots with walking and flight locomotion [4, 5]

34.3 Aerial-Aquatic Robots with Multimodal Locomotion

The aerial-aquatic multimodal mobile robots move quickly in the air using flight. Flying robots can easily and quickly reach/enter aquatic environments.

In Fig. 34.6 the main aerial-aquatic locomotion methods used in mobile robotics are presented.

Figure 34.7 presents the 3D model and the prototype of the NEZHA robot with multimodal locomotion realized by flying, swimming and floating on the water surface. It can perform two locomotion methods for the aerial locomotion: horizontal flight and vertical flight. The swimming locomotion is achieved using the expansion/compression of an air chamber bladder filled with high-pressure gas [8, 9].

In Fig. 34.8 is shown an aerial-aquatic mobile robot which plunge-dive into the water with folded wings. The robot moves in the air by flight and gliding. The transition process between air and water is bioinspired from the plunge process of the gannet seabirds (morus bassanus) [10].

Fig. 34.6 Aerial-aquatic multimodal locomotion methods used in mobile robotics

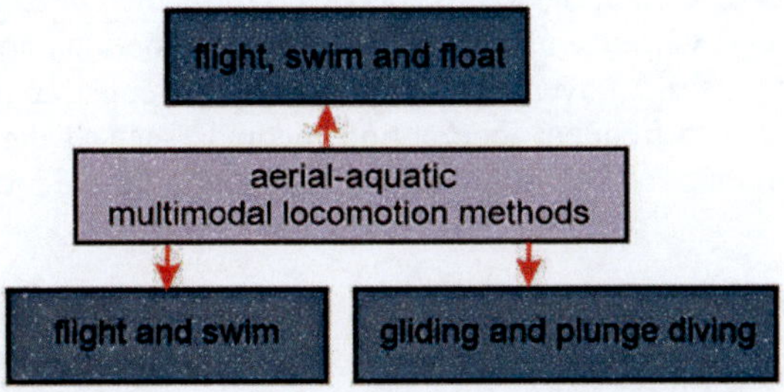

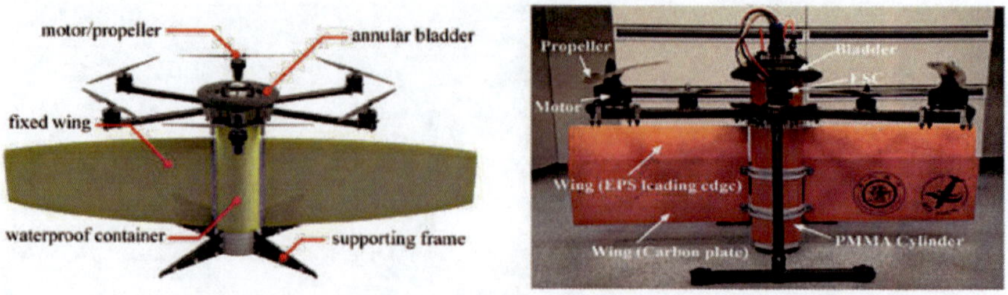

Fig. 34.7 The NEZHA aerial-aquatic robot with flying, swimming and floating locomotion [8, 9]

Fig. 34.8 Flying robot with foldable wings which plunge-dives and swim [10]

34.4 Terrestrial-Aquatic Mobile Robots with Multimodal Locomotion

The terrestrial-aquatic (amphibious) mobile robots with multimodal locomotion moves efficiently and quickly on the soils surface, in water or at the separation limit of the two environments, using at least two different locomotion methods. The terrestrial mobile robots perform their locomotion process commonly using legs, wheels and tracks, depending on the locomotion surfaces particularity's. The terrestrial movement is restricted by the constraints of the locomotion system used.

The frequent locomotion methods seen at the terrestrial-aquatic (amphibious) mobile robots with multimodal locomotion are presented in Fig. 34.9.

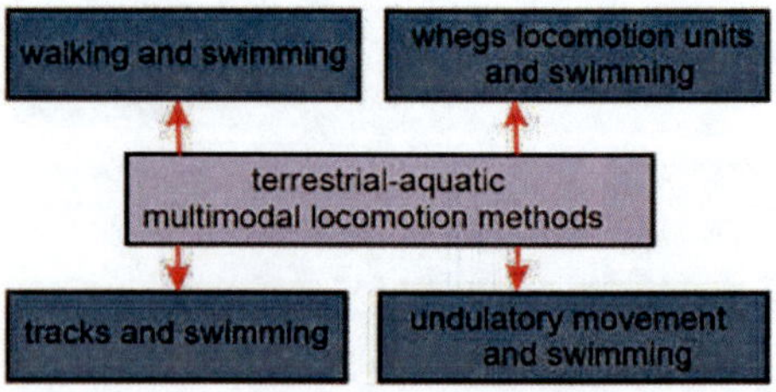

Fig. 34.9 Multimodal terrestrial-aquatic locomotion methods frequently used by mobile robots

Fig. 34.10 Terrestrial-aquatic mobile robots with whegs units locomotion and swim [11, 12]

In Fig. 34.10 mobile robot variants capable of multimodal locomotion are presented. They move on the soils surface using whegs locomotion units, respectively swim.

The first one is equipped with reconfigurable wheels, which act as whegs units.

Figure3434...34.10a shows both configuration of the locomotion units used by the amphibious robot: whegs units and wheels. The reconfigurable structure of the locomotion units allows the diversification of the robot's terrestrial locomotion methods. The robot can float on the water surface or swim using the units as paddles [11].

Figure 34.10b presents images from the aquatic and terrestrial locomotion process of the Whegs™ robot equipped with sealed structure [12].

34.5 Terrestrial Robots with Multimodal Locomotion

The terrestrial mobile robots with multimodal locomotion, also known as hybrids, use a combination of two or more conventional locomotion methods in order to adapt to the particularities of the locomotion surfaces as efficiently as possible. The conventional terrestrial locomotion methods are legged, wheeled and tracked.

Figure 34.11 presents the main terrestrial locomotion methods used by the multimodal locomotion capable mobile robots.

Within the Department of Mechatronics and Machine Dynamics (MDM) of the Technical University of Cluj-Napoca were developed two mobile robot models with terrestrial multimodal locomotion. They are equipped with hybrid leg-wheel locomotion units. The **combined locomotion** means both locomotion methods are used at the same time: wheeled and legged.

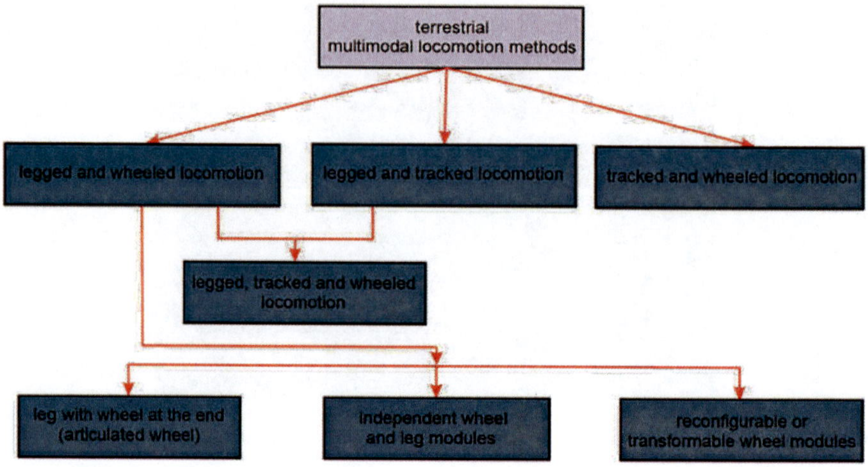

Fig. 34.11 Terrestrial multimodal locomotion methods in mobile robotics

The left side of Fig. 34.12 shows a quadruped mobile robot prototype with conventional wheels attached to the extremity of the legs. The legs have 3 and the wheels 2° of freedom. The cable powered robot moves using stepping, wheeled and combined locomotion [13].

The right side of the figure presents a three-legged mobile robot prototype with conventional wheels attached to the extremity of the legs. The legs and the wheels have only 1° of freedom each of them. The Li-Po battery powered robot moves using wheeled and combined locomotion [14].

The structural diagram of the quadrupedal mobile robot prototype with leg-wheel locomotion units [14] is presented in Fig. 34.13.

Figure 34.14 shows the kinematic diagram of the three-legged mobile robot prototype equipped with hybrid leg-wheel type locomotion units [14].

The locomotion strategy concept of the hybrid leg-wheel three-legged robot for overpassing obstacles is presented in the following figures. The locomotion

Fig. 34.12 Mobile robots with multimodal legged and wheeled locomotion [13, 14]

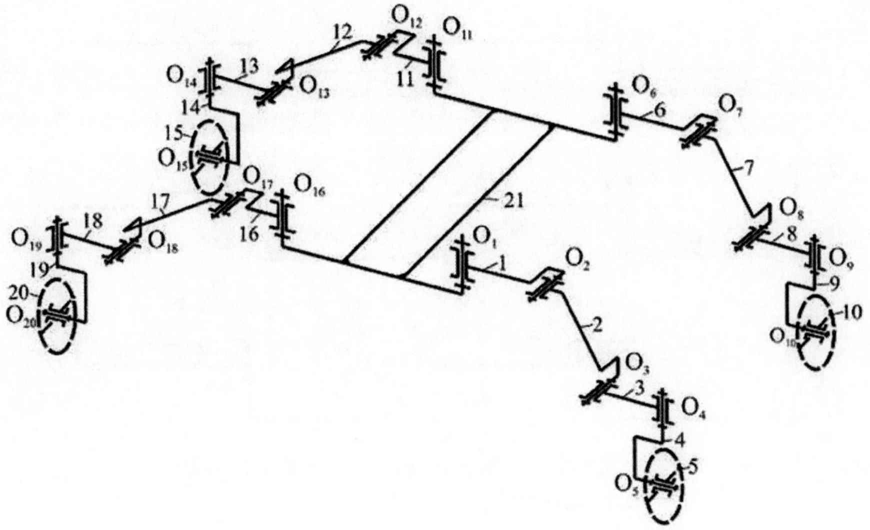

Fig. 34.13 Structural diagram of the quadruped hybrid leg-wheel locomotion unit robot [14]

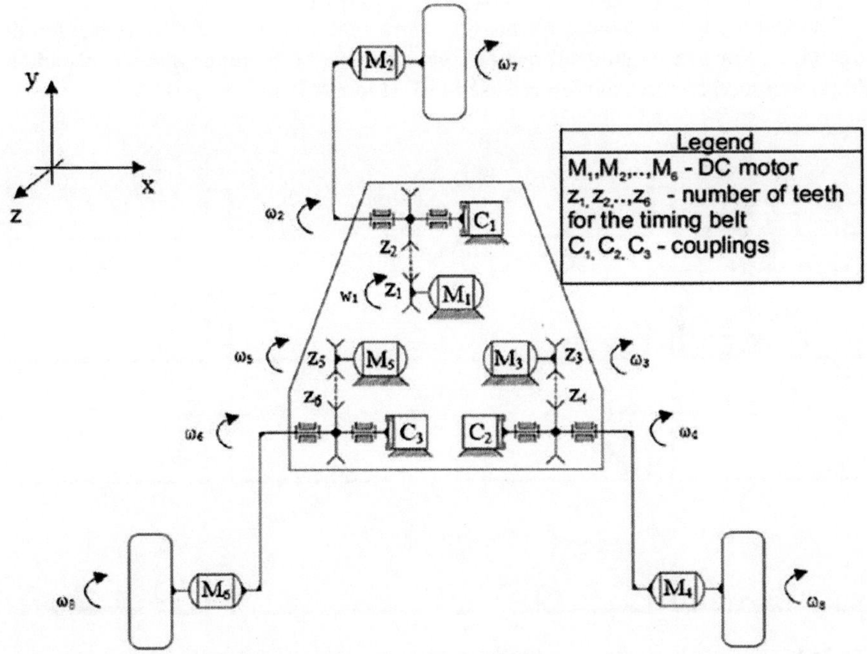

Fig. 34.14 Kinematic diagram of the hybrid leg-wheel locomotion unit robot prototype [14]

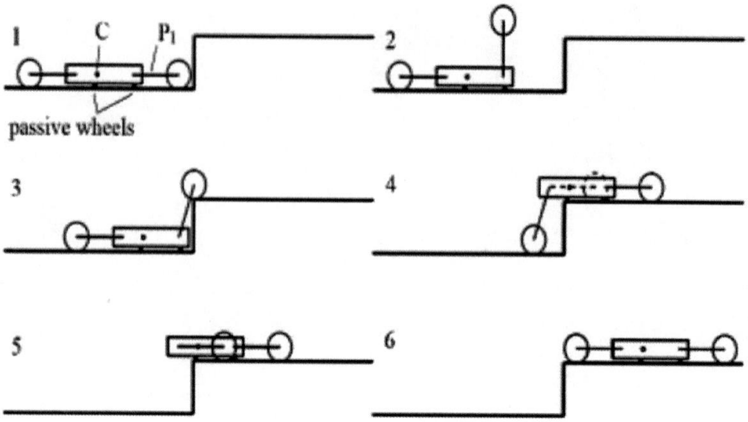

Fig. 34.15 Locomotion strategy concept for passing a step type obstacle [15]

sequences for passing a step like obstacle are presented in Fig. 34.15. In this case the passive wheels attached underneath the chassis of the robot are used for stability during the climbing process of the step type obstacle [15].

The locomotion sequences for passing a threshold like obstacle are presented in Fig. 34.15. For overpassing a threshold type obstacle both the passive wheels and a shield mounted on the front leg are used [15] (Fig. 34.16).

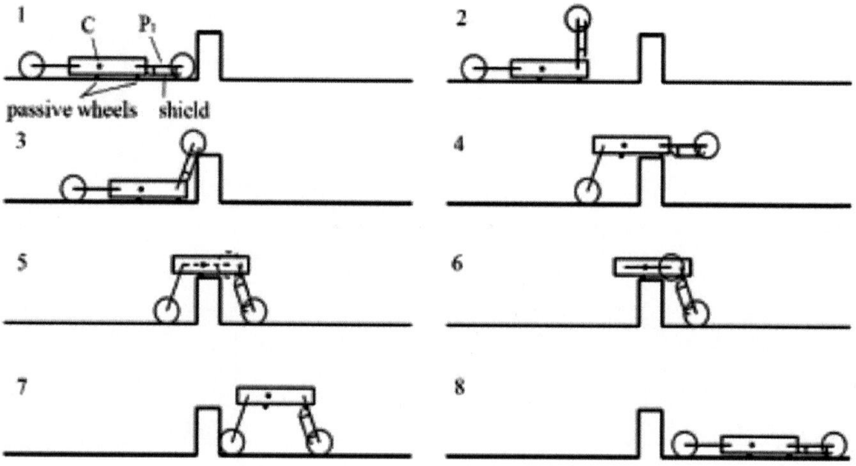

Fig. 34.16 Locomotion strategy concept for passing a threshold type obstacle [15]

34.6 Design Proposal for a Multi-environment Mobile Robot with Multimodal Locomotion

The authors propose a mobile robot design capable of moving using multiple locomotion methods: flying, floating and whegs unit locomotion for swim and crawling. The robot can be used in aerial-aquatic-terrestrial environment [16].

The current multimodal locomotion mobile robot variants are mainly designed to be efficiently used in only one or at most two work environments. We identified the necessity of developing a multimodal robot which can operate in more than two environments. The proposed mobile robot will be successfully used in aerial-aquatic-terrestrial environment The robot will move on the soil surface using the crawling method and the big obstacles will be overpassed by flight, using the propellers. The robot can successfully swim in the water using the whegs units. To increase the operating time of the robot and reduce the energy consumption, the robot can passively float on the water's surface.

Figure 34.17 presents the proposed mobile robot protype with multimodal locomotion and the structure of the whegs locomotion units. The robot is lightweight and hermetically sealed. The proposed locomotion units allows both aquatic and terrestrial locomotion. To keep the robot on the water's surface, sponges are placed inside the robot's structure [16].

The proposed multimodal robot is suitable for search and rescue operations. Video cameras are mounted both in the front and back of the robot to real-time monitor the work environment. The robot will be remotely controlled, equipped with sensors, such as GPS sensor for localization. As energy source Li-Po batteries will be used [16].

The proposed multimodal robot design eliminates the constraints of the work environments using its best fitting locomotion methods and is suitable for a wide range of application. The main challenge will be the optimization of the energy consumption without reducing the locomotion performances of the robot.

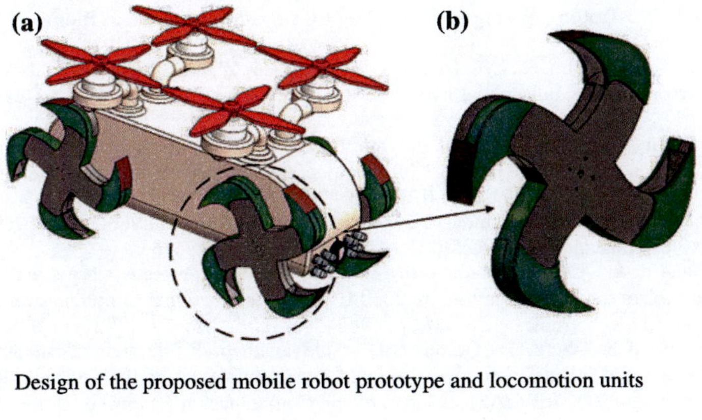

Fig. 34.17 Design of the proposed mobile robot prototype and locomotion units

34.7 Conclusions

This paper presents a short state of the art on the existing mobile robot variants capable of multimodal locomotion. Contains a classification based on the working environment and relevant examples of multimodal mobile robots are briefly presented for each category. Terrestrial robot prototypes, with hybrid leg-wheel locomotion units, developed in the Mechatronics and Machine Dynamics Department at the Technical University of Cluj-Napoca are briefly described.

The paper presents the variety of locomotion systems and methods used by the mobile robots with multimodal locomotion. The mobile robot's locomotion systems and methods are chosen according to the specific/constraints of the operating environments and locomotion surfaces in order to optimize the energy consumption and the locomotion process.

References

1. Zhang, X., Zhou, F., Xu, X., Chen, H.: Configuration design and analysis of a multimodal wheel with deformable rim. In: 2019 IEEE/ASME International Conference on Advanced Intelligent Mechatronics (AIM), pp. 772–777, July 2019
2. Zhou, F., Xu, X., Xu, H., Zhang, X.: A multimodal hybrid robot with transformable wheels. In: 2017 IEEE International Conference on Real-time Computing and Robotics (RCAR), pp. 139–144, July 2017
3. Kalouche, S., Rollinson, D., Choset, H.: Modularity for maximum mobility and manipulation: control of a reconfigurable legged robot with series-elastic actuators. In: IEEE International Symposium on Safety, Security, and Rescue Robotics (SSRR), pp. 1–8, October 2015
4. Peterson, K., Fearing, R.S.: Experimental dynamics of wing assisted running for a bipedal ornithopter. In: 2011 IEEE/RSJ International Conference on Intelligent Robots and Systems, pp. 5080–5086, September 2011. IEEE (2011)
5. Transform Quadcopter—Multi Legged Robot (2017). https://m.blog.naver.com/PostView.naver?isHttpsRedirect=true&blogId=ohryan77&logNo=220912529077
6. Siegwart, R., Nourbakhsh, I. R., Scaramuzza, D.: Introduction to Autonomous Mobile Robots. MIT Press (2011)
7. Raibert, M.H.: Trotting, pacing and bounding by a quadruped robot. J. Biomech. **23**, 79–98 (1990)
8. Lu, D., Xiong, C., Zhou, H., Lyu, C., Hu, R., Yu, C., Zeng, Z., Lian, L.: Design, fabrication, and characterization of a multimodal hybrid aerial underwater vehicle. Ocean Eng. **219**, 108324 (2021)
9. Lu, D., Xiong, C., Zeng, Z., Lian, L.: A multimodal aerial underwater vehicle with extended endurance and capabilities. In: 2019 International Conference on Robotics and Automation (ICRA), pp. 4674–4680, May 2019. IEEE (2019)
10. Hybrid Aerial Underwater Vehicle, SCOPE Project, MIT Lincoln Laboratory (2018). https://phoenixfiles.olin.edu/do/23bf82e5-0d2c-471c-a8d2-62a51d619b16
11. Vijaychandra, A., Alex, C., Mathews, M., Sathar, A.A.: Amphibious wheels with a passive slip mechanism for transformation. In: 2019 IEEE 4th International Conference on Advanced Robotics and Mechatronics (ICARM), pp. 960–965, July 2019. IEEE (2019)
12. Boxerbaum, A.S., Werk, P., Quinn, R.D., Vaidyanathan, R.: Design of an autonomous amphibious robot for surf zone operation: part 1 mechanical design for multi-mode mobility. In: Proceedings, 2005 IEEE/ASME International Conference on Advanced Intelligent Mechatronics, pp. 1459–1464, USA, July 2005

13. Tătar, M.O., Cirebea, C.: Mobile minirobot with hybrid locomotion. Solid State Phenomena **260**, 51–58 (2017)
14. Tătar M.O., Barbu P., Teutan E.: Design and development of the hybrid mobile robots. In: International Conference on Automation, Quality and Testing, Robotics, AQRT 18 TETA 21, May 24–26, Cluj-Napoca, Romania, pp. 1–6 (2018)
15. Barbu, P., Tatar, M.O.: Development and testing of a mobile robot with hybrid legged-wheeled locomotion. Robotica Manag. 22(1) (2017)
16. Gyarmati M., Kadar F., Guțu H., Tătar MO.: Contributions to the search and rescue robots' development with TriStar and whegs units. In Acta Technica Napocensis-Series: Applied Mathematics, Mechanics, and Engineering, vol. 65, no. 2, June 2022

Chapter 35
Mobile Robots on Adapted Wheels for Stair Climbing

O. Antonescu, D. Valeanu, and P. Antonescu

Abstract The paper presents several types of mobile robots on 2, 4 and 6 wheels special adapted for climbing the stairs and uneven terrains. In the case of the mobile robots on 2 and 4 noncircular wheels, each wheel is designed to have three curved or roller arms which make possible the stair climbing. In the classical case of the mobile robot on 4 circular wheels, their diameter is inferiorly limited, the body being in a high position because of the high ground clearance, over the stairs. In the case of the mobile robot on 4 and 6 circular wheels for uneven terrain, the body is linked to the wheels through articulated bars as suspension and variable geometry.

Keywords Mobile robot · Noncircular wheel · Arc-shape wheel · Planetary driving unit · Wheeled leg

35.1 Introduction

The constructive structure [1] of the mobile robots has diversified especially regarding the transmission and rolling system on uneven terrains [2, 3] and stair climbing [4–6].

The mobile robots are using driving wheels for moving on flat or uneven terrain, including the stair climbing. The most used wheels are circular driven by simple transmissions with differential or by an actuator (motor) on each wheel offering an overall low-cost.

In the case of 2-wheel mobile robots, the balance is the most critical issue for staying in vertical position, being utilized as vacuum cleaners. These mobile robots have the advantage of having a high maneuverability, spinning very easily.

The mobile robots having 3 wheels use the differential transmission on two driving wheels, the third wheel being for steering.

The propulsion system that drives most of the mobile robots consists of an electric motor or several motors, the control being insured by means of a software.

O. Antonescu (✉) · D. Valeanu · P. Antonescu
University Politehnica of Bucharest, Bucharest, Romania
e-mail: oval33@hotmail.com

I. Doroftei et al. (eds.), *Proceedings of SYROM 2022 & ROBOTICS 2022*, Mechanisms and Machine Science 127, https://doi.org/10.1007/978-3-031-25655-4_35

The scientific research on the mobile robot domain in Romania has been driven through the establishing in Iasi city of a laboratory of mobile robots being endowed with some functional models of educational mobile robots [7].

35.2 Mobile Robot on Two Noncircular Driving Wheels

The mobile robot on two noncircular driving wheels has three elastic curved profiles on each wheel (Fig. 35.1a) and it uses two support rollers mounted at the rear of the mobile robots [8]. Each noncircular wheel consists of three thin arc-shape profiles which are angularly equidistant and fixed in the wheel hub (Fig. 35.1b).

The two driving wheels are fixed on a shaft which is driven by means of a worm gear [1].

The electric motor is fixed on a metallic frame which has two rollers as rear axle. The transmission is very simple consisting of a driving shaft which transmit the torque to wheel shaft through a worm gear.

35.3 Mobile Robots on Four Noncircular Driving Wheels

Figures 35.2 and 35.3 show two models of mobile robots on noncircular driving wheels. The first mobile robot (Fig. 35.2) has four elastic arc-shape wheels of the same special type as we have seen at Sect. 35.2 (Fig. 35.1).

The second mobile robot (Fig. 35.3) has a different type of noncircular wheels i.e. each wheel has three rigid arms with two rollers through which the contact with the stair surface is kept.

To be noticed that in both variants of the mobile robot, a single electric motor drives the two axles through chain transmissions.

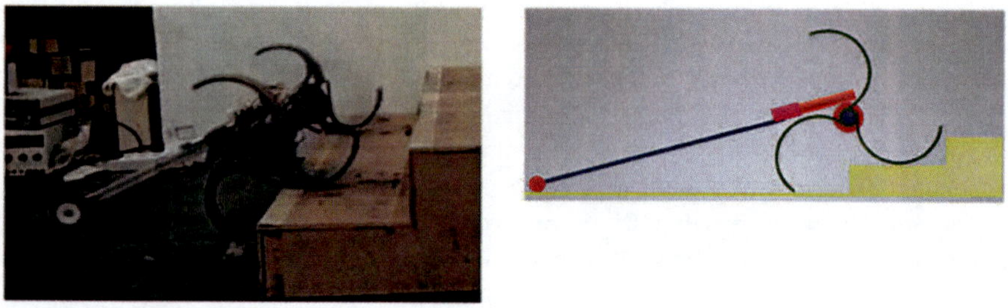

(a) (b)

Fig. 35.1 Mobile robot on special arc-shape wheels: **a** — prototype [8]; **b** — simulation

Fig. 35.2 Arc-shape wheel mobile robot [8]

Fig. 35.3 Three-arm wheel mobile robot [9]

35.4 Mobile Robot on Four Circular Driving Wheels

In Fig. 35.4 we present a mobile robot on four circular driving wheels with a metallic rigid frame having the ground clearance high enough to overcome obstacles and climbing the stairs.

First, the front driving wheels roll on the vertical surface of the stair, and then on the horizontal surface of it (Fig. 35.4a). Also, when the rear driving wheels reach the vertical surface of the stair, they begin to climb it (Fig. 35.4b).

This mobile robot has four mobilities, all driving wheels being driven by means of an electric motor.

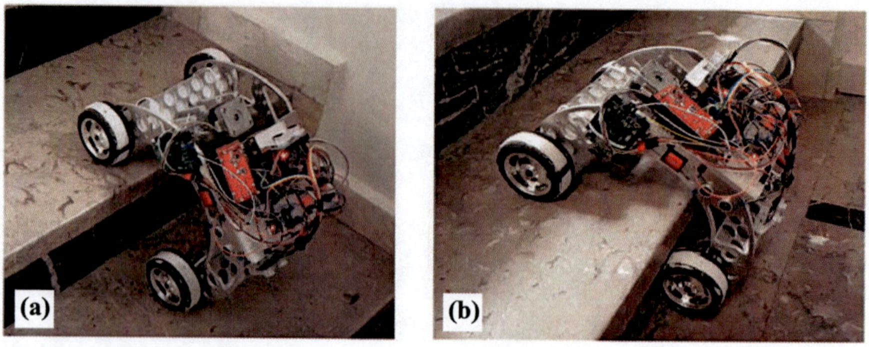

Fig. 35.4 Mobile robot on four circular driving wheels [10]

35.5 Mobile Robot with Four Planetary Driving Units

The mobile robot with four planetary driving units is rolling using eight driving wheels on horizontal or inclined terrain (Fig. 35.5a), and it is changing the three driving wheels of each unit if necessary in stair climbing (Fig. 35.5b).

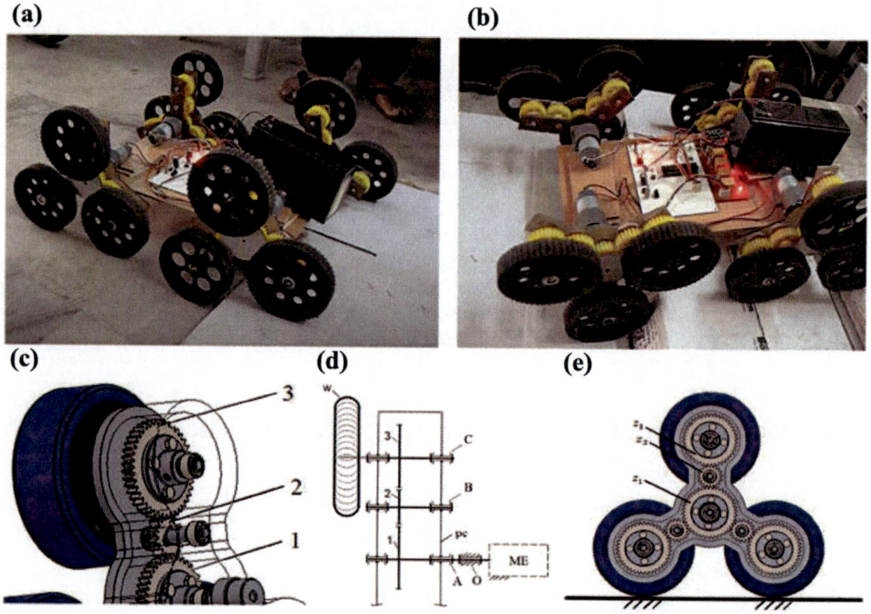

Fig. 35.5 **a** Mobile robot with four planetary driving units on horizontal terrain. **b** Mobile robot with four planetary driving units climbing the stairs. **c** The three gears of one arm of the planetary carrier [11]. **d** Kinematic scheme of one arm of the planetary carrier [11]. **e** Planetary driving unit (three-arm driving wheel) [12]

To be noticed that in each planetary unit the driving wheel is solidary with the planetary gear 3 (Fig. 35.5c) and the sun gear 1 is solidary with the electric motor (ME) driving shaft (Fig. 35.5d).

According with the cylindrical mechanism structure of each unit with three-arm planetary carrier (Fig. 35.5e), each electric motor actuates three driving wheels with the same angular velocity [1, 13].

Although the cylindrical planetary mechanism has two degrees of freedom, only one electric motor is necessary to drive the unit. Of course, we need a driving torque high enough to rotate the whole planetary unit with respect to the driving wheel that is in contact with a stair or other obstacle (Fig. 35.5e). During the rolling on a flat terrain, the angular velocity (rpm) of the driving wheel 3 n_w is:

$$n_w = n_{ME} \cdot \frac{z_1}{z_3} \tag{35.1}$$

where n_{ME} is the speed of the electric motor, z_1 and z_3 are the teeth numbers of sun and planet gears. During the stair climbing, from the condition of the movement with respect to the planetary carrier (pc) [1], the following equation results:

$$\frac{n_1 - n_{pc}}{n_w - n_{pc}} = \frac{z_3}{z_1} \tag{35.2}$$

where n_1 and n_{PC} are the speeds of the sun gear and the planetary carrier.

In Eq. (35.2) we impose that the driving wheel to be stopped with non-sliding hypothesis during the obstacle climbing ($n_w = 0$). Therefore, we obtain

$$n_{pc} = n_1 \frac{z_1}{z_1 - z_3} \tag{35.3}$$

with the condition: $z_1 \neq z_3$.

35.6 Mobile Robot on Six Circular Driving Wheels

This mobile robot has six circular wheels driven independently by means of an electric motor (Fig. 35.6). When the mobile robot encounters a stair, each driving wheel rolls on the vertical surface and then on the horizontal surface. Figure 35.6 show the mobile robot climbing up (a) and down (b) on stairs.

We propose an improved kinematic scheme (Fig. 35.7a) inspired from the famous Sojourner rover [15]. It consists of an open kinematic chain articulated, in vertical plane, directly to the body and indirectly through two shock absorbers.

The body 7 is articulated in A to the rocker 1 which has two arms of which the wheel r_3 and bar 2 are articulated in C and B respectively.

The wheels r_1 and r_2 are articulated in E and D to the bar 2 which has two arms BD and BE.

Fig. 35.6 Mobile robot on six circular driving wheels climbing up (**a**) and down (**b**) on stairs [14]

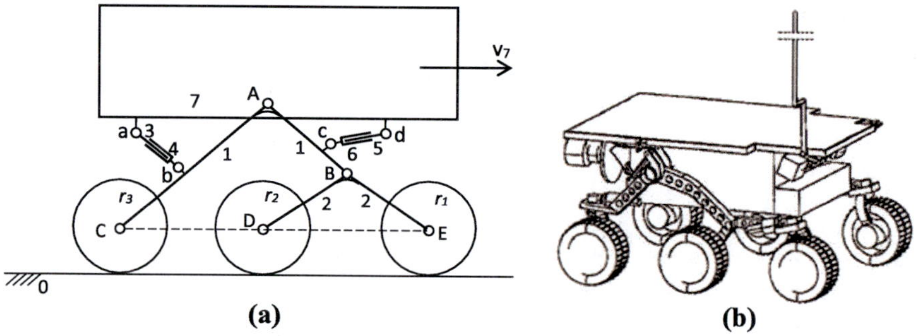

Fig. 35.7 **a** Kinematic scheme of the six-wheel mobile robot (right side-view). **b** Sojourner rover [15]

The body 7 is also linked to the rocker 1 through two kinematic Dyad Chains DC(3, 4) and DC(5, 6) articulated in the points a and b, and in the points c and d respectively.

The Kinematic Chain KC(1, 2) allows the independent rolling of each wheel on steep or vertical surfaces, making possible to overcome some obstacles encountered by the mobile robot on a rough terrain (Fig. 35.7a).

As the analyzed kinematic scheme corresponds to the right side of the mobile robot, on the left side, in symmetrical position, there is a similar linkage with the wheels r_4, r_5 and r_6.

The Dyadic kinematic Chains DC(3, 4) and DC(5, 6) type RTR (Rotation-Translation-Rotation) can also have helicoidal springs as suspension.

The constructive layout of the Sojourner robotic Mars rover is displayed in Fig. 35.7b.

Fig. 35.8 Mobile robot on four wheeled legs [16]: **a** on a flat terrain; **b** in stair climbing

35.7 Mobile Robot on Four Wheeled Legs

The mobile robot on four wheeled legs (Fig. 35.8) has three degrees of freedom on each leg in the hip, knee and wheel.

The kinematic chains of the legs can insure a variable ground clearance of the body which allows the stair climbing (Fig. 35.8b) or to overcome large obstacles.

The wheeled legs consist of a thigh linked to the body through a rotate joint (hip) and a shank which has a driving wheel articulated in the end. Thus, the mobile robot can maintain the balance when encounters an obstacle.

35.8 Conclusions

In the paper several types of mobile robots on wheels have been analyzed. The mobile robots are designed to overcome obstacles and climb stairs.

The mobile robots on wheels, capable to climb stairs, have a wide variety of technical solutions for the structure and geometrical configuration of the driving wheels.

The advantage of using the mobile robots on circular wheels is that the contact between the wheel and the terrain is continuous giving a better stability of the rover. But they have the disadvantage of requiring a high-performance control system to maintain this contact even on vertical surfaces.

The disadvantage of using the mobile robots on noncircular wheels is that the ride is not so smooth, especially in stair climbing.

In the case of the last analyzed type of mobile robot on four wheeled legs, the stair climbing is even smoother due to variation of the length between the hip and the contact point with the rolling surface.

This analysis represents a base for a future work on design of a new type of rolling system including the mechanisms for driving a mobile robot.

References

1. Antonescu, O., Antonescu, P.: Mechanism and Machine Science — course book. Politehnica Press, Bucharest (2016)
2. Wang, Z., Gu, H.: A review of locomotion mechanisms of urban search and rescue robot. Indus. Robot: Int. J. **34**(5), 400–411 (2007)
3. Faruk Kececi, E.: Design and prototype of mobile robots for rescue operations. Robotics **27**, 729–737 (2009)
4. Lim, S.H., Teo J.: Recent advances on locomotion mechanisms of hybrid mobile robots. In: vol. 14, pp. 11–25 (2015)
5. Jakimovski, B., Horenz, M., Kotke, M., Maehle, E.: Design of a hybrid wheeled-legged robot — WheeHy. In: ISR/Robotik, pp. 1233–1238 (2010)
6. Nitulescu, M.: Mobile Robots (in Romanian). Sitech Press, Craiova (1999)
7. Ivanescu, M., Staretu, I., Nitulescu, M., Joni, N.: Robotics History in Romania (in Romanian). AGIR Press, Bucharest (2020)
8. https://rodel.snu.ac.kr/research/current-research/High-Speed-Mobile-Robot-with-Transformable-Wheels/. Accessed 10 Oct 2022
9. https://www.unusuallocomotion.com/pages/more-vehicles-iv/142-wheeled-robots-and-lunar-cars-4.html/. Accessed 10 Oct 2022
10. Smith, L.M., Quinn, R.D., Johnson, K.A., Tuck, W.R.: The tri-wheel: a novel wheel-leg mobility concept. In: Proceedings of the 2015 IEEE/RSJ International Conference on Intelligent Robots and Systems (IROS), pp. 4146–4152 (2015)
11. Gyarmati, M. Tatar, M.O.: Locomotion systems for search and rescue robots. Robotica & Management **24**(1), 8–13 (2019)
12. Antonescu, O., Robu (Nan), C., Antonescu, P.: Structure and analysis of linkages for quadruped bio-robot legs. In: Proceeding of the 26-th Working Meeting on Standardization of Terminology, Bucharest, Romania, pp. 25–30 (2016)
13. Tatar, M.O., Gyarmati, M.: Locomotion unit for mobile robots. In: IOP Conference Series: Materials Science and Engineering, vol. 444 (2018)
14. Moosavian, S., Ali, A., Semsarilar, H., Kalantari, A.: Design and manufacturing of a mobile rescue robot. In: IEEE/RSJ International Conference on Intelligent Robots and Systems (IROS), Beijing, China, pp. 3982–3987 (2006)
15. https://en.wikipedia.org/wiki/Sojourner_(rover)/. Accessed 10 Oct 2022
16. https://rsl.ethz.ch/. Accessed 10 Oct 2022
17. https://www.makeuseof.com/what-you-need-to-build-your-own-autonomous-robot/. Accessed 10 Oct 2022
18. Gyarmati, M. Tatar, M.O.: Mathematical modelling of a search and rescue robot with TriSTAR locomotion units. In: IOP Conference Series: Materials Science and Engineering, vol. 997, no. 1, p. 012077
19. https://www.pinterest.com/pin/113934040570446430 9/. Accessed 10 Oct 2022

Chapter 36
Kinematics and Design of a 12-DOF Quadrupedal Robot

D. G. Buleandră, C. Rusu, A. O. Băneasă, and R. C. Donca

Abstract The following paper aims to model the movement of a quadrupedal robotic platform, using the Denavit Hartenberg approach for direct kinematic analysis and a geometric approach for inverse kinematics. The final aim is to use these equations in order to control a robotic platform. The platform was designed and built using widely available materials and technologies.

Keywords Robotics · Legged-robotics · Inverse-kinematics · Paralel-platform

36.1 Introduction

The agility of quadrupeds represents the inspiration for engineers who want to design legged robots. However, one cannot simply copy the mechanical and neural control systems from animals to robotic models. Rather, properties must be selected that are critical to certain behaviors that the engineer wishes to capture with a particular robot. Most robotic vehicles have identical legs, rigid bodies, and rudimentary sensors for environment interaction [1].

Examples of possible applications of walking robots are in 3-D environments (Dirty, Doll, Dangerous), exploration of remote locations: in space or on other planets [2], at the bottom of the sea, in hostile or hazardous environments: nuclear power plants or in places with high levels of radiation, mining prospecting and exploration

D. G. Buleandră · C. Rusu · A. O. Băneasă · R. C. Donca (✉)
Department of Mechatronics and Machine Dynamics, Technical University of Cluj-Napoca, Cluj-Napoca, Romania
e-mail: radu.donca@mdm.utcluj.ro

D. G. Buleandră
e-mail: buleandra.fl.debora@student.utcluj.ro

C. Rusu
e-mail: calin.rusu@mdm.utcluj.ro

A. O. Băneasă
e-mail: Baneasa.Eu.Andrei@student.utcluj.ro

I. Doroftei et al. (eds.), *Proceedings of SYROM 2022 & ROBOTICS 2022*, Mechanisms and Machine Science 127, https://doi.org/10.1007/978-3-031-25655-4_36

[3], demining and unexploded munition recovery applications, natural disaster areas or other catastrophic situations, search and rescue operations, military operations.

This paper has the goal to present the kinematic analysis done in order to control the movement of a quadrupedal robotic platform. The general presentation of the state of the art is followed by an analysis of both the direct and inverse kinematic models. The last section highlights the design and construction of a quadrupedal robot using widely available materials and technologies. The control of the robot is obtained using the inverse kinematic equations.

36.2 State of the Art

We are going to mention in this section the current stage of development in the study of quadrupedal robots, along with the main projects in the field.

The *Boston Dynamics* team of researchers have continued to bring innovations in the field of legged robots. Using a 15 HP internal combustion engine to power hydraulic pumps, *BigDog* is a robot designed to traverse rough terrain by walking, running, climbing and capable of carrying heavy loads [4].

The *ANYbotics* team from Zurich, Switzerland, has developed another quadrupedal robot, with applications in inspection, supervision, search and rescue operations in industrial settings. The designed robot, *ANYmal*, can perform this type of work autonomously, capable of traversing difficult environments [5].

The final robot to be briefly mentioned is the MIT *Cheetah 3*. *Cheetah 3* uses a new leg design that includes proprioceptive action on abduction/adduction in addition to a greater range of motion at the hip and knee joints [6].

36.3 Kinematic Analysis

Analyzing the structure of a dog's skeleton, we observe the kinematic chain corresponding to a leg. After removing redundant kinematic joints, we can use three simple rotation joints for each leg. The total number of degrees of freedom is determined by using the following relation:

$$M = 6 * (n - 1) - \sum_{i=1}^{5} iC_i \tag{36.1}$$

where: n represents the number of elements, and C_i represents the number of joints of type i. In our case, the only joints used are rotation joints with a single degree of freedom. Figure 36.1 shows the kinematic chain of the robot.

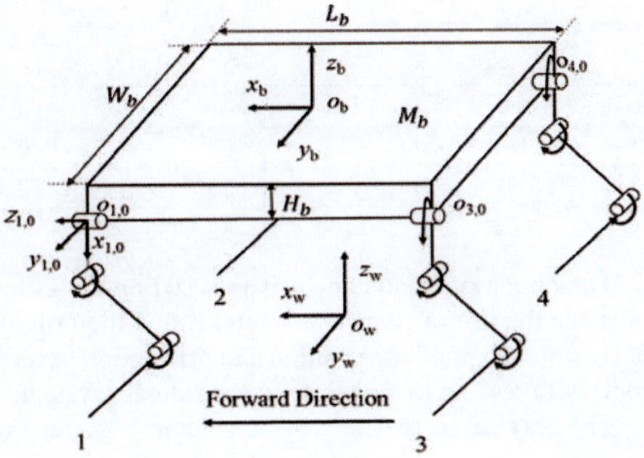

Fig. 36.1 The kinematic chain of the robot [7]

$$M = 6 * (12 - 1) - 5 * 12 = 6 * 11 - 5 * 12 = 66 - 60 = 6 \tag{36.2}$$

We can interpret this result as representing the ability of the robot to move its body in all three directions of the x, y and z axis and also to rotate around these axes. Some performed movements may be composed.

In order to generate the required trajectories, it is necessary to determine a mathematical model. This model is based on the notations from the Fig. 36.2.

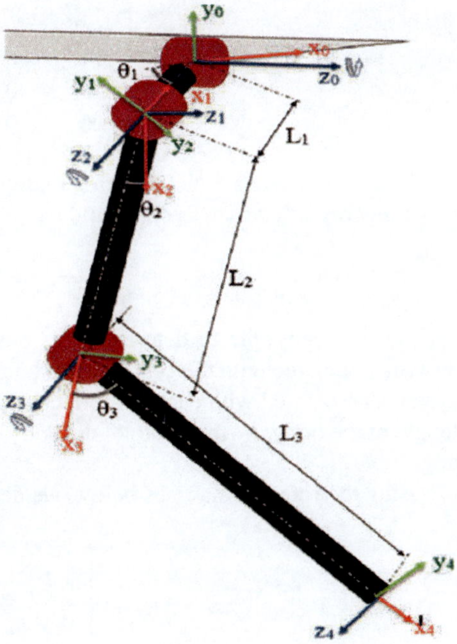

Fig. 36.2 The coordinate systems associated with the leg's joints [8]

Table 36.1 D-H parameters

	α	a	d	θ
0–1	0	L1	0	θ_1
1–2	0	L2	0	θ_2
2–3	0	L3	0	θ_3

The direct kinematic analysis is based on the Denavit-Hartenberg method. This involves the use of four parameters, called the D-H parameters presented in Table 36.1, which express the position and orientation of the final effector (in our case, the foot) with respect to the reference coordinate system.

The next step is to determine the elementary transformation matrices, as follows:

$$T_0^1 = \begin{bmatrix} cos\theta_1 & -sin\theta_1 & 0 & L_1 cos\theta_1 \\ sin\theta_1 & cos\theta_1 & 0 & L_1 sin\theta_1 \\ 0 & 0 & 1 & 0 \\ 0 & 0 & 0 & 1 \end{bmatrix} \tag{36.3}$$

$$T_2^3 = \begin{bmatrix} cos\theta_2 & -sin\theta_2 & 0 & L_2 cos\theta_2 \\ sin\theta_2 & cos\theta_2 & 0 & L_2 sin\theta_2 \\ 0 & 0 & 1 & 0 \\ 0 & 0 & 0 & 1 \end{bmatrix} \tag{36.4}$$

$$T_3^4 = \begin{bmatrix} cos\theta_3 & -sin\theta_3 & 0 & L_3 cos\theta_3 \\ sin\theta_3 & cos\theta_3 & 0 & L_3 sin\theta_3 \\ 0 & 0 & 1 & 0 \\ 0 & 0 & 0 & 1 \end{bmatrix} \tag{36.5}$$

The homogeneous transformation matrix results by multiplying the four elementary transformation matrices, as follows:

$$T_0^4 = T_0^1 * T_2^3 * T_3^4 \tag{36.6}$$

Thus it is possible to determine the position and orientation of the coordinate system associated with the center of the body x_m, y_m and z_m, knowing the coordinate systems associated with each end effector. We used the following notations: L—the length of the body, l—the width of the body, ϕ—yaw angle, Ψ—pitch angle, ω—roll angle.

Using the rotation matrices below, we determined the transformation matrix T_M:

$$R_x = \begin{bmatrix} 1 & 0 & 0 & 0 \\ 0 & cos\omega & -sin\omega & 0 \\ 0 & sin\omega & cos\omega & 0 \\ 0 & 0 & 0 & 1 \end{bmatrix} \tag{36.7}$$

$$R_y = \begin{bmatrix} cos\phi & 0 & sin\phi & 0 \\ 0 & 1 & 0 & 0 \\ -sin\phi & 0 & cos\phi & 0 \\ 0 & 0 & 0 & 1 \end{bmatrix} \quad (36.8)$$

$$R_z = \begin{bmatrix} cos\Psi & -sin\Psi & 0 & 0 \\ sin\Psi & cos\Psi & 0 & 0 \\ 0 & 0 & 1 & 0 \\ 0 & 0 & 0 & 1 \end{bmatrix} \quad (36.9)$$

$$T_M = R_x * R_y * R_z * \begin{bmatrix} 1 & 0 & 0 & x_m \\ 0 & 1 & 0 & y_m \\ 0 & 0 & 1 & z_m \\ 0 & 0 & 0 & 1 \end{bmatrix} \quad (36.10)$$

The kinematic equations that describe the relationship between the position and orientation of the center of the body and the positions and orientations of the first leg is given by the following transformation matrix:

$$T_1 = T_M * = \begin{bmatrix} cos(\frac{\pi}{2}) & 0 & sin(\frac{\pi}{2}) & L/2 \\ 0 & 1 & 0 & 0 \\ -sin(\frac{\pi}{2}) & 0 & cos(\frac{\pi}{2}) & l/2 \\ 0 & 0 & 0 & 1 \end{bmatrix} \quad (36.11)$$

The same approach can be used to determine the transformation matrices for the other three legs.

For the inverse kinematic analysis of the leg, we first considered a front view of it as presented in Fig. 36.3a. The hip joint and the thigh joint are coaxial, also the thigh and leg are coplanar.

In the following figures, we denote with l_f the length of the femur, l_t the length of the tibia and with h the distance between the body and the ground.

$$h_2 = \sqrt{h^2 + y^2} \quad (36.12)$$

$$\theta_3 = arctan\left(\frac{y}{h}\right) \quad (36.13)$$

Next, to determine the remaining two angles, we considered an inclined side view as shown in Fig. 36.3b. The inclination of the lateral plane is given by θ_3.

$$\theta_2 = arctg\left(\frac{\pm\sqrt{1 - a^2}}{a}\right) \quad (36.14)$$

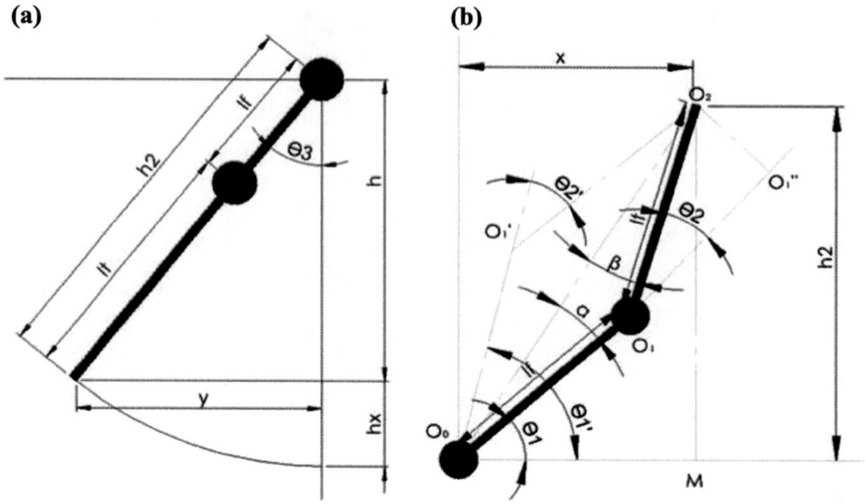

Fig. 36.3 **a** Front view of the leg, **b** tilted lateral view of the leg

The value of θ_1 is given by:

$$\theta_1 = arctg\left(\frac{h_2}{x}\right) - arctg\left(\frac{l_t sin\theta_2}{l_f + l_t cos\theta_2}\right) \tag{36.15}$$

For positioning in the correct configuration, we chose the value of θ_2 angle from the two obtained solutions:

$$\theta_2 = arccos\left(\frac{x^2 + h_2^2 - l_t^2 - l_f^2}{2l_t l_f}\right) \tag{36.16}$$

Knowing the x, y and h values we can determine the values of the three angles from the kinematic joints of each leg. However, even if the legs are identical, they are placed symmetrically. For this reason, the angles obtained from inverse kinematics are not directly translated to the real angles of the motors. We determined the actual angles of the motors. For example, the angles we used for the first leg are:

$$\varphi_1 = \theta_3 \tag{36.17}$$

$$\varphi_2 = 90 - \theta_1 \tag{36.18}$$

$$\varphi_3 = \theta_2 \tag{36.19}$$

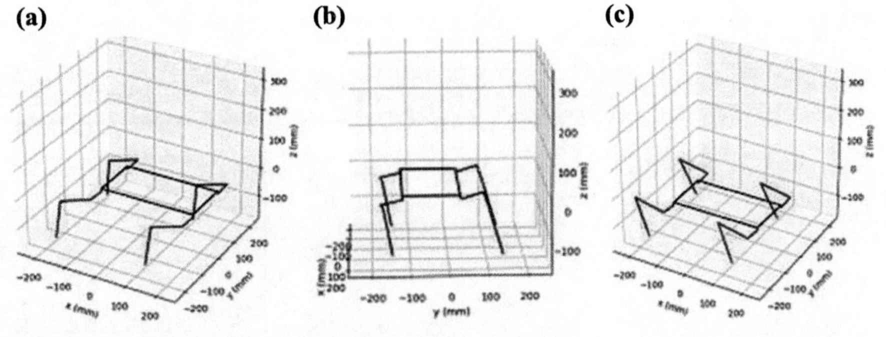

Fig. 36.4 Translations on **a** x axis, **b** y axis and **c** z axis [10]

Further, we analyzed the six movements that the center of the body can perform, with the aim of determining the movement performed by each leg separately.

The translations of the center of the body are relatively trivial, they can be obtained by increasing or decreasing one of the parameters *x, y* or *h,* as shown in Fig. 36.4.

The three elementary rotations present a greater difficulty because two of the parameters that determine the position of the legs must be calculated. The geometry of the movements are presented in Fig. 36.5.

The first of the elementary rotations addressed is the rotation around the z axis. We can visualize the trajectory of the foot as being described by an arc of a circle with the center in the center of the body.

We know the initial *x* and *y* parameters, as well as the rotation angle α. We want to determine the new x' and y' values. The positive sense of rotation will be considered the trigonometric sense. We obtained:

$$x' = d * sin\alpha_f - \frac{L}{2} \tag{36.20}$$

$$y' = d * cos\alpha_f - \frac{l}{2} \tag{36.21}$$

We used the same approach for the other two rotations.

36.4 Quadrupedal Robot Design

The mechanical design of the robot's components was carried out with the help of SolidWorks CAD design software. We chose 12 identical servo motors.

The main elements of the structure were made of plexiglass, by laser cutting. This technology allows the creation of complex and precise 2D shapes in a short period of time.

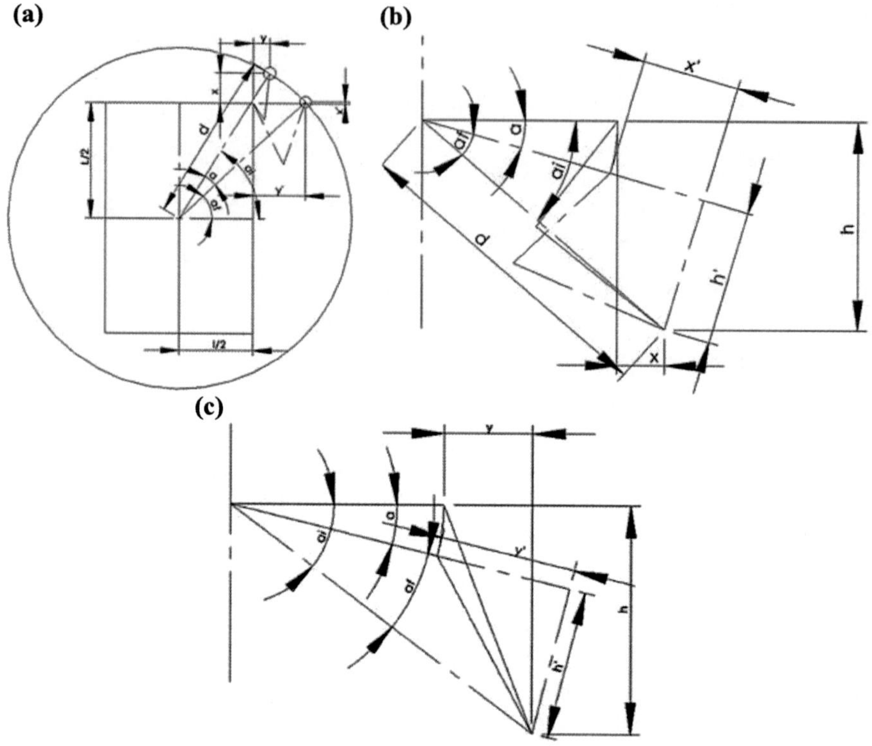

Fig. 36.5 **a** Rotation around the z axis (top view), **b** rotation around the y axis (side view), **c** rotation around the x axis (front view)

The motors that actuate the femurs are directly attached to the motors that actuate the robot's hips. A hemispherical element representing the foot is assembled at the end of the tibia. Therefore, the contact between the leg and the ground will always be in one point. This simplifies the kinematic analysis of the robot leg. Figure 36.6 shows the CAD model of the robot.

The algorithm we used is based on the inverse kinematic equations we obtained in the previous chapter. Figure 36.7 presents the final version of the robot.

36.5 Conclusions

Considering the large number of constructive solutions for walking with artificial legs we can conclude that, at the moment, these systems have some shortcomings. Among these are reduced autonomy and low processing power.

However, legged robots have many practical applications. Compared to wheeled robots, legged robots have the advantage of being able to move over rough terrain.

Fig. 36.6 The complete assembly of the robot

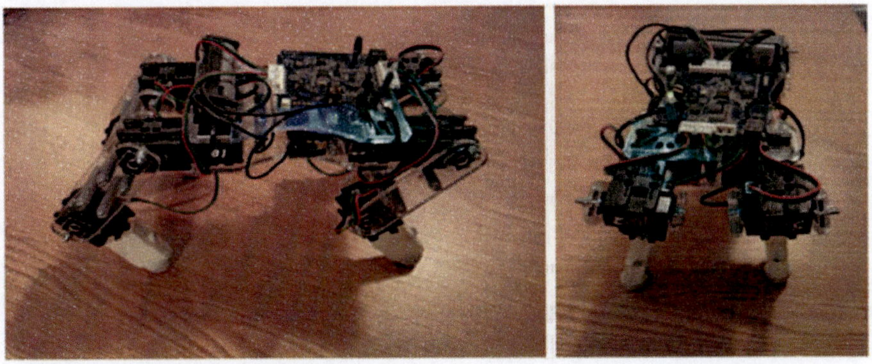

Fig. 36.7 The quadrupedal robot

They are much more adaptable to their environment, being able to avoid a wide range of obstacles.

We took inspiration from the leg of a dog. Analysing its structure, we chose a configuration that we designed and built. Adding the control elements, the result of these actions is a four-legged robot capable of walking.

References

1. Ritzmann, R.E., Quinn, R.D., Fischer, M.S.: Convergent evolution and locomotion through complex terrain by insects, vertebrates and robots (2004)
2. Kubota, T., Takahashi, H.: Micro walking robot design for planetary exploration. In: Proceedings CLAWAR'2003—6th International Conference on Climbing and Walking Robots; pp. 357–364 (2003)

3. Roberts, J.M., Corke, P.I., Winstanley, G.J.: Development of a 3500-tonne field robot. Int. J. Robot. Res. **18**(7), 739–752 (1999)
4. BigDog—The most advanced rough-terrain robot on earth. Boston Dyn. (2017)
5. Fankhauser, P., Hutter, M.: ANYmal: a unique quadruped robot conquering harsh environments (2018). https://doi.org/10.3929/ethz-b-000262484
6. Bledt, G, Powell, M.J., Katz, B., Di Carlo, J., Wensing, P.M., Kim, S.: MIT Cheetah 3: design and control of a robust, dynamic quadruped robot (2018)
7. Sun, W., Tian, X., Song, Y., Pang, B., Yuan, X., Xu, Q.: Balance control of a quadruped robot based on foot fall adjustment (2022)
8. Sen, M.A., Bakircioglu, V., Kalyoncu, M.: Inverse kinematic of a quadruped robot (2017)
9. Elarabawy, A.: 12-DOF quadrupedal robot: inverse kinematics (2020)
10. Sugahara, Y., Lim, H.-O., Hosobata, T., Mikuriya, Y., Sunazuka, H., Takanishi, A.: Realization of dynamic human-carrying walking by a biped locomotor. In: Proceedings of the 2004 IEEE International Conference on Robotics and Automation, pp. 3055–3060 (2004)

Part VIII
Other Robotic and Automation Areas

Chapter 37
Soft Robotics Modeling and Simulation

F.-A. Braşoveanu

Abstract Soft robots are systems designed to have mechanical properties similar to those of living tissues. To design and evaluate soft robotic structures, mechanical-oriented software environments are often used, software able to describe the design, material structure, and dynamics. In the control-oriented field, the design of the controller must be managed according to a specific model. With respect to this, the present research shows a procedure for the SoRoSim toolbox that can use both the kinematic and dynamic models of a classic soft robot and provide a realistic simulation environment. A comparison study centered on the dynamic performance, between the developed robot and others with similar designs/structures found in literature, was carried out, proving the SoRoSim toolbox capabilities in simulating pneumatic structures-based soft robots.

Keywords Soft robots · SoRoSim · PneuNet · Simulation

37.1 Introduction

Soft robots are a new way of building and controlling robotic actuators. The general advantages of this type of robot came from two considerations: safety of usage and speed vs mass. While rigid robotic systems are known for their energetic usage and massive bodies, soft robotic systems tend to be made lightweight and easy to actuate (using pressure, hydraulic oils and fluids, or even electricity—in small voltages and currents). The safety component is possible due to the characteristics of the building materials, such as soft polymers and rubber-silicone composites. The resulting robots are easily deformable and, during actuation, a surface or object does not tend to be deformed when in contact with them. Since the beginning of the'90s, the scientific world has received a first approach [5], even if not related to soft robotics, but to path-planning theory.

F.-A. Braşoveanu (✉)
"Gheorghe Asachi" Technical University of Iasi, Iasi, Romania
e-mail: florian-alexandru.brasoveanu@academic.tuiasi.ro

I. Doroftei et al. (eds.), *Proceedings of SYROM 2022 & ROBOTICS 2022*, Mechanisms and Machine Science 127, https://doi.org/10.1007/978-3-031-25655-4_37

In more recent papers, the area of soft actuators grew, varying the composition of materials and the mechanical structure. Most of the developed actuators are nature-inspired shapes and/or dynamics, aquatic animals [14], invertebrate and jelly-like bodies [2]. The obstacle is that, in aquatic environments, the density of water is providing an almost gravity-free space, and so the designed actuator has to be light to simulate accurately the underwater conditions.

One of the most popular designs of soft actuators is known as pneumatic networks (PneuNets) [11, 12, 16, 20]. As described in [7], the idea of this robot came from a precedent interpretation of the inflatable flexible microactuator (FMA) - one single chamber with various numbers of constraints. The evolution of the PneuNets involved design changes regarding dimensions, the material used, and mostly the shape of the composing chambers. The actuator is represented as a series of interconnected consecutive chambers inside a homogenous elastic material.

The PneuNet has 3 basic components: a superior layer containing the consecutive chambers and the interconnecting channel, a homogenous inferior layer and a stretch-resistant piece of material, thin enough to be placed inside the inferior layer. The last two elements form a passive constraint layer that ensures a known movement. The design of the superior layer, with separated chambers, ensures a rapid and uniform deformation along the actuator (Fig. 37.1 (left)).

The PneuNets are known today as having multiple types of designs. Around the idea of PneuNets, many research groups are using combinations of these actuators to build purpose-oriented complex actuators. For example, a star-fish design is presented in [7] or [2], other groups are using PneuNets to build muscle-enhancement devices in [6, 15], and even a design for a soft pneumatic hand [22]. In the last two examples, the controlled data is the pressure in the actuator to ensure a sturdy grip and a position according to the wanted one.

From the control perspective, recently, soft robotic systems control is mostly done using some basic principles (PID controller) or predictive control [1], while other approaches involve even neural networks and machine learning processes [4]. The controlled quantity can vary from force to position and liquid/air pressure.

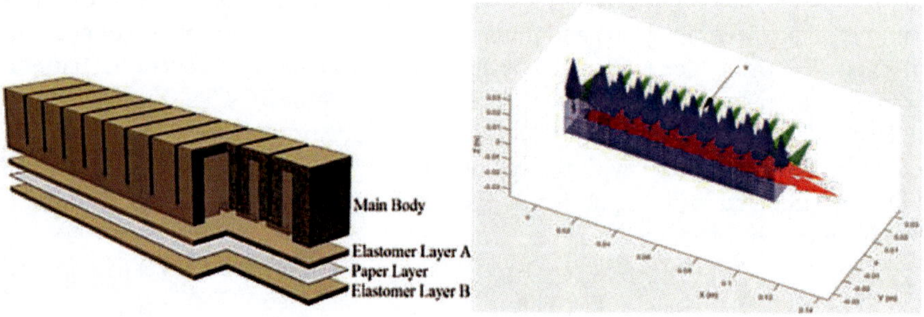

Fig. 37.1 Sectioned representation of the fast PneuNet [17] (left) and the structural form obtained with the toolbox (right)

The method used in the SoRoSim toolbox, which will be presented in the next section, is called Geometric Variable-strain (GVS). Other papers are presenting different approaches [3, 9, 19] one example being the Polynomial Curvature Case in [18] and with a variation in [8]. It is a method that involves an alternative version of modeling in a "sequence space of functions" that is equivalent to the special model and involves a configuration variable defined in the mentioned space.

The main contributions of our work are:

- model development for a pneumatic structure with a non-homogenous body (including channels and chambers)
- developing a pneumatical actuation mimic using string-based actuation.
- validation on using homogenous body simulation environments for non-homogenous body integration.

These contributions prove that soft-robotic models have similar behavior regardless of the actuation. Therefore, a general modeling method can be developed.

The paper is structured as followings: next, the mathematical preliminaries used for soft robots' modeling are discussed, also various types of actuators are presented, the main target being the one chosen to be analyzed in simulation. The third section presents the procedure developed for Fast PneuNets integration in SoRoSim. The fourth section presents the simulation results and the comparison of state-of-the-art alternatives. In the end, conclusions and future work are revealed.

37.2 Mathematical Preliminaries

The specific actuator considered for this study is the generic fast PneuNet having the chambers connected via a common channel, in which the chambers are separated in the main body. For the explored case, existing in literature, the actuation is done using the pneumatic input that determines the bending of the actuator around the constrained bottom wall (having incorporated a paper piece that blocks the extension of the wall).

The chosen fast PneuNet actuator will be reconstructed in a specialized toolbox for soft robots, called SoRoSim, developed for the MATLAB environment. As the toolbox is presented in [10], the mathematical description of soft robots is made using the GVS approach, a method generally used for modeling uniform body robots (with no chambers or/and channels, cuts, etc.).

As described in [10], the GVS has its origins in the modeling of rigid robots. The principal idea is that a continuous infinite DOF soft robot can be modeled in the same manner as the rigid case. This is done by discretizing the continuous body frame and finding a correlation between various movements and their position and orientation. The technique used to map a shape reconfiguration of the robot uses the Cosserat roads description theory. This is known as micropolar elasticity, which incorporates the elastic behavior (twisting, bending, etc.) and defines the dynamics of a continuous segment. From this theory, the authors in [10] propose the interpretation of a continuous soft link as being from the initial configuration, or the Cosserat road base, ($X_i = 0$) to the final configuration at the end of the road ($X_i = L_i$).

Starting from a rigid body transformation, where g_i is an element in the Special Euclidian group (SE(3)), the configuration of a soft or rigid body with respect to the predecessor is defined as a curve mapping the body-frame X_i, which also includes joint constraints. A specific g_i in Eq. 37.1 can be used to express the relation between a soft or rigid body segment and the previous chain of segments. For a certain configuration X_i, a transformation is in form Eq. 37.2.

For the exponential form of g, the partial derivatives must be introduced. An SE(3) element is defined with references in time and space, giving the two possible derivatives, where ξ_i is the strain twist that contains the angular and linear strains and η_i^r is the velocity twist that contains the angular and linear velocity of the previous body frame as in Eqs. 37.3 and 37.4.

$$g_i(\cdot) : X_i \in [0, L_i] \tag{37.1}$$

$$g_i(X_i) = \begin{pmatrix} R_i & r_i \\ 0 & 1 \end{pmatrix} \subseteq SE(3) \tag{37.2}$$

$$g_i'(X_i) = g_i \begin{pmatrix} \widetilde{k_i} & p_i \\ 0 & 0 \end{pmatrix} = g_i \widehat{\xi_i} \tag{37.3}$$

$$\dot{g}_i(X_i) = g_i \begin{pmatrix} \widetilde{w_i^r} & v_i^r \\ 0 & 0 \end{pmatrix} = g_i \widehat{\eta_i^r} \tag{37.4}$$

In this manner, the product of exponentials describing the configuration of link i for the spatial frame can be developed using the Magnus expansion of $\widehat{\xi_i}$ at $X_i = L_i$, an integral series of Lie Brackets of $\widehat{\xi_i}(X_i)$.. Determining a relation between the time derivative of the strain twist and the links velocity needs to consider the equality of the mixed partial derivatives and the usage of the adjoint operator and adjoint map of SE(3) (see [10] for the detailed equations).

Following, the dynamics based on forces action is considered. The basis of the mathematical form is found in the d'Alembert principle and it is including in equations the screw inertia density matrix of the cross-section $\overline{M}_i(X_I) \in R^{6x6}$ and the boundary conditions in the local frame. $\overline{M}_i$ is also containing the mass density, the cross-section area and the second moment of area J_i. . These mathematical expressions are also considering the external load and the internal wrenches (F_{e_I} and respectively F_{i_i} and F_{a_i}). From the soft robotics perspective, the following forms of the classic dynamic matrices (Eqs. 37.5, 37.6, 37.7, 37.8, 37.9, 37.10) are the soft dynamic behavior.

$$M(q) = \sum_{i=1}^{N} \int_0^{L_i} J_i^T \overline{M} J_i dX_I \tag{37.5}$$

$$C(q,\dot{q}) = \sum_{i=1}^{N} \int_{0}^{L_i} J_i^T \left(ad_{\eta i}^{*} \overline{M} J_i + \overline{M} \dot{J}_i\right) dX_I \quad (37.6)$$

$$D = \underset{i=0}{\overset{N}{diag}} \left(\int_{0}^{L_i} B_{\eta i}^T \Upsilon_i B_{\xi i} dX_i \right) \quad (37.7)$$

$$K = \underset{i=0}{\overset{N}{diag}} \left(\int_{0}^{L_i} B_{\eta i}^T \Sigma_i B_{\xi i} dX_i \right) \quad (37.8)$$

$$B(q) = \underset{i=0}{\overset{N}{diag}} \left({}^{i}S_i^T \, or \int_{0}^{L_i} B_{\eta i}^T B_{\tau i} dX_i \right) \quad (37.9)$$

$$F(q,\dot{q}) = \sum_{i=1}^{N} \int_{0}^{L_i} J_i^T \overline{F}_{e_i} dX_I \quad (37.10)$$

37.3 Fast PneuNet Model Integration in SoRoSim

The considered PneuNet is a structure with 11 chambers interconnected via a common channel. The chambers are separated by intermediary transversal cuts. The paper layer represents the bending resistive layer (Fig. 37.1 (left)). To develop a similar structure, the toolbox requires specific physical characteristics. For the dimensional design, the data of the robot can be found in [13]. These data elements are collected from [1], the specific dimension data given for their simulation is given in Table 37.1 (first half). For completing the physical specification, the considered soft robot must be described with the composing material specifications. As collected from the literature, various research groups used various materials for fabrication, giving a rich number of material specifications. For the presented simulation and development, the physical data of the considered material is presented in Table 37.1 (second half).

In the toolbox, the procedure to create a dynamic structure is developed to offer the possibility of creating rigid, soft or mixed structures. The procedure involves creating separate links (segments) that can have different geometrical and mechanical/material characteristics. With the created segments, one or more linkages can be assembled and configured to actuate and their interaction can be established.

For the development of a simulated structure similar to the fast PneuNet from Fig. 37.1 (left), the SoRoSim toolbox provides an interactive graphic interface. In the development process, a number of iterations were carried out: one using a multi-link robot design, each link representing a chamber of the robot and a link corresponding

Table 37.1 Soft robot physical dimensions from [13] and material characteristics

Physical dimensions		Material characteristics	
Data measurement	Value (mm)	Data measurement	Value
PneuNet length	103	Young modulus	1,000,000 Pa
Rigid layer length	110	Sheer modulus	3.33 $*10^5$ Pa
Chamber interior dimensions (L ǀl ǀh)	3 ǀ 14 ǀ 13	Poison ratio	0.5
Chamber exterior dimensions (L ǀl ǀh)	8 ǀ 20 ǀ 15	Robot density	500 $\div$ 1000 kg/m^3
Common channel dimensions (L ǀl ǀh)	65 ǀ 2 ǀ 2	Material damping	0.112$*10^5$ Pa·s
Inter-chambers distance	6		
Wall thickness (top ǀ lateral)	2 ǀ 3		
Rigid layer thickness	5		

to the common channel, connected in a linkage with one actuation; another idea was to develop a similar multi-link linkage with consecutive chambers representing different consecutive segments interconnected. For these two ideas, the weak point was the actuation relation, which was only between the robot origin that models a rigid connection to a virtual wall, and the final link, the other links being unactuated and having their dynamics given only by the material proprieties.

Finally, the approach that gives the wanted behavior of the simulation was to develop a single link with interior segments, similar to the ideas described before, the difference being the actuation done between consecutive segments of the link that gives the linkage.

In the toolbox pop-up interface, we can decide the type of link constructed, namely a soft one. The material characteristics (Table 37.1 second half) are then introduced, to describe the elastic characteristics of the robot body according to the mathematical model data individually for each link specified, 21 links in this case.

The next step is done by introducing the dimensional data and the positioning that can be customized in space. All 21 segments have to be individually dimensioned and positioned relative to the previous segment. Once the link is created, the linkage is generated using a similar interface. Starting with the singular link, the linkage must be developed using an actuated structure with 21 actuators. Each actuator is placed between two divisions to fully actuate the entire robot body.

The link and linkage respectively can be altered lately to provide different behavior by modifying and/or eliminating some physical properties, such as gravity, damping and actuator position.

One important aspect is that the simulator does not offer the possibility to develop a dedicated pneumatic actuation and the designed robot does not have internal chambers. The effect might influence the stiffness factor or damping factor. That is why these factors have been slightly altered to simulate the overall properties of the real soft actuator presented in the literature.

37.4 Simulation Results

Based on the described actuator, the general structure obtained is presented in Fig. 37.1 (right). It is important to note that the gravitational axis is y, z being the axis around which the movement is executed and analyzed.

The analyzed dynamics are considered in two different conditions, with or without gravitational force. The toolbox also allows a static analysis of the system. The result of this analysis is a final stance for the desired configuration input. Once the initial force is introduced, the robot starts bending on the direction of the rigid layer.

SoRoSim toolbox allows the implementation of different types of inputs to our system. Depending on the target, our dynamics can differ. For the first classical input analysis, the robot was considered in a non-gravitational space and with no object of interaction. Note that the actuation is not of the pneumatic type, but it is using the above-described method (force on the commanded string).

Initially, the analysis was done when the actuator was not actioned. The measuring unit for our application is Newton (N). The PneuNet is simulated with 0 N force given as input and without the presence of gravity; in this case, the robot is not moving. Introducing the gravity factor, the obtained behavior in a 3 s time span is presented in Fig. 37.2 (left).

A generic step function can be generated in the form of a constant value. The simulation was done considering the presence of gravity, for a 0.1 N input on each actuator, over a one-second time span. The PneuNet is designed to have a fast dynamic. For the first 0.16 s, the pose of the robot is presented in Fig. 37.2 (center). The motion over this short time interval shows a dislocation on the free end of the actuator. The fast response is due to the gravitational influence added to the actuation force. After the complete simulation is done, the final pose of the actuator is presented in Fig. 37.2 (right). The simulation actuated all the components, resulting in a circular configuration. The actuated segments present also resistance to gravity and, in case of physical interaction with an object in space, the force will try to dislocate the position of that object.

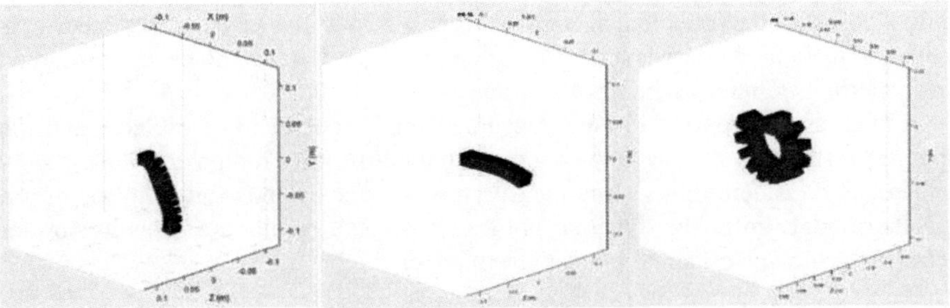

Fig. 37.2 Effect of the gravitational force on an unactuated PneuNet for a 3 s time span (left), PneuNet pose after 0.16 s of actuation at 0.1 N on each actuator (center), Final pose after 1-s actuation at 0.1 N on each internal actuator (right)

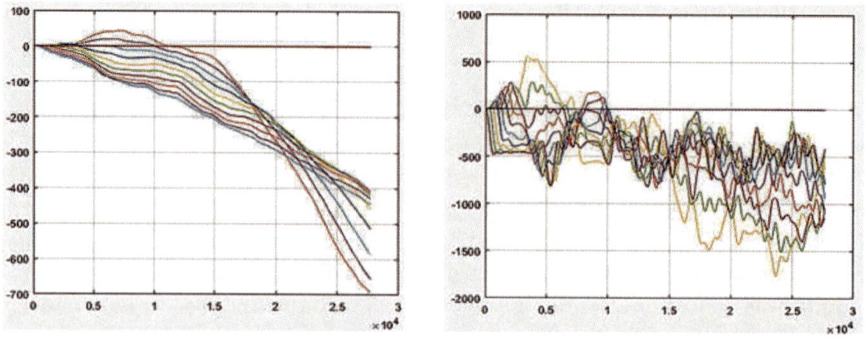

Fig. 37.3 The variation of the joint coordinates (left), Derivative of the joint coordinates (right)

The presented simulation results are to be compared with those found in literature, both theoretical/simulation results and that obtained by the physical implementation. Note that, as mentioned, the paper does not intend to mimic the results in the literature, but to obtain similar behaviors. The purpose is to propose a way for designing and simulating PneuNets in the SoRoSim toolbox.

As a more control-oriented remark, the toolbox is using the mathematical procedures presented in the previous sections to calculate the joint coordinates and their derivatives. The joint coordinates variation, respectively the derivatives of these coordinates are presented in Fig. 37.3. The data is collected for the actuation in Fig. 37.2.

Recalling the experiments from the literature, the simulations are realized in specific environments, specialized in physical proprieties and material characteristics, and in this manner describe more accurately the elastic behavior. In [3], one of the models is approximating our considered PneuNet; it presents a type of dynamic that can be used for our comparison. A similar configuration of a PneuNet is shown, with a similar response to input pressure as in Fig. 37.3 (right).

In [12], the PneuNet starts from a different pose, and the gravitational field is used in opposition to the movement. The results for different pressure inputs determine similar poses for the robot like the one in this paper. In [20], a physical 3D PneuNet is built, and its behavior at constant pressure input is presented. The bending returned on actuation is similar to the results of this paper.

In [11], as well as in [15], the fast PneuNet is specifically simulated and the obtained results are similar to the ones got in the SoRoSim toolbox. In [21], a slow actuator is considered and some measurements offer the possibility to identify a specific model. From the dynamic point of view, the results are showing similar behavior to the one obtained in the present paper.

37.5 Conclusions

In the presented study, the SoRoSim toolbox has been used in order to implement a PneuNet, a soft robot well-known in the field. The focus of this paper was on dynamic behavior and robot type compatibility as well as maintaining an accurate representation of the physical proprieties.

The methods and considerations used to overcome the problems in development were specifically done to maintain the natural model of the robot and describe an identical structure in the simulation environment. This was accomplished without limiting the versatility of the simulation or denaturizing the proprieties of the robot. We presented a method to mimic the pneumatic actuation using the model of a string one, with consideration of the actual behavior of a pressure input on the PneuNet's chambers, the elasticity and eventual resistance.

Finally, we proved that a typical pneumatic actuator, with a wide range of applications and specific body proprieties and restrictions, can be simulated using an environment nonspecialized on pneumatic actuation.

This study provides new possibilities for developing various soft structures and opens a control possibility. Ideas can vary from simple actuation and positioning of the robot on a simple circular trajectory to different types of actuation. According to the presented simulation method, controlling the robot with precision can invoke various control strategies. It can open ways even to unknown control theories.

References

1. Bruder, D., Gillespie, B., Remy, C.D., Vasudevan, R.: Modeling and control of soft robots using the Koopman operator and model predictive control (2019). arXiv: abs/1902.02827
2. Choi, W., Mazzeo, A.D.: Multigait soft robot. Proc. Nat. Acad. Sci. **108**(51), 20400–20403 (2011)
3. Fang, G. Matte, C.-D., et al.: Kinematics of soft robots by geometric computing. IEEE Trans. Robot. **36**(4), 1272–1286 (2020)
4. Gillespie, M.T, Best, C.M. et al.: Learning nonlinear dynamic models of soft robots for model predictive control with neural networks. In: IEEE ICSR (RoboSoft), pp. 39–45, Livorno (2018)
5. Gregory, J.W.B., Chirikjian, S.: An obstacle avoidance algorithm for hyper-redundant manipulators. In: IEEE ICRA, vol. 1, pp. 625–631 (1990)
6. Heung, K.H., Tong, R.K., Lau, A.T. & Li, Z.: Robotic glove with soft-elastic composite actuators for assisting activities of daily living. Soft Robot. **6**(2), 289–304 (2019)
7. Ilievski, F., Mazzeo, A.D. et al.: Soft robotics for chemists. In: Angew. Chem. Int. Ed., No. 50, pp. 1890–1895 (2011)
8. Katzschmann, R.K., Santina, C.D. et al.: Dynamic motion control of multi-segment soft robots using piecewise constant curvature matched with an augmented rigid body model. In: 2nd IEEE ICSR (RoboSoft), pp. 454–461. Seoul (2019)
9. Li, W.-Y., Nabae, H., Endo, G., Suzumori, K.: New soft robot hand configuration with combined biotensegrity and thin artificial muscle. IEEE RA-L **5**(3), 4345–4351 (2020)
10. Mathew, A.T., Hmida, I.B. et al.: SoRoSim: a MATLAB® toolbox for soft robotics based on the geometric variable-strain approach. In: IEEE Robotics & Automation Magazine, pp. 1–45 (2021). arxiv: 2107.05494

11. Mosadebgh, B., Panagiotis, P. et al.: Pneumatic networks for soft robotics that actuate rapidly. Adv. Funct. Mater. **24**(12), 2163–2170 (2014)
12. Ogura, K., Wakimoto, S., Suzumori, K., Nishioka, Y.: Micro pneumatic curling actuator - nematode actuator. In: IEEE ICRB, pp. 462–467. Bangkok (2008)
13. Panagiotis Polygerinos, B.M.A.C.: Soft robotics toolkit (2013). https://softroboticstoolkit.com/book/pneunets-design. [Accessed February 2022]
14. Pinskier, J., Howard, D.: From bioinspiration to computer generation: developments in autonomous soft robot design. Adv. Intell. Syst. **4**(1), 1–20 (2022)
15. Polygerinos, P. et al.: Towards a soft pneumatic glove for hand rehabilitation. In: IEEE International Conference on Intelligent Robots and Systems, pp. 1512–1517. Tokyo (2013)
16. Rus, D, Tolley, M.: Design, fabrication and control of soft robots. In: Nature Publishing Group, vol.. 521, pp. 467–475 (2015)
17. Santina, C.D., Katzschmann, R.K., Bicchi, A., Rus, D.: Dynamic control of soft robots interacting with the environment. In: IEEE ICSR (RoboSoft), pp. 46–53. Livorno (2018)
18. Santina, C.D., Rus, D.: Control oriented modeling of soft robots: the polynomial curvature case. IEEE Robot. Autom. Lett. **5**(2), 290–298 (2020)
19. Shiva, A., Sadati, S. et al.: Elasticity versus hyperelasticity considerations in quasi-static modelling of a soft finger-like robotic appendage for real-time position and force estimation. Soft Robot. **6**(2), 228–249 (2019)
20. Souhail, A., Vessakosol, P.: PneuNets bending actuator design and fabrication using low cost silicones. In: The 9th TSME ICoME, No. December, pp. 1–7 (2018)
21. Sun, Y., Song, Y.S., Paik, J.: Characterization of silicone rubber based soft pneumatic actuators. In: IEEE IROS, pp. 4446–4453. Tokyo (2013)
22. Zhang, Y., Xu, Z., Pan, L., Bao, G.: Research on humanoid pneumatic soft robot hand. In: Proceedings of the 2018, IEEE ICRB (ROBIO), pp. 355–360. Kuala Lumpur (2018)

Chapter 38
Smart City Concepts and Urban Service Robots

Y. K. Jawad and M. Nitulescu

Abstract In recent years, the new concepts of smart cities (SC) and urban service robots (USR) have become more important in our daily lives. In SC, information is gathered from a diversity of sensors placed all over the place, creating centralised data hubs, which are then used to make all the city systems work better, giving better services to the citizens. Integrating the big steps conquered in computer vision and artificial intelligence, the USR systems have gone from being able to do only simple tasks to more complex ones, like helping hotel concierges, museum guides, waiters in cafes and restaurants, home assistants, drones that deliver packages on their own, and more. As digital agents, all USRs can access the SC data to get the information they need to succeed in a specific task. On the other hand, the SC can also use the USR agents as mobile sensors and actuators. Traffic jams and the lack of parking solutions are problems in most big cities around the world, which can make driving stressful and a waste of time. We propose here a conceptual solution to solve the parking problem based on the intelligent management of parking lots assured by a USR system and a blockchain architecture type.

Keywords Smart city · Urban service robotics · Artificial intelligence · Intelligent parking solutions

38.1 Introduction

Information and communications technology (ICT) systems and the many other recent technical innovations offer new ways to run our cities better and more efficiently in many aspects and to transform them finally toward the new concept of "smart cities" (SC). The actual strategies implemented for SC are a good way to learn about the future of smart development. These days, cities are becoming more

Y. K. Jawad (✉) · M. Nitulescu
University of Craiova, Craiova, Romania
e-mail: yahya.joad@gmail.com; jawad.yahya.q7n@student.ucv.ro

M. Nitulescu
e-mail: nitulescu@robotics.ucv.ro

I. Doroftei et al. (eds.), *Proceedings of SYROM 2022 & ROBOTICS 2022*, Mechanisms and Machine Science 127, https://doi.org/10.1007/978-3-031-25655-4_38

and more dependent on technologies like high-speed internet, 5G mobile networks, the Internet of Things (IoT), and big data [2]. Artificial intelligence (AI) and urban service robots (USR), mobile or not, are now currently used in many applications, and they are undoubtedly necessary support for increasing the level of intelligence and automation. Digital communication and other cutting-edge infrastructure technologies change many parts of social life, such as how people work and act, what they produce and consume, and how they interact with each other. These USR machines have become a part of our everyday lives, and recent technological advances, such as better vision components and associated software technologies, more accurate localisation in space, optimised energy consumption, and better machine learning, have made it possible to move from the closed environment of a house to the more open and complex environment of a city for service robotics [10]. It is clear that AI and USR are being used in SC development more in the Global North right now [3], but more and more countries in the Global South are also getting involved. Several cities in Europe, America, Australia, and Asia are working to make AI and robotics better in their cities. To remember that one of the first service robots for the home came into our lives almost 20 years ago. There are a lot of different USRs out and about in our cities right now. People are especially interested in automating logistics (which includes both the transportation of people and goods), transportation (traffic control and advanced driver assistance systems), public safety and surveillance (facial people recognition), manufacturing (process control), and online retail. All of this makes AI very important for building SC. Self-driving platforms for customised public transportation have been tested in many places around the world. Self-driving ground delivery robots are now already in use in several cities and college campuses, and companies are trying to get flying robots for fast deliveries off the ground [4]. Cities also have put-up security robots that can work on their own. They do the same thing as security cameras, but they also have a way for people to get in touch with emergency services. The ongoing healthcare crisis has also shown that robots can help public officials manage and contain emergencies. Some examples are robots that clean large public spaces and robots that remind people about rules for their health care [6].

Nowadays, city dwellers frequently "fight" to obtain a parking location that suits their needs and preferences. In this paper, we will address the development of an effective parking solution as a USR type system, which contributes significantly to time savings, energy conservation, security increase, pollution decrease, and less traffic in an SC. Therefore, this strategy aids in both achieving smart traffic and helping vehicles find the best parking spots. By utilising passing sensors and new adaptable payment methods, smart parking is one of the breakthroughs that offer simple-to-use parking services to urban commuters. Actual technology can offer individualised parking facility information, such as parking availability, cost, distance, etc. So, it is possible to solve this both difficult and delicate issue by combining the knowledge about all of the SC parking spaces available into a single platform (blockchain) and creating a so-called management of the USR parking lots in the entire SC.

38.2 Blockchain

Blockchain is a peer-to-peer distributed ledger technology that keeps track of sales, contracts, and other business transactions. Blockchain technology, which was initially designed to enable transactions, may be used directly for any kind of transaction. The advantage of blockchain is that an attacker needs to control 51% of the systems in order to outperform the target network's hashing power. As a result, launching an attack against the blockchain network is computationally impractical. The workings of blockchain technology are shown in the example below. This strategy can ensure ownership and control over the data streams the devices produce. Ownership of data is crucial. Additionally, the blockchain can improve security in the SC and IoT network's device communication. Consider A and B as two separate entities in a parking system powered by blockchain. A is paying a parking fee to B, the parking authority. This transaction is shown online as a block that contains details about it, including the block number, proof of work, previous block, and transaction records. This block is broadcast to all network nodes. The block is verified by the other entities, and the transaction is validated and added to the chain if the block is approved by more than 50% of the entities. Following that, entity A transfers the fee to entity B's account. These are some of the features.

Control and Authority. You will see how authority functions if we compare blockchain with databases. While databases are always centralised, the blockchain is designed to operate in a decentralised manner. Blockchain has the advantage it needs to advance into the next technology generation because of this distinctive trait. It eliminates the need for centralised control and enables networks to operate autonomously.

Architecturally. Uses a network of distributed ledgers as its architecture. Each peer on the peer-to-peer network can connect to another peer using safe cryptographic methods. Since there is no central node, the consensus algorithm can be worked on by all nodes simultaneously. An architecture of client and server underpins a database. It is an extremely effective architecture that can operate in both small- and large-scale settings. In this case, the client serves as the receiver, and the server functions as the centralised processing hub. A secure connection is used to keep clients and servers in constant communication.

Data management and immutability. Data may be easily saved and accessed in a database. Because databases lack immutability, which is necessary for ensuring the effective operation of the program, they are more vulnerable to manipulation by rogue administrators or outside hackers. Even though it operates differently when it comes to data storage, blockchain supports immutability, which implies that once written, data cannot be changed or wiped. Since the network is immutable, no data manipulation is conceivable.

38.3 Innovations

Parking lifts are an idea that is rapidly becoming more popular. It is planned to add an application (using Java or Python) like Waze on smart devices linked with parking lifts to facilitate obtaining parking spaces by specifying the number of vacant places and the speed of access to parking lifts in SC. The use is not limited to vehicles and bicycles. With the development of the era of electric vehicles, "Electric vehicle sales may increase from 2% of the global share in 2016 to 30% by 2030" [12]. An inevitable result of the increase in the number of electric vehicles shows the need to increase the locations of charging stations for cars and electric bikes inside parking lifts spaces to reduce the waiting time for charging vehicles. An electric vehicle can be charged in as little as 30 min or as long as 12 h. In order to make the application available, it is possibly linked to cloud computing.

38.4 Concepts

In this section, we will introduce some basic concepts necessary to develop a proposed general architecture [5] for the management of SC parking lots using a USR solution.

Management of the USR parking lots (MUSRP). The SC has parking lots that are each run by an automated ticketing machine, or (USR), at the entrance and exit of the lot. Drivers get a ticket with a QR code and a timestamp when they use this USR. The system can also be used for surveillance, which is done by taking a picture of the vehicle to find out its make, model, colour, and vehicle number. So, this system can work to collect data, information, and images. All this information is also stored on a server, and the right algorithms are used to analyse and evaluate these [7]. If the vehicle stays longer than the free period, the payment system will be set up through MUSRP. The primary objective of the integrated smart parking system that we suggest is to provide a single connection between all parking service providers' platforms in the SC. While most of the current solutions focus on the centralisation of the smart parking issue, transparency is produced by our system, and it is primarily due to the decentralised infrastructure, which is more invulnerable.

Transportation and vehicular traffic. In SC, traffic information and data can be analysed to help both city officials and residents. For example, residents can figure out how long a trip will take based on this data. There are already TV camera-based traffic monitoring systems in place in a lot of cities. Generally, deep integration of IoT for SC means using different sensor categories, like the GPS systems on vehicles, the sensors that measure air quality, the sensors for detecting noise levels on certain streets, etc. For the best online traffic control solutions, such as automatic traffic light control or sending human traffic agents to the key spots for re-planning congested routes [9], there is a need for a permanent, strong communication infrastructure in the SC.

38.5 Principles, Architecture and Solutions Implemented

We now present the general architecture and principles used for the suggested integrated auto parking system, based on blockchain [11, 12], as a part of the application of the USR concept in an SC. Our proposed solution (Fig. 38.1) has three parts: a parking service provider, a blockchain network (MUSRP), and a user. Each part has a unique application interface for interfacing with the integrated smart parking system.

Assume that the city offers a variety of smart car parking options from several car parking service providers. For the sake of simplicity, assume each smart parking lot is managed by a single service provider. A blockchain-based integrated smart parking system is connected to every parking space.

In the system, there can be two different kinds of transactions. First, data outputs from the parking sensors. Assume that every parking space in a smart parking lot is fitted with an IoT device (such as a parking sensor or a TV camera) that may indicate the availability of parking as a transaction. To generate a transaction, each automobile parking service provider has a smart contract and internal control (decrease/increased number of places). The associated IoT device creates a transaction when a parking space changes from "empty" to "occupied". Like this, when the parking spot is changed from "occupied" to "empty," the IoT device generates a new transaction. The MUSRP processing server receives the transaction first. To verify this transaction, the MUSRP transmits responses 1 and 2 to the control system, the user interface,

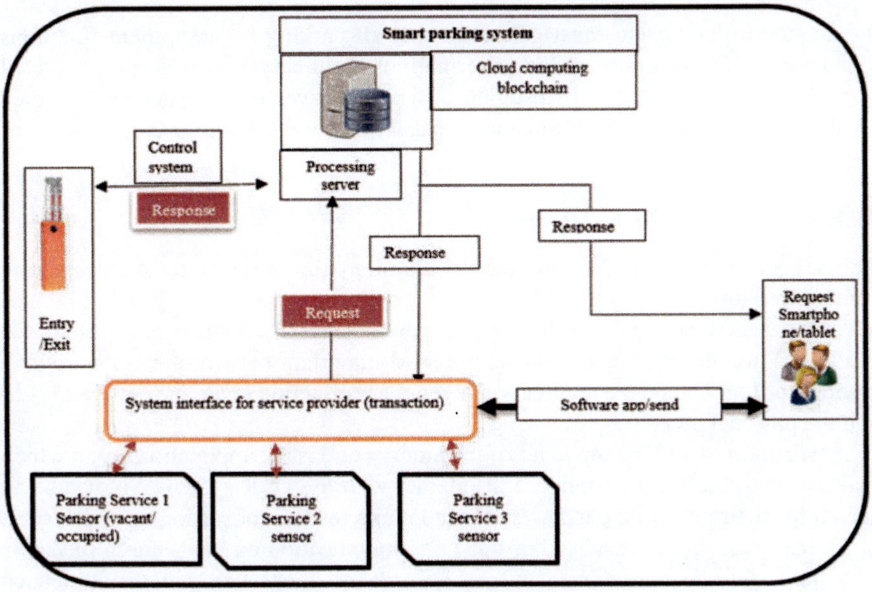

Fig. 38.1 The block diagram of the general proposed architecture

and the blockchain network. Secondly, it provides information about parking costs. Assume that the service of parking is determined by the amount of time. Smart contracts are used to set parking rates. The parking price smart contract is moved to the blockchain network. A transaction is produced each time when the parking price is adjusted dynamically based on the current parking time used. The blockchain network receives the transaction for verification. Next, the blockchain network uses a consensus process to verify the transaction.

A. USR Management Software

This USR software tries to make the parking procedure simpler for everyone by enabling communication between all parties and the ability to request assistance from the system administrator when needed. The steps of collecting, processing, and consulting are all crucial. After gathering parking-related data on the processing server level (such as licence plates, exit and entry times, congested areas, etc.), this data is stored as a database on the server level to be used next to the Web server level. This level interacts with the system administrator and drivers via a Web application to present the parking offer in real-time and provide possible other private services. The following elements are the foundation for this architecture:

- The users interact with the online application; after logging in or registering, they may perform a variety of tasks like monitoring the parking lot's status (to see whether a spot is available), booking it in advance, or submitting a complaint.
- The system administrator interface must make sure that the application's data is updated and has visibility into the entire system; he also handles users' special demands.
- The controller system must guarantee that all parking lot equipment functions flawlessly, that no uncontrolled abnormalities occur, and that no alert is generated without being thoroughly investigated. The users have one thing in common: they all take advantage of the application's capabilities while having varying levels of access permission.

B. Multilayers USR Architecture

This section presents how our proposed technology for an integrated smart parking solution is built up in layers. The layered architecture that was suggested defines and standardises the typical architecture of smart parking systems that are integrated designs. Also, the architecture shows the most important parts of the system and is made up of four layers: the application layer, the transaction layer, the network layer, and the physical layer.

Application layer. The top layer of the architecture is the application layer, which makes it possible for a participant to interact with the USR system. So, users can search for their preferred parking spots and make reservations using an application on their mobile phone (Android or iOS) or an application on the Web. In the same way, parking service providers can send information about their own services, such as the availability of parking spaces and special deals. The integrated MUSRP system suggests a good place to park based on what the user wants and what's available.

Since users interact directly with the integrated MUSRP system, this layer gives the end users the final parking service.

Network layer. The network layer makes sure that users, parking centres, and the integrated USR system can all interact with each other. The information from users and from the parking lots will be sent through this layer, and the system will work. On this layer, different ways to communicate are implemented, such as Local Area Network (LAN) and Wide Area Network (WAN), which will be used by drivers, parking system providers and other IoT devices (sensors for parking, security cameras, etc.). This layer includes several wireless communication technologies that are already in use, such as Bluetooth, Wi-Fi, GSM, 4G and 5G networking communications etc. Also, this layer assures data security and possible future growth for the USR system [5].

Transaction layer. It is the way to make sure that all participants (both business and technical) in each environment get the same result. It is responsible for decentralisation. The network layer must first concur on a transaction's legitimacy before settling it. If the system has a lot of users, the agreement could take some time. As a result, we can demonstrate that when security criteria are the same, scalability is inversely proportional to decentralisation.

Assuming that the degree of decentralisation of two proof-of-work blockchains is the same and defining security as the hash rate of the blockchain. Scalability increases as security increases, and confirmation time lowers as the hash rate increases. Scalability and security, therefore, increase with ongoing decentralisation.

Physical layer. Also called layer number one, this oversees sending bits from one device to another. It also lets the electrical or mechanical interfaces connect to the physical medium so that devices can communicate by a synchronised strategy. This layer handles most of the network's physical connections, such as wireless transmission, cabling standards and types, connectors, network interface cards accepted, and more, depending on the network's particularities.

38.6 Conclusions

Developing the SC concept needs to figure out how to take advantage of new ideas and technologies like IoT, which can add important features to any new SC project. Also, the USR concept can add specific features to any SC project, including modern parking systems and their optimal management proposed here.

The goal of this paper was to develop and describe a general model that can be integrated into each SC project. Then, developers and manufacturers of future autonomous vehicles will have to adopt this model and its requirements.

This paper proposes an architecture for a USR management parking system in SC, a big and strong problem nowadays in the majority of important cities all over the world. The developed architecture is, in fact, a blockchain-based infrastructure and information system to offer smart parking solutions for drivers. So, for smart parking, we propose here a layered integrated architecture that assures data integrity, trust among users and an open structure for possible developments and evolutions. For emphasis, the new idea of employing parking lifts in parking management has already generated some interest and can be added to our proposed MUSRP as a new USR system. In fact, this is a general robotic procedure for storing, where an automobile is piled on the available space, allowing for the use of two, three, or even more layers of stacking to accommodate vertically several vehicles in a single parking place.

As a result, we can conclude that the parking system of the future will be a smart parking system that incorporates various actors in various industries, as well as the delivery of varied services integrated into the SC complex system and control. By doing so, it will be possible to consider the current drivers' customs for parking their cars and ensure the transfer to contemporary and intelligent parking systems based on USR.

References

1. Biswas, K., Vallipuram M.: Securing smart cities using blockchain technology. In: 18th IEEE International Conference on High Performance Computing and Communications (2016)
2. Dameri, R.P.: Searching for smart city definition: a comprehensive proposal. Int. J. Comput. Technol. 2544–2551 (2013)
3. Golubchikov, O., Thornbush, M.: Artificial intelligence and robotics in smart city strategies and planned smart development (2020)
4. Jeelani, S., Al, E.: Robotics and medicine: a scientific rainbow in hospital. Int. J. Pharmacy Bio Allied Sci. (2015)
5. Nikitas A. et al.: How can autonomous and connected vehicles, electro mobility, BRT, Hyperloop, shared use mobility and mobility-as-a-service shape transport futures for the context of smart cities (2017)
6. Nitulescu, M., Jawad, Y.K.: Smart city and internet of things technologies. In: 25th IEEE International Conference on System Theory, Control and Computing (ICSTCC), Romania, pp. 606–612 (2021)
7. Panagiotis, H., Fourlas, G.K.: Reliability and maintainability analysis of a robotic system for industrial applications. Int. J. Performab. Eng. 453 (2015)
8. Quang-Vinh, D., Nguyen, C.T. and Rudová, H.: Scheduling of mobile robots for transportation and manufacturing tasks. J. Heurist. 175–213 (2019)
9. Sabbir, A., Rahman, M.S., Rahaman, M.S.: A block-chain-based architecture for integrated smart parking systems. In: Proceedings of 2019 IEEE International Conference on Pervasive Computing and Communications Workshops (PerCom workshops). IEEE (2019)
10. Tiddi, I., et al.: Robot city interaction: mapping the research landscape a survey of the interactions between robots and modern cities. Int. J. Soc. Robot. 299–324 (2020)
11. Xie, J., Tang, H., Huang, T., Yu, F.R., Xie, R., Liu, J., Liu, Y.: A survey of blockchain technology applied to smart cities: Research issues and challenges. IEEE Commun. Surv. Tutor. **21**(3), 2794–2830 (2019)
12. https://en.wikipedia.org/wiki/Electric_vehicle

Chapter 39
Synthesis of an Automation System for the PCR (Polymerase Chain Reaction) Samples Preparation Process

M.-O. Sandu, C.-M. Gruescu, E.-C. Lovasz, and V. Ciupe

Abstract Robots are widely used in all activities, including medical research and practice. The paper shows a preliminary design of an automation system for samples preparation process in view of PCR (Polymerase Chain Reaction) analysis. The current manually samples preparation takes a long time to go through many steps, involving manipulation of microliters (μl) of genetic material and reagents. The process can be subject of human errors, because it is highly repetitive and demands constant and long focus. The proposed automated samples preparation process significantly reduces the pipetting time as well as increases the validity of samples for PCR analysis and eliminate the human errors. The robotic cell is based on a robot UR3, which may be used in various laboratory analyses, with proper programming.

Keywords DNA samples manipulation · Automated system · Robotic application

39.1 Introduction

Genetic analyses involving DNA (deoxyribonucleic acid) require relatively high amounts of high purity DNA. Practically, the samples of DNA, which can be taken are very small and, thus, multiplication of DNA is necessary. The multiplication of DNA is achieved with the PCR (Polymerase Chain Reaction) technique (Fig. 39.1).

The objective of this technique is to make a very large number of copies of a portion of a DNA molecule, under the conditions of knowing the nucleotide sequences, adjacent to the portion to be amplified.

M.-O. Sandu · C.-M. Gruescu · E.-C. Lovasz (✉) · V. Ciupe
Politehnica University of Timişoara, Timişoara, Romania
e-mail: erwin.lovasz@upt.ro

M.-O. Sandu
e-mail: melania.sandu@upt.ro

C.-M. Gruescu
e-mail: corina.gruescu@upt.ro

V. Ciupe
e-mail: valentin.ciupe@upt.ro

I. Doroftei et al. (eds.), *Proceedings of SYROM 2022 & ROBOTICS 2022*, Mechanisms and Machine Science 127, https://doi.org/10.1007/978-3-031-25655-4_39

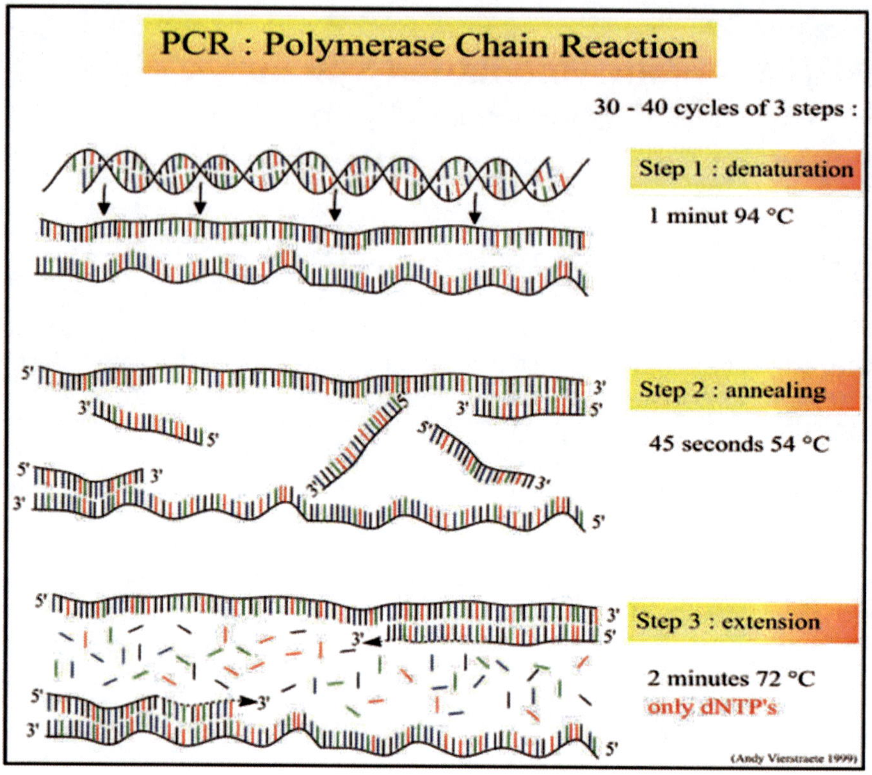

Fig. 39.1 PCR reaction diagram [1]

The first step is to denature the DNA by heating (Fig. 39.1), resulting in two free polynucleotide strands, which come into contact with an excess of two oligonucleotides (primers). The two oligonucleotides, attached to the two ends of the DNA fragment to be copied, will be primers from which a DNA polymerase will copy the segment from the primer to the end of the strand [1–4].

In this way, two pairs of chains are obtained from the initial pair of chains. These will be subjected again to the cycle of denaturation, primer attachment, polymerization, obtaining four pairs of chains and so on exponentially, so that at 20 cycles are obtained approximately 10^6 copies (Fig. 39.2). Because the strands synthesized in one cycle serve as a template in the next, a million-fold increase in the DNA amount is achieved in just 20 cycles [1–4].

The PCR technique is performed in the biochemistry laboratories and needs medical instrumentation and specific apparatuses consisting of different pipettes and adequate tips, racks with wells for samples, well boxes, elution tubes, reagent kits, PCR thermocycle, incubator, extractor, refrigerators, computers etc. Figure 39.3 shows several images of devices in the biochemistry laboratory of Victor Babeş University of Medicine and Pharmacy Timişoara.

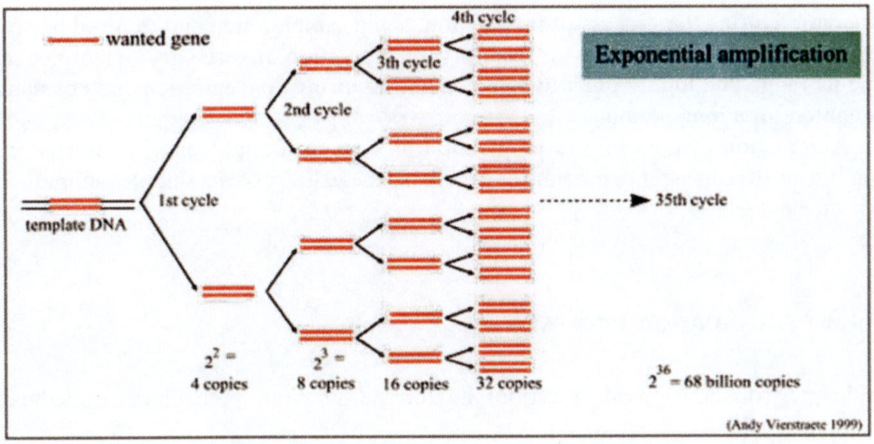

Fig. 39.2 Exponential amplification [1]

Fig. 39.3 Laboratory devices: Vortex (**a**), Pasteur pipette (**b**), Incubator (**c**), Maxwell Extractor (**d**), Eppendorf pipette (**e**), 96 rack wells (**f**), PCR thermocycle (**g**), Graphical and numerical results (**h**)

The preparation of samples with DNA follows standard procedures, consisting of a big number of steps, performed in a logical sequence. The primers, the buffer solution, and the biological material are fed to the wells, which are transported to apparatuses performing mixing, heating etc. The present working method is based on manual preparation of a complete 96 wells rack, taking about 2 h for a well-trained operator. During the process of preparation, some samples are compromised due to operator errors, because of poor visibility, manipulation of very small quantities in the μl range and highly repetitive procedure, requiring concentration and constant attention for a long period.

Automation of the samples preparation process in the laboratory is desired to significantly reduce pipetting time and increase the validity of the samples submitted for medical analysis.

39.2 Automation Process

Industrial robots are used in various flexible automation applications in medical laboratories [5–8].

The medical analysis equipment contains of a workstation in a thermally insulated niche, where the reagent containers and the rack with wells are positioned. The flexible robot cell is composed minimally of an industrial robot (IR), racks to support containers (RSC), reagents containers (RC) and the biological samples, medical instrumentation (MI) and medical analysis equipment.

The robot which is going to be used for the application is type UR3 (Fig. 39.4), working in the laboratories of the Politehnica University of Timişoara.

The UR3 [9] collaborative robot weighs 11 kg, has a payload of 3 kg, ± 360 degree rotation on all wrist joints, and a reach of 500 mm, which makes it adequate for the use in a laboratory.

The handling operation requires the matching of movements and positions of all components of the kinematic chains formed by the industrial robot, reagent container, medical instrumentation and medical analysis equipment. In the case of automatic operation of the flexible cell, the correlation is carried out by the control system using programs developed on the basis of suited algorithms. The calculation algorithms are related to reference systems attached to each component of the flexible laboratory cell and are obtained by successive coordinate transformations between the different

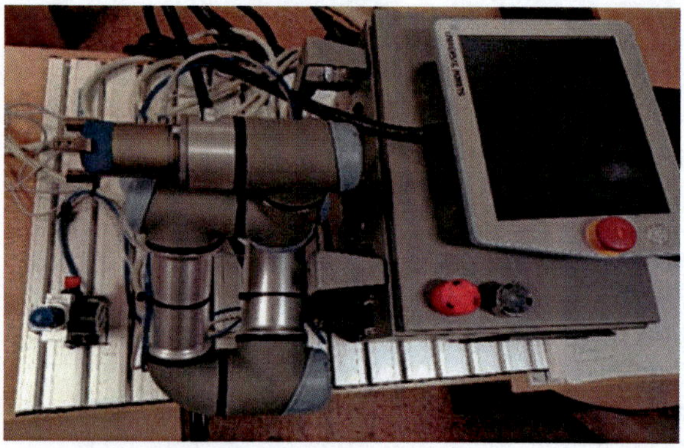

Fig. 39.4 Robot UR3 (Laboratories of Politehnica University of Timişoara)

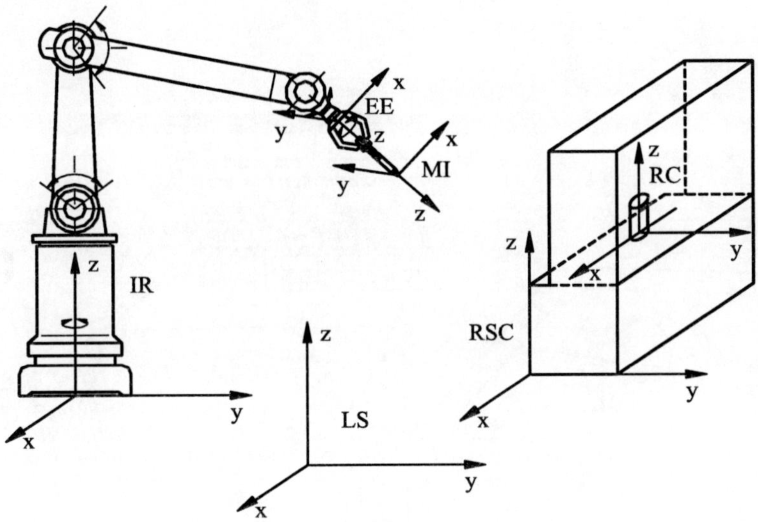

Fig. 39.5 Robotic cell (IR—industrial robot, EE—enf-effector, MI—medical instrumentation, RC—reagent container, LS—laboratory space, RSC—rack to support containers

reference systems. Figure 39.5 presents a generic view of the robotic cell, with reference systems attached to the main components of the cell.

Figure 39.6 shows an intuitive model of the robotic cell, suggesting the movements of the end-effector, which travels within the laboratory space to grip pipettes and suited tips from reagent (primer) containers to the wells, where samples are prepared. The movements are conceived so that the distances and time of displacement to be minimum.

In this regard, the first reagent is pipetted inside all 96 wells (12 rows × 8 columns), then the second reagent and so on. For each reagent is needed a different pipette and a different tip, so travels to a recycle bin is planned after each sequence.

In order to program the movement of the end-effector, the dimensions of all medical instrumentation and distances between them must be known (as shown in Fig. 39.7).

The transmission control protocol (TCP) of the robot should refer to the parameters pipettes (i), peaks (j), reagents (k), wells (g), rows of wells ($n = 8$), number of reagents ($p = 4$).

The workflow on which the programming is based on, is presented in Fig. 39.8.

39.3 Conclusions

An automated robotic cell designed for a genetic analysis laboratory is prepared to handle genetic material and different reagents in sequencies, which are programmed

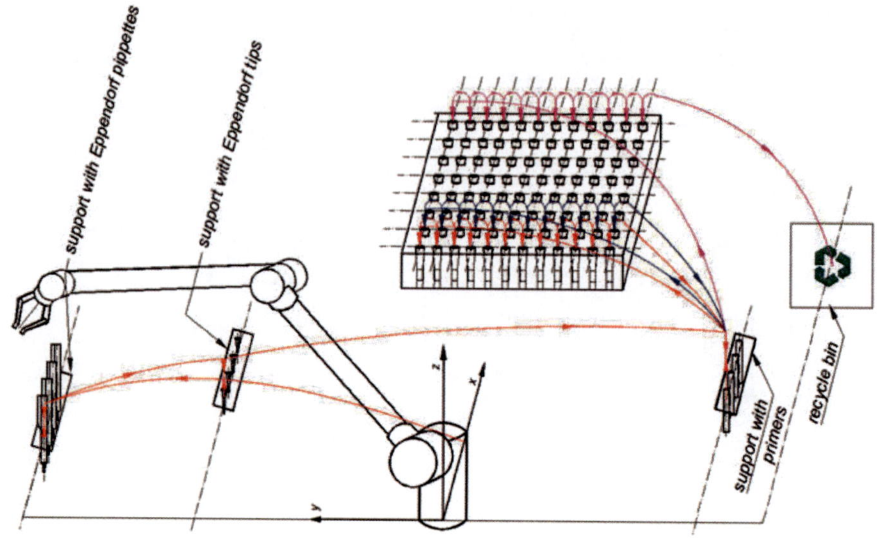

Fig. 39.6 Simplified model of trajectories and sequencies of movement

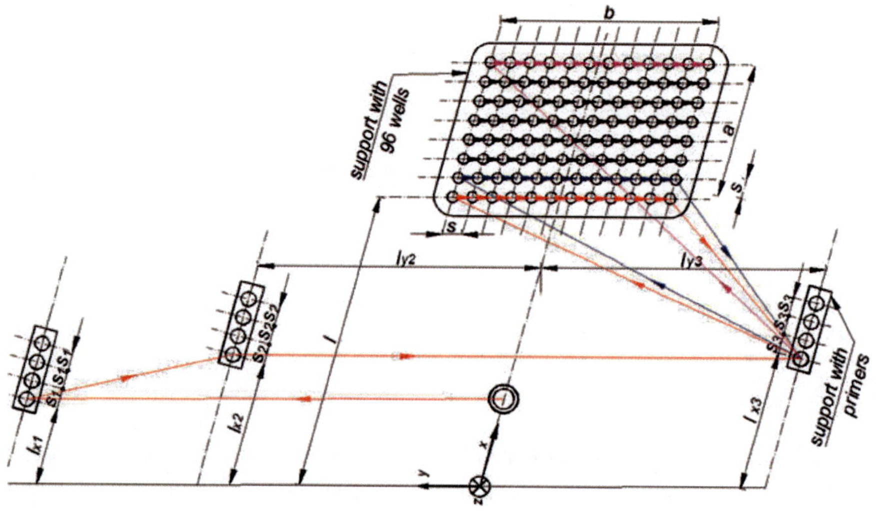

Fig. 39.7 Simplified scheme of the robotic cell emphasizing the dimensions to be taken into account in programming the movements

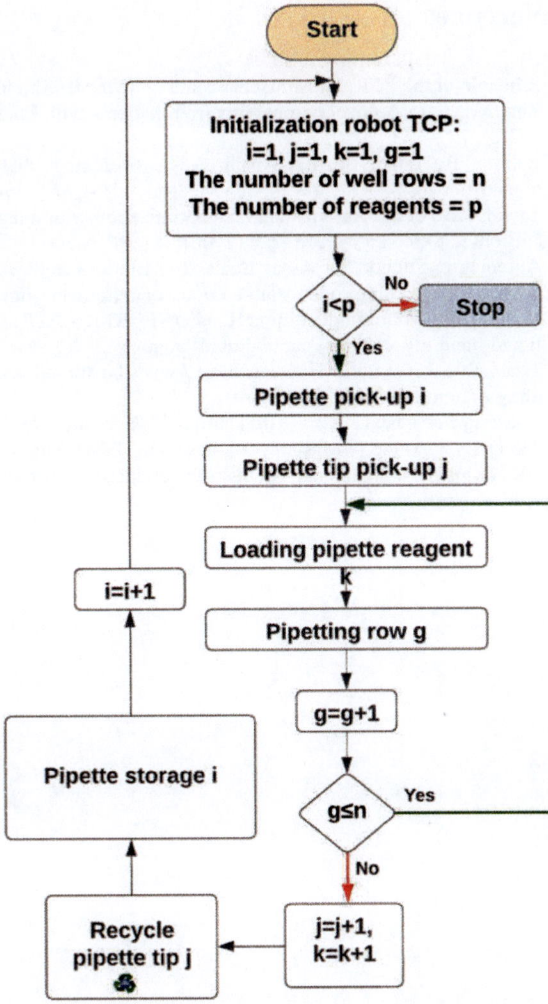

Fig. 39.8 Workflow of the preparation process of samples

to go through standard procedure protocols. The robot foreseen with adequate grippers can be used in various analyses, considering the standard dimensions of medical instrumentation, such as pipettes, containers and wells.

The use of a robotic cell significantly shortens the preparation time of samples and eliminates the human errors, which occur during manual handling.

The application described in this paper may be of interest for small and medium laboratories.

References

1. Principle of the PCR. https://users.ugent.be/~avierstr/principles/pcr.html
2. Zhu, N., et al.: A novel coronavirus from patients with Pneumonia in China. N. Engl. J. Med. **382**(8), 727–733 (2020)
3. Rahimi, F., Talebi Bezmin Abadi, A.: Case-finding: fast, available and efficient font-line diagnostics for SARS-CoV-2. Arch MedRes. 51(5), 453–454 (2020)
4. Taylor, P.B., et al.: Automated assay optimization with integrated statistics and smart robotics. J. Biomol. Screen **5**(4), 213–226 (2000)
5. Automation Support for Assay Scale-Up I Liquid Handler Assay Automation (2021). https://www.promega.com/products/lab-automation/automated-liquid-handler-assay-support/
6. MANTIS® - Microfluidic Liquid Handler - FORMULATRIX® (2021). https://formulatrix.com/liquid-handling-systems/mantis-liquid-handler/
7. Tecan Evo 150 Liquid Handler. https://www.bostonind.com/laboratory/automated-liquid-handling-systems-and-robotics
8. Mastercycler® nexus X2 - PCR Thermal Cycler. https://online-shop.eppendorf.at/SP-en/PCR-44553/Cyclers-44554/Mastercycler-nexus-X2-PF-82586.html
9. UR3. https://www.universal-robots.com/products/ur3-robot/

Chapter 40
Algorithm and Software for Achromatic Doublet Synthesis

S.-L. Oprean and C.-M. Gruescu

Abstract Robotics uses imaging applications on large scale. The optical components associated with cameras must provide high quality images, using assemblies which are compact, small in size and light in weight. Optical design is not published in national or international standards, but it is mostly developed by companies, which design and manufacture optical devices. The paper proposes an original general algorithm for imaging applications and then an algorithm and appropriate software to synthetize the achromatic doublet as the simplest compound entity to provide very high quality of image. The algorithm is designed to minimize the spherochromatism of the assembly. The solutions obtained with the algorithm and software were tested in the program OSLO, which performs complex analysis of optical systems. The doublets resulted from the proposed algorithm and software were compared with catalogue components existing on the market. The original doublets are very good regarding the basic optical characteristics and image quality, so that the algorithm and software may be of interest for optical companies.

Keywords Imaging applications · Achromatic doublet · Algorithm and software

40.1 Introduction. Imaging Applications

Imaging is used in robotics in a large variety of applications, such as sorting by color or shape, detection of errors in dimension or position, guiding of an end-effector and many others [1–3].

Image taking and processing uses visible range (the image can be seen directly by the human operator) or infrared range (thermal images need conversion to visible in order to be seen).

S.-L. Oprean (✉) · C.-M. Gruescu
University Politehnica of Timisoara, Timisoara, Romania
e-mail: sonia.oprean@student.upt.ro

C.-M. Gruescu
e-mail: corina.gruescu@upt.ro

I. Doroftei et al. (eds.), *Proceedings of SYROM 2022 & ROBOTICS 2022*, Mechanisms and Machine Science 127, https://doi.org/10.1007/978-3-031-25655-4_40

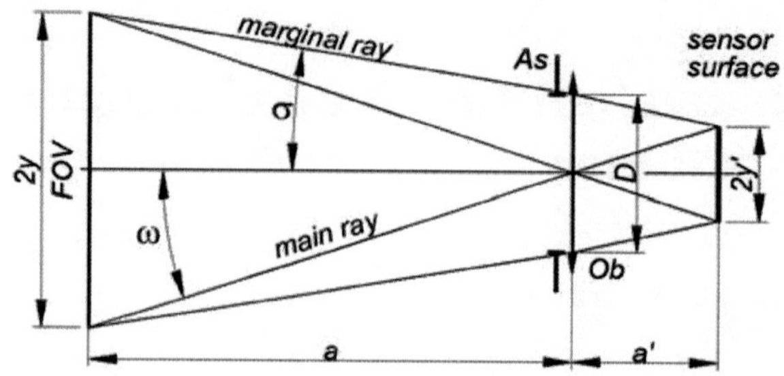

Fig. 40.1 General scheme of an imaging application (2y—object field or field of view (FOV), a, a—object, image distance, As—aperture stop, D—diameter of the objective, 2y—image field, ω—half angle of the object field, σ—half angle of the marginal rays, Ob—objective)

Depending on the application, the parameters are very variable, regarding the size of the camera, the working distance, the resolution etc. However, the optical parameters, illustrated in Fig. 40.1, are not independent, but in a complex relationship.

The general optical scheme proposed for an imaging application in Fig. 40.1, considers the camera chosen from a catalogue. The specifications needed for the optical design are the sensor size (HxV) or the number of pixels and the pixel size.

There are no standard algorithms to calculate the optical and geometrical parameters suited for an application, with a given camera. The authors propose an algorithm based on Eq. (40.7). The calculus considers as input data the object size 2y, the object distance a, and a model of camera from a catalogue, from which the sensor size and the pixel size are known.

The characteristics of the objective result from the following steps:

- Camera resolution:

$$R_{CCD} = \frac{1000}{2\ \text{Pixel Size}[\mu m]}\left[\frac{\text{cycles}}{\text{mm}}\right], \tag{40.1}$$

- Primary magnification:

$$\text{PMAG} = \frac{\text{Sensor Size [Horiz, mm]}}{2y\ [\text{mm}]}, \tag{40.2}$$

- Objective resolution:

$$R_{ob} = \frac{R_{CCD}}{\text{PMAG}}\left[\frac{\text{cycles}}{\text{mm}}\right], \tag{40.3}$$

– Image distance:

$$a' = \mathrm{PMAG} \cdot a[\mathrm{mm}], \tag{40.4}$$

– Effective focal length:

$$f' = \frac{aa'}{a + a'}[\mathrm{mm}], \tag{40.5}$$

– Numerical aperture (choice for f-number):

$$D = f' \cdot f - \mathrm{number} \tag{40.6}$$

– Angle of the marginal ray:

$$\omega = \mathrm{arctg}\left[\frac{y}{a}\right], \tag{40.7}$$

The objective can be chosen from a catalogue on basis of sizes f′, D, a, resolution and image quality. However, catalogues list assemblies with $a = \infty$, limited range of focal lengths and diameters, and do not provide information on resolution and image quality. Thus, the synthesis of the objective is the choice for a better component. In this case, all parameters of interest can be taken into account.

Considering the requirements regarding the axial dimensions and the weight, which should be as small as possible, the first choice for the objective is an achromatic doublet or a pair of symmetrically set achromatic doublets. The doublet is the first compound optical entity, which can provide a very good quality of the image.

40.2 Algorithm and Software for the Synthesis of the Doublet

The achromatic doublet is an assembly made of two lenses with a common radius, on which the elements are cemented. Figure 40.2 shows a cemented doublet, consisting of lenses L1 and L2, in two usual mountings: using a threaded lock-ring (Fig. 40.2a), for diameters bigger than 6 mm and a crimping solution (Fig. 40.2b), for small diameters.

In optical engineering there are no standard methods or procedures for the synthesis of optical assemblies to provide good quality solutions [4–10].

The authors propose an algorithm based on contribution theory, developed by Kingslake [6]. Lenses L1 and L2 are made of a pair Crown-Flint of glasses. The first lens is positive and the second one is negative. There are three degrees of freedom (three radii), which are assigned to three conditions: effective focal length of the assembly and correction of two aberrations (spherical and chromatic).

The synthesis of the cemented doublet follows the steps:

- total curvatures for the lenses (as a convention index, "a" refers to the first lens of the doublet and index "b" to the second one):

$$c_a = \frac{1}{f'(\nu_a - \nu_b)dn_a} \tag{40.8}$$

$$c_b = \frac{1}{f'(\nu_b - \nu_a)dn_b} \tag{40.9}$$

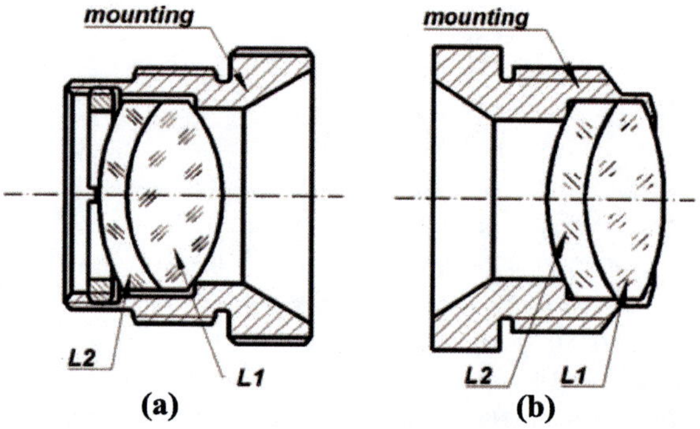

Fig. 40.2 Mounting solutions. a With threaded lock-ring; b Crimped

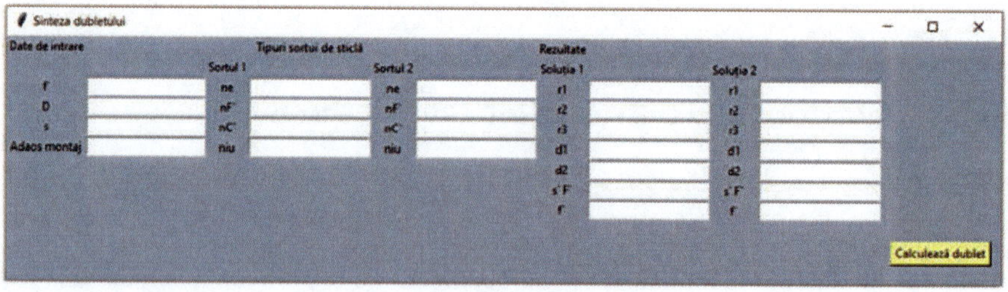

Fig. 40.3 Interface of the software application for the synthesis of the cemented doublet

where n_F, n_C, n_e are the refractive indices for the blue, red and green spectral lines, V_e Abbe number.

- G-sums (for both lenses):

$$G_1 = n_e^2(n_e - 1)/2;\ G_2 = (2n_e + 1)(n_e - 1)/2;\ G_3 = (3n_e + 1)(n_e - 1)/2 \tag{40.10}$$

$$G_4 = (n_e + 2)(n_e - 1)/2/n_e;\ G_5 = 2\left(n_e^2 - 1\right)/n_e;\ G_6 = (3n_e + 2)(n_e - 1)/n_e/2. \tag{40.11}$$

- condition to minimize of the primary spherical aberration, with the unknown variable c_1 (the first curvature of the first lens):

$$k_1c_1^2 + k_2c_1 + k_3 = 0, \tag{40.12}$$

$$k_1 = G_{4a}c_a + G_{4b}c_b, \tag{40.13}$$

$$k_2 = -G_{2a}c_a^2 - G_{5a}c_ap_a - G_{2b}c_b^2 - 2G_{4b}c_ac_b - G_{5b}c_bp_b \tag{40.14}$$

$$k_3 = G_{1a}c_a^3 + G_{3a}c_a^2p_a + G_{6a}c_ap_a^2 + G_{1b}c_b^3 + G_{2b}c_ac_b^2 + G_3bc_b^2p_b + G_4bc_a^2c_b + G_5bc_ac_bp_b + G_6bc_bp_b^2, \tag{40.15}$$

with $p_{a,b}$—inverse of object distance of surfaces 1 and 2.

- the second curvature of the first lens:

$$c_2 = c_1 - c_a, \tag{40.16}$$

- condition of marginal achromatization:

$$\sum\left(\tilde{d} - d\right)dn = 0 \text{ or } \left(\tilde{d}_a - d_a\right)dn_a + \left(\tilde{d}_b - d_b\right)dn_b = 0, \tag{40.17}$$

where $\tilde{d}$—length of the marginal ray, d—length of the paraxial ray

- center thickness of lenses.
- third radius (intermediate data in Eqs. (40.18)–(40.20) needed):

$$\begin{cases} x_j = r_j\left[1 - \cos\left(\tilde{\sigma}_j - \tilde{i}_j\right)\right], \\ y_j = r_j \sin\left(\tilde{\sigma}_j - \tilde{i}_j\right). \end{cases} \quad (40.18)$$

$$\tilde{d}_a = (d_a + x_2 - x_1)/\cos\tilde{\sigma}_1' \quad (40.19)$$

$$\begin{cases} x_3 = \tilde{d}_b \widetilde{\cos\sigma_2'} + x_2 - d_b \\ y_3 = y_2 - \tilde{d}_b \widetilde{\sin\sigma_2'} \end{cases} \quad (40.20)$$

$$r_3 = \frac{x_3^2 + y_3^2}{2x_3}. \quad (40.21)$$

with x_3, y_3 the coordinates of the third incidence point, resulted from a ray tracing.

Considering Eq. (40.12) is a second-degree equation, there are two solutions resulting from the algorithm.

To accomplish the design process with speed and accuracy a software application specially created based on the above methodology was written in Python. The interface of the program is presented in Fig. 40.3.

40.3 Analysis of the Solutions

The solutions provided by the original software were analyzed with OSLO [11].

Figures 40.4 and 40.5 shows synthetically the image quality parameters for a doublet made of N-BK7/N-SF2, with $f' = 8$ mm. For geometrical parameters, Fig. 40.4a illustrates the spherochromatism. Wavefront analysis provides the values of PV OPD, RMS OPD and Strehl Ratio (Fig. 40.4b). The Fourier analysis estimates the resolution (MTF) and distortion/contrast issues (PTF) Fig. 40.5. All images and values indicate a diffraction limited component, which means the image quality of the doublet is very good.

In order to validate the algorithm and software, the solutions were compared with products listed in the catalogue of an important provider of optical components, Thorlabs [12].

The comparison procedure was as follows:

1. One doublet from the catalogue was chosen and with the input data (f′, D and pair of glasses—from the catalogue Schott [13]), the assembly was redesigned using the algorithm and software proposed in this paper.
2. The component from the catalogue and the designed doublet were compared in OSLO, on basic optical characteristics and image quality.
3. Notices and conclusions were formulated regarding the proposed algorithm and software.

From the THORLABS catalog were selected 10 cemented doublets Crown-Flint with different values of apertures and various material type combinations to test

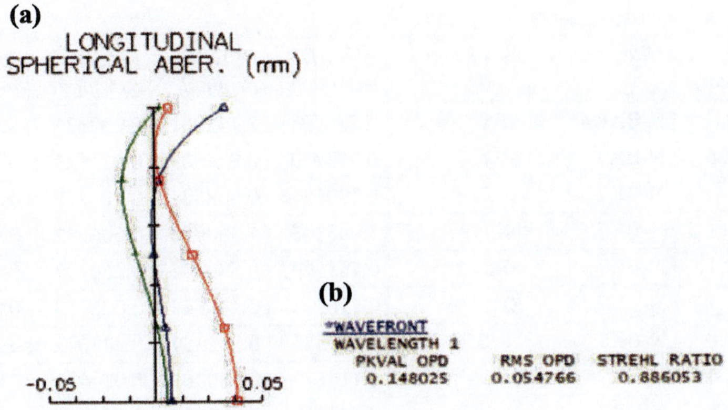

Fig. 40.4 Sperochromatism (**a**) and wavefront parameters (**b**)

Fig. 40.5 MTF and PTF

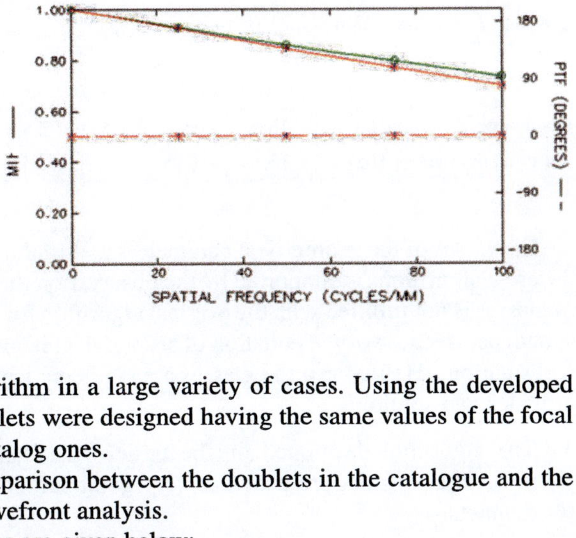

the performance of the algorithm in a large variety of cases. Using the developed software application 10 doublets were designed having the same values of the focal length and aperture as the catalog ones.

Table 40.1 shows the comparison between the doublets in the catalogue and the original ones, in terms of wavefront analysis.

The most important notices are given below:

1. The components in the catalogue and the original doublets are very similar in image quality, which is very good (PV OPD < 0.2, RMS OPD < 0.07 for diffraction limited class). The parameters are slightly lower for wide openings (the cemented doublet is recommended to work with f-number < f/2.5). The f-number for the first two lines correspond to f/2.
2. The algorithm used by the company Thorlabs is different from the one proposed in this paper.

Table 40.2 presents the data in the catalogue and the solution provided by the software for $f' = 30$ mm, $D = 8$ mm, glasses N-BK7/N-SF2.

Table 40.1 Comparison results

f' *[mm]*	*D* *[mm]*	*Glass lens 1*	*Glass lens 2*	*PVOPD catalogue*	*PVOPD original*	*RMSOPD catalogue*	*RMSOPD original*
10	5	N-BAK4	SF5	0.551795	2.143951	0.152833	0.605385
13	6.35	N-BK7	SF2	0.313920	0.377603	0.111621	0.116715
20	8	N-BK7	SF2	0.299654	0.308520	0.107278	0.109561
30	8	N-BK7	N-SF2	1.608105	1.308950	0.506942	0.501657
30	12.7	N-BK7	SF2	0.788199	0.589334	0.264787	0.205823
50	12.7	N-BK7	SF2	0.216307	0.218643	0.076358	0.079362
200	50.8	N-BK7	N-SF2	0.975525	0.982359	0.343647	0.345178
400	25.4	N-BK7	SF2	0.116172	0.130061	0.034611	0.038720
250	50.8	N-BK7	N-SF2	0.834433	0.832975	0.266896	0.266817
1000	50.8	N-BK7	N-SF2	0.218445	0.218063	0.065005	0.064890

Table 40.2 Comparison in geometry of components (catalogues versus original design)

	R1 *(mm)*	*R2* *(mm)*	*R3* *(mm)*	*d1* *(mm)*	*d2* *(mm)*	*Glass lens 1*	*Glass lens 2*
Catalogue	16.0	−13.5	−59.4	2.5	1.5	N-BK7	N-SF2
Original doublet	15.9	−13.8	−1.95	2.3	1.5	N-BK7	N-SF2

The values of the geometrical parameters are similar, but not identical. The idea of different algorithms is supported by the observation that mathematical compatibility of glasses is not fulfilled with the original algorithm for some pairs of materials in the catalogue. Because of the equation of second degree, not any pair of glasses leads to real solutions. However, if the glasses are mathematically compatible, the algorithm provides two solutions.

3. The algorithm developed on the basis of fundamentals in literature, written in a working software provides very good solutions for designing achromatic doublets.

40.4 Conclusions

The subject of this paper is imaging applications, which may be of interest in a wide range of technical application, including robotics. The authors propose a complete design of an imaging application, starting with input data, including geometrical and optical requirements. First, an algorithm to find optical power of the objective and its position to the object and camera is proposed. To complete the design, it is necessary to have an appropriate objective, whose original design, with adequate software are presented. Beginning from fundamentals stated by Kingslake, the synthesis of doublets and a software to perform the calculus are described in detail. The solutions provided by this software are as good as the best products on the market.

The software product may be of interest for the designers of optical apparatuses or companies which manufacture optical components.

References

1. Dewi, T., Risma, P., Oktarina, Y.: Fruit sorting robot based on color and size for an agricultural product packaging system. Bull. Electr. Eng. Inf. **9**(4), 1438–1445 (2020)
2. Zhang, W., Zhang, C., Li, C., Zhang, H.: Object color recognition and sorting robot based on OpenCV and machine vision. In: IEEE 11th International Conference on Mechanical and Intelligent Manufacturing Technologies (ICMIMT), pp. 125–129 (2020)
3. Hazrat Ali, M., Aizat, K., Yerkhan, K., Zhandos, T., Anuar, O.: Vision-based robot manipulator for industrial applications. Proc. Comput. Sci. **133**, 205–212 (2018)
4. Smith, W.J.: Modern Optical Engineering, 3rd edn. McGraw Hill, New York (2000)
5. Fischer, R.E., Tadic, B., Yoder, P.R., Jr.: Optical System Design. McGraw Hill, New York (2008)
6. Kingslake, R., Johnson, B.: Lens Design Fundamentals, 2nd edn. Elsevier, Burlington US (2010)
7. Bass, M. et al, Handbook of Optics. Fundamentals. Techniques. Design, McGraw Hill, NY (1995)
8. O–Shea, D.C.: Elements of Modern Optical Design. Wiley, New York (1985)
9. Hecht, E.: Optics, 5th edn. Eddison Wesley Longman Inc., New York (2015)
10. Gruescu, C.M.: Ingineria optica, Ed. Politehnica, Timisoara (2012)
11. OSLO. https://lambdares.com/oslo/
12. Unmounted Achromatic Doublets, AR Coated: 400–700 nm. https://www.thorlabs.com/newgrouppage9.cfm?objectgroup_id=120
13. Optical glass. https://www.schott.com/en-ro/products/optical-glass-p1000267/downloads

Chapter 41
An Experimental Evaluation of the Performance and Key Factors of Intelligent Recommendation System

G. Rus, M. Deriaz, C. Vaida, L. Nae, B. Gherman, E. Oprea, A. Pisla, L. Stulens, and D. Pisla

Abstract The implementation of recommendation systems on different websites, to offer a satisfying experience to user (related both to work and pleasure time), became a vital component for various platforms. Such a system must consider a variety of needs that are unique to the domain for which it is utilized, making its development difficult. The aim of this paper is to analyse the requirements for a recommendation system that can be implemented in applications designed act as a matchmaker between demand and supply in human resources, connecting companies with senior mentors, and to develop a system that satisfies these requirements. The proposed hybrid system is implemented in a knowledge transfer platform, dedicated to elderly people, where the recommendation is using artificial intelligence agents.

G. Rus · C. Vaida · B. Gherman (✉) · A. Pisla · D. Pisla
CESTER, Technical University of Cluj-Napoca, Cluj-Napoca, Romania
e-mail: Bogdan.Gherman@mep.utcluj.ro

G. Rus
e-mail: Gabriela.Rus@mep.utcluj.ro

C. Vaida
e-mail: Calin.Vaida@mep.utcluj.ro

A. Pisla
e-mail: Adrian.Pisla@mep.utcluj.ro

D. Pisla
e-mail: Doina.Pisla@mep.utcluj.ro

M. Deriaz
38 Route d'Ambilly, 1226 Thonex, Switzerland
e-mail: michel.deriaz@unige.ch

L. Nae · E. Oprea
Digital Twin Ltd., Bucharest, Romania
e-mail: laurentiu.nae@digitaltwin.ro

E. Oprea
e-mail: eduard.oprea@digitaltwin.ro

L. Stulens
In4Care, Happy Aging, Leuven, Belgium
e-mail: leen.stulens@in4care.be

I. Doroftei et al. (eds.), *Proceedings of SYROM 2022 & ROBOTICS 2022*, Mechanisms and Machine Science 127, https://doi.org/10.1007/978-3-031-25655-4_41

The system has been tested on different scenarios to estimate the accuracy of the matchmaking process.

Keywords Recommendation system · Matchmaking · Semantic similarity · Machine learning

41.1 Introduction

The most well-known recommendation systems (RS) are found in e-commerce, e-learning, movie and music streaming websites, and job search websites. These platforms use recommendation algorithms, also referred to as recommendation engines, to provide recommendations to users based on their individual needs.

The systems are designed to examine a profile of user to offer an appropriate recommendation based on characteristics such as interests or similarities between people, who share related interests. RS creates suggestions utilizing a variety of methods, the most well-known of which are: Content based recommendations (CB), where the suggestions are made using the previous choices of the user; Collaborative Filtering (CF), which recommend new items based on clustering approach, where the suggestion is made to a user based on previous choices of another users with similar profiles and Hybrid Systems, which imply the combination of different systems to develop a robust framework [1].

Although recommendation systems are frequently studied for streaming or e-commerce platforms [7], where the suggestions are essentially content or purchases, these systems can also be used to assist recruiters on job platforms or to recommend an appropriate person for vocational or training platforms, where the suggestion is a person. The major challenge in this case is to build a system which can recognize the features of involved parts: the profile of company/students and the profile of candidate/mentor, in order create the best pair between them.

Because these types of recommendation systems must recommend a person suitable for a specific task, the main challenge for a system like this, is to choose why a person is more suitable than others for a task or a project. In addition, the degree of accuracy of a system used to recommend a person must be much higher than the degree of accuracy for recommendations based on products or content.

The knowledge transfer platform presented in [3], is a response to a well-known problem in Europe: reintegration of elderly people in professional activity, considering that a sedentary and aimless life, often leads toward problems like social isolation, depression, physical degradation and so on. In addition to this, the platform has the potential to respond to another problem of our society: the scarcity of people with experience in engineering. The recommendation system task is to pair a company (mentee) with a senior (mentor), considering the task of the company and the skills of the mentor.

An analysis of the system is performed based on the main characteristics and challenges specific to this type of system. The paper is structured as follows: Sect. 41.2

evaluates different architectures for recommendation systems and analyzes the characteristics for a recommendation system used on a knowledge transfer platform, Sect. 41.3 describes the proposed recommendation system, followed by the Sect. 41.4 in which testing results made on system is presented. Section 41.5 offer a perspective about future work and presents the conclusion.

41.2 Possible Architectures of Recommendation Systems

The architecture of recommendation systems used in job search websites or knowledge transfer platforms differs depending on the specific requirements of each assignment.

In [8] the authors present a recommendation system for Research and development (R&D) projects, in which companies obtain recommendations for potential researchers to collaborate on various projects. The proposed system is consisting of two parts: an offline module for selection of candidates and an online module for context-aware recommendation (recommendation based on the context of users). Another perspective for a recommendation system used in a knowledge transfer platform is described in [4], where AI agents based on Machine Learning (ML), are used to recommend a mentor to a mentee. They proposed a system based on historical data of the user, assisted by an Apriori algorithm based on extra association rules. In [2] the authors present an overview of e-Recruitment recommender systems, where the main challenges of this type of systems development, are highlighted.

An analysis of the primary characteristics of recommendation systems used in e-recruitment or mentoring platforms was conducted based on the articles mentioned above, to assess the value of these qualities for system development. For this analysis, Analytical Hierarchy Process (AHP) method was used, utilizing the following characteristics:

Matchmaking the job/task with the candidate: represents the process that achieves the matching with the available task or job, based on different characteristics.

Bidirectional process: represents the concept through which the system is built to be bidirectional, combining both the candidate/mentor and the recruiter/mentee needs.

Recommendation is made considering the attributes of candidates: refers to the need of make a recommendation based on the attributes of the candidates, in order to provide a relevant recommendation.

Large Database: refers to the amount of data that a system must be trained on in order to extract useful information.

Usability of AI agents: refers to the introduction of Artificial intelligence agents in recommendation process in order to obtain valid results in a short time.

Trust in the expertise of candidates: refers to the trust in candidates and how a system can manage to provide the most reliable candidates based on an honest expertise.

The implication of human factor in the process refers to the implication of a human factor during the entire process, to verify and validate the final result.

Predisposition for bias: refer to the possibility through a system can be exploited to provide a bias for a certain candidate.

Based on AHP analysis presented in Fig. 41.1 a hybrid system for recommendations was developed, considering the trust in the mentor expertise and bias vulnerability, vital elements for platform. The AI agents are integrated throughout the system, allowing for optimal matchmaking based on candidate characteristics and company requirements.

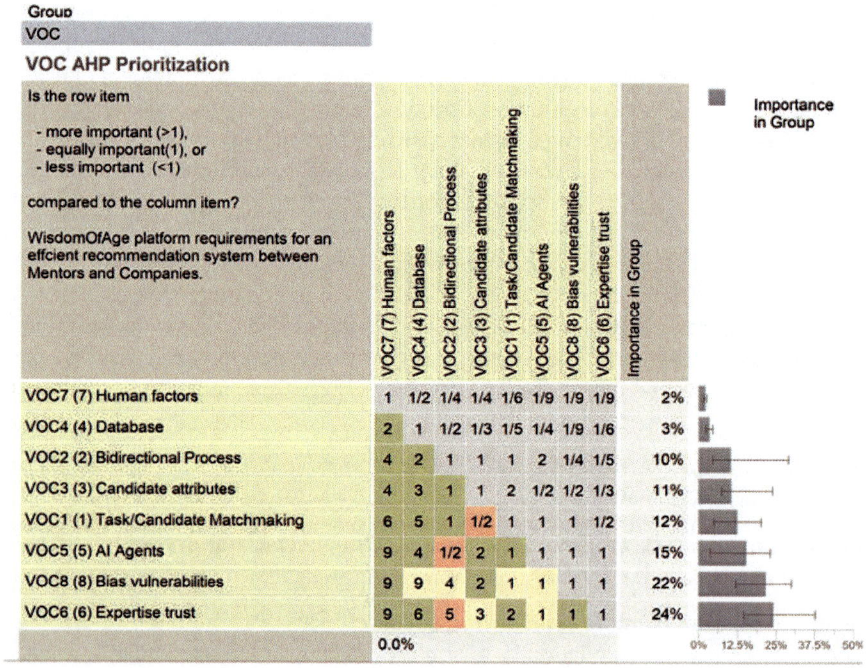

Fig. 41.1 AHP analysis for an efficient recommendation system

41.3 General Architecture of the WoA Recommendation System

The platform was designed to respond to a request very often met in present, the reintegration of elderly people in a healthy work environment, where the seniors (mentors) can share their knowledge and experience to a company (mentee).

The recommendation system is designed to offer the maximum comfort for mentors, and for this, the recommendations are made towards companies. A schematic representation of the system can be seen in Fig. 41.2.

The recommendation system architecture is based on two major components: the match-making system and the collaborative filtering.

The Matchmaking system has the role to match the mentor with the company. To make a valid match between company and candidate, a bidirectional system was developed that calculates both the semantic similarity between the keywords of the company's task and the keywords of the mentor, as well as the semantic similarity between the mentor's description and the description of the company's task. The keywords and the description for mentors and companies are required when the profiles are completed, and this information is saved in a Jason database. The process starts comparing every word from the list of keywords of the company saved in database, with every word from keywords list of every candidate, to calculate the similarity degree between them. The semantic similarity for keywords. is calculated using Word2Vec natural language processing technique. Considering that words with similar meaning are situated in close vicinity, the distance between them can be calculated, yelding a semantic similarity degree [5].

One of the major obstacles to train the Word2Vec algorithm was to find a suitable database considering that the algorithm has to calculate the semantic similarity

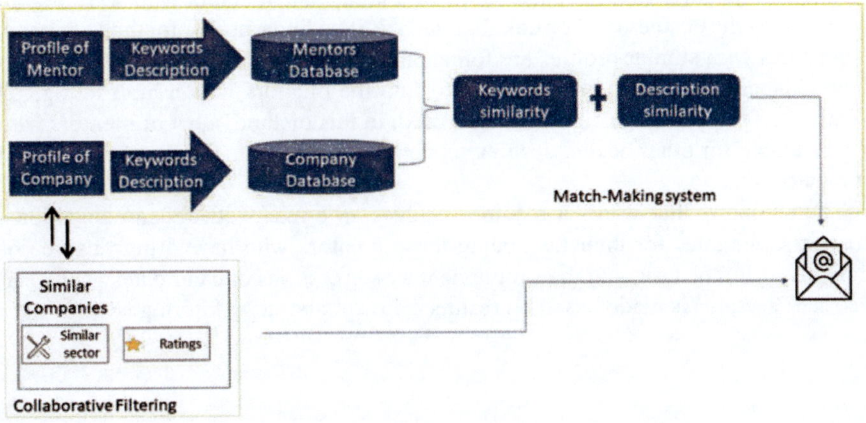

Fig. 41.2 Schematic representation of the system proposed

between specific words (eg Mechanical-Autocad), and the databases from the specialized sites are too generic. To solve this problem, was created a database composed from jobs description, in this way, the algorithm being able to learn with a suitable database. The descriptions were collected from 5 different sites (Ejobs, Linkedin, Indeed, Monster and Glassdor). The first version of database was composed from 20.000 descriptions, but has been proved to be insufficient, the algorithm presenting just 75% accuracy. After this result, to the database was added more descriptions, the final database having over 35.000 descriptions, which led to a superior accuracy, respectively 93%. Both databases were trained using the same parameters:

$$Sample = 6e - 5;$$
$$Window : 2, 3, 4, 5;$$
$$Epochs : 100, 200, 300;$$

At beginning the database was trained using window 2 and different number of epochs (started with 100 and increased to 300), at the second step the window parameter was set to 3 and the number of epochs was variated as in the previous example. The high accuracy was achieved with Sample = *6e-5,* Window = 3 and No of epochs = 300.

Based on a similar principle, the semantic similarity between descriptions is calculated using a pre-trained framework, named Bidirectional Encoder Representations from Transformers (BERT), using a Machine Learning (ML) model for natural language processing [6].

Collaborative filtering was used to complete the recommendation process. Besides the matchmaking system, which provides to the company the most suitable mentors for their task, was also introduced a collaborative filtering which makes new recommendations based on the preferences of other companies from similar domains. The system uses positive previous selections of a company (mentors who were highly evaluated by the businesses they collaborated with), to find a profile for that company. Companies with similar profiles are found, and the rating of the mentor with whom they collaborated is analyzed. In order to find the mentors with a high rating, K-Nearest Neighbor (KNN) algorithm was used. In this method, a list of mentors who were suitable for other businesses in a related sector is sent to the company in need of mentoring.

The company that wants a mentor, receives an e-mail with recommendations, the best candidates for their task being those mentors who have a high degree of similarity with the task, and the supplementary suggestions of candidates ("You can also see" section) is made based on results from collaborative filtering.

Test 1 – Order of Keywords
Description: verify if similar keywords a different order have the same similarity
Inputs:
Company-Keywords: „electrical, engineer"
Candidate 1: Keywords: "electrical, engineer"
Candidate 2: Keywords: "engineer, electrical"
Outputs: Candidate 1: 0.9323, Candidate 2: 0.9323
Observations: Candidates have the same description.
Interpretation: The system is not affected by the order of keywords.

Fig. 41.3 Representation of Test 1 with related results

Test 2 – Extra keywords
Description: verify if an extra keyword for a candidate can influences the result
Inputs:
Company-Keywords: „electrical "
Candidate 1: Keywords: "electrical, engineer"
Candidate 2: Keywords: "electrical"
Outputs: Candidate 1: 0.8312, Candidate 2: 0.8988
Observations: Candidates have the same description.
Interpretation: Because of the extra keyword (engineer), the system offers a lower similarity to the Candidate 1.

Fig. 41.4 Representation of Test 2 with related results

41.4 Experimental Evaluation of the Robust Recommendation System

In purpose of developing a reliable recommendation system for WisomOfAge platform, a series of trials were conducted. The potential exploits of the system and the ability to recommend a person with respect to the bidirectional process, were the main studied aspects. These tests are presented in Figs. 41.3, 41.4 and 41.5.

41.5 Conclusions

Recommendation systems became a very useful tool for a large range of domains, where the interests and experiences of user are important. Although these systems are common in e-commerce and streaming applications, they have become very popular in applications where the recommendation is a person, not a product. Comparing to streaming applications, where the accuracy of recommendations is not necessarily a goal, the e-recruitment platforms require extensive testing, considering the hidden character of the process. The challenges for recommendation systems used in E-learning, job sites or mentoring platforms are closely related to the bidirectional process (in which all participating parts receive a solution for their needs) and the bias exploit. This paper has examined the possibility to offer a solution for these problems,

Test 3– Detailed description

Description: verify in what degree a detailed description of mentor, influences the final result.
Inputs:

Company -Description: "In this role, the Electrical Design Engineer will complete all hardware design activities in compliance with all Project Management Office processes and procedures. He/she will be responsible for the hardware part of the automation system, including system architecture, cabinet design, network and field in system interfaces. Also, this person will work closely with the project manager and the hardware lead engineer, as well as with the supply chain assembly center."

Candidate 1: "I am an open-minded engineer who has a good ability to solve tasks that require creativity, learning, which leads me to constant progression."

Candidate 2:" Electrical Engineer with 7+ years of experience working with blue-chip companies in the consumer electronics industry, primarily focusing on implementing and debugging electro-mechanical systems. Possess excellent interpersonal abilities and a wide range of technical skills. Enjoy being part of a team, as well as managing, motivating, and training a productive team, and thrive in high-pressure and challenging working environments. "

Outputs: Candidate 1: 0.8075, Candidate 2: 0.6853.

Observations: Candidates have the same list of keywords.

Interpretation: The system showed that detailed descriptions can increase the chances of collaborating with companies. Based on this fact, the platform will recommend to users to write detailed description.

Fig. 41.5 Representation of Test 3 with related results

and a hybrid system composed of a matchmaking system and a collaborative filter was developed.

The AHP analysis effectuated to underlines the main characteristics for a system which must recommend the most suitable person for a specific task, revealed that elements such as bidirectional process, bias exploits, and implications of AI agents in recommendation process have a high significance. Considering these aspects, the recommendation system was tested based in different scenarios meant to highlight the possible weakness of the system.

Future work will concentrate on eliminating weaknesses and improving the recommendation system, introducing an algorithm able to detect discrepancy between keywords and descriptions if exists, detecting in this way possible exploits of the system or a misunderstood of the mentor/company.

Acknowledgements This work was supported by a grant of the Romanian Ministry of Research and In-novation, CCCDI—UEFISCDI and of the AAL Programme with co-funding from the European Union's Horizon 2020 research and innovation programme project number AAL-CP-AAL-2020-7-83-CP-WisdomOfAge within PNCDI III.

References

1. Das, D., Sahoo, L., Datta, S.: A survey on recommendation system. Int J Comput Appl **160**(7), 6–10 (2017)
2. Freire, M.N., de Castro, L.N.: e-Recruitment recommender systems: a systematic review. Knowl Inf Syst **63**, 1–20 (2021)

3. Gherman, B., Nae, L., Pisla, A., Oprea, E., Vaida, C., Pisla, D.: WisdomOfAge: designing a platform for active and healthy ageing of senior experts in engineering. In: ICT for Health, Accessibility and Wellbeing. IHAW 2021. Communications in Computer and Information Science, vol. 1538
4. Kostopoulos, G., Neureiter, K., Papatoiu, D., Tscheligi, M., Chrysoulas, C.: ProMe: a mentoring platform for older adults using machine learning techniques for supporting the "live and learn" concept". Mob. Inf. Syst. (2018)
5. Lilleberg, J., Zhu, Y., Zhang, Y.: Support vector machines and Word2vec for text classification with semantic features. In: 2015 IEEE 14th International Conference on Cognitive Informatics and Cognitive Computing (ICCI*CC), pp. 136–140 (2015)
6. Reimers, N., Gurevych, I.: Sentence-BERT: sentence embeddings using siamese BERT-networks. In: Proceedings of the 2019 Conference on Empirical Methods in NLP (2019)
7. Singh, P., Pijush, D.P., Choudhury, P.: Collaborative filtering in recommender systems: technicalities, challenges, applications, and research trends, new age analytics: transforming the internet through machine learning. IoT, Trust Model. **220**
8. Wang, Qi., Ma, J., Liao, X., Du, W.: A context-aware researcher recommendation system for university-industry collaboration on R&D projects. Decis. Support Syst. **103**(2017), 46–57 (2017)

Chapter 42
Optimization of Marker Design in Garment Industry on the Criterion of Utility Coefficient

S. Hora, C.-M. Gruescu, C. Bungau, and R. Bodea

Abstract Garment industry consumes very large amounts of fabric and, thus, this industry is one of the most important waste generators. The transition from linear to circular economy is based mainly on reducing the waste. In this regard, the solutions must be found in all stages of manufacturing, starting with the design of production. The operation that allows the control on the efficiency of fabric used in garment industry is the design of the marker, which, nowadays, partially benefits of computer aid. The operator brings his skills in the process, based on personal experience. One of the problems, which is not totally solved by CAD is the marker design. The operator chooses the number of garment sizes assigned to one marker, which strongly influences the coefficient of utility of fabric. By using a multi-objective method of optimization, it is demonstrated that the optimum number of sizes is two, which saves up to 10% of the fabric in comparison with the variants 1, 3, 4 or more sizes on one marker. The mathematical approach in the paper is original, as there are no references in the literature on the subject.

Keywords Garment industry · Marker design · Pareto optimization

42.1 Introduction

At present, one of the defining features of the economy is the transition from the linear model (based on a take-make-consume-throw away pattern) to the circular one (involving sharing, leasing, reusing, repairing, refurbishing and recycling existing materials and products as long as possible). Practically, the circular economy implies reducing waste to a minimum [7].

S. Hora · C. Bungau · R. Bodea
University of Oradea, Oradea, Romania

C.-M. Gruescu (✉)
Politehnica University of Timisoara, Timisoara, Romania
e-mail: corina.gruescu@upt.ro

I. Doroftei et al. (eds.), *Proceedings of SYROM 2022 & ROBOTICS 2022*, Mechanisms and Machine Science 127, https://doi.org/10.1007/978-3-031-25655-4_42

The European Green Deal was issued by the European Commission in 2020, as the new circular economy action plan that includes proposals on more sustainable product design, reducing waste and empowering consumers. The European Parliament adopted in 2021 a resolution on the new circular economy action plan demanding additional measures to achieve a carbon–neutral, environmentally sustainable, toxic-free and fully circular economy by 2050 [7].

The most amounts of waste result from sectors such as electronics, plastics, textiles and construction. Documents issued at the European level and different research carried out by universities and companies provide data regarding the actual environmental impact of textile industry [3, 5, 8].

The solutions for reducing the waste must be found in new concepts of design and technology.

The paper introduces the coefficient of utility of fabric and the influence of sizes number on it, in Sect. 2.1 and describes the results of a research based on simulation. The mathematical approach, which validates the simulations follows in Sect. 2.2.

42.2 the Marker Design and the Coefficient of Utility of Fabric

42.2.1 The Influence of Sizes Number on the Coefficient of Utility of Fabric—Simulation

A garment model is generated by a fashion designer. The next steps are performed by an engineer, who designs the production.

The garment piece is going to be manufactured from several plan pieces sewn together. The pattern for these pieces can be generated in electronic format and further processed in specific software applications.

All simulations described below were performed in Gemini CAD [9].

Considering the patterns are already generated and graded for the common range of sizes 38–40–42–44-46–48-50–52-54, the following stage is the design of the markers. The marker contains the patterns of different sizes in a combination that is provided by the user. The software application is able to optimize the position of the patterns on the marker and to compute the necessary length of fabric and the coefficient of utility of fabric for the marker.

Figure 42.1 shows a marker containing patterns for sizes 38 and 46. There are 9 patterns necessary for the model. They are highlighted in a different color for each size. The CAD program carries out the optimization in patterns arrangement on the fabric surface. In Fig. 42.1, the patterns are denoted 1…18 and the necessary length is L. The width of the fabric, l, is a standard value within a small range (usually 1.2, 1.38 and 1.6 m). The program displays L and CU (coefficient of utility).

CU is the ratio between the sum of the patterns area (useful surface Su) and available area, L x l (S):

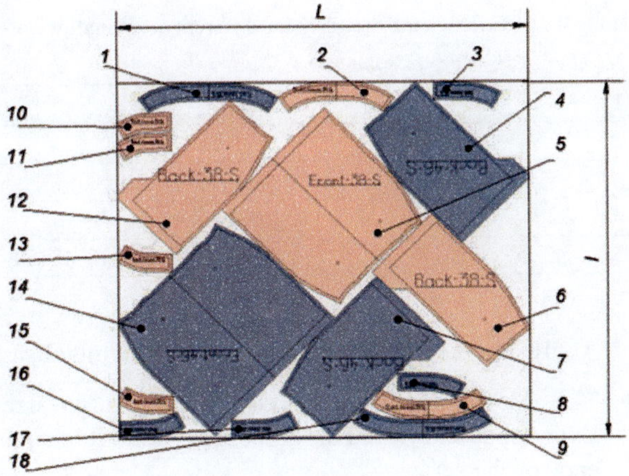

Fig. 42.1 Marker and elements to illustrate the definition of coefficient of use of fabric

$$CU = Su/S * 100\,[\%]. \quad (42.1)$$

Considering all sizes are required for a model, which is a common situation, the problem is how many sizes should be assigned to one marker and what is the best combination of sizes so that the fabric is most rationally used, and waste is minimized. The marker is further used in tailoring the patterns in a number of layers equal to the number of products required in every size. Thus, the coefficient of utility on one marker is decisive for the efficiency of fabric use for the entire production.

The literature does not provide any recommendations in this important matter [2, 4, 6, 9]. One method to find the best combination is to make simulations for each model and specific range of sizes required.

The analysis is developed for a mid-complexity model (Fig. 42.2), considering all sizes 38…54 required.

In Table 42.1 are presented the results of simulations for placing one, two, three, four and all nine sizes on one marker.

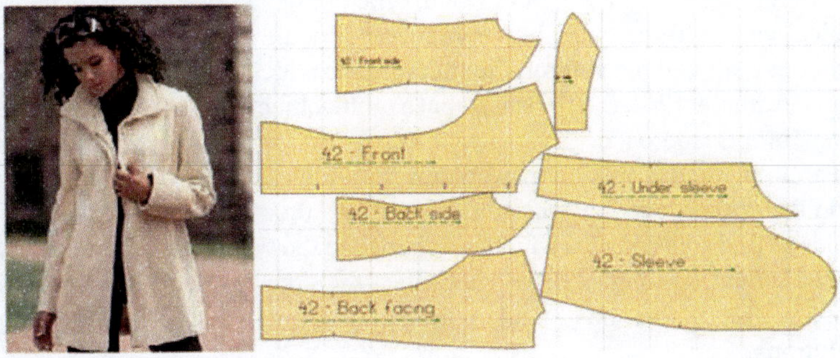

Fig. 42.2 The model and necessary patterns

Table 42.1 Coefficients of utility for different combinations of sizes

Number of sizes per one marker	*Sizes*	*CU [–]*	*L [m]*
1	38	64.3	1.123
2	38–54	81.2	2.037
3	38–46–54	71.7	3.458
4	38–40–46–54	75.6	4.272
5	38–40–42–44–46–48–50–52–54	78.4	9.490

Analyzing the above data result the following observations:

- The best framing is obtained by combining two sizes, which are not consecutive. Experience shows that the combination should contain two sizes from the opposite ends of the series
- In general, the even number of sizes is more favorable in terms of material use efficiency
- The most unfavorable is the placement of a single size and, in general, the odd number leads to lower coefficients of utility
- Combining of all sizes could seem the best choice, as the number of layers is minimum and the waste on the margins of the layers is smaller, accordingly. However, the arrangement of all sizes may result in lengths that exceed the length of the cutting board, which make this concept unfeasible.

42.2.2 *Optimization of Sizes Number on One Marker*

Optimizing the coefficient of utility of fabric is a complex problem, which involves defining relationships to quantify a wide range of parameters of influence, counting to allow the analytical writing of criteria. Some influence parameters, such as the size of the patterns depending on the model, the inclination of the patterns in relation to the thread direction of the fabric and the presence of geometric or printed pattern can be characterized only by logical value, type applies / does not apply. However, their logical value will be quantitatively reflected by the size of the marker.

In order to optimize the marker, the literature provides general recommendations and simple examples [2, 4, 6, 9] or states that a multi-objective method of optimization would be a proper approach.

The Pareto method is applied to problems for which several criteria can be formulated, and the criteria may be contradictory. The method states that the idea of optimization is to find solutions for which the change in the value of a criterion should be taken into account, as long as it does not affect the others. The result is not a single solution, but a set of solutions found on the Pareto front, which contains the equally good solutions.

Table 42.2 Numerical data for the optimization

Length of the marker L [m]	*CU [–]*	*Number of sizes on the marker n [–]*	*Total length for the batch Lt [m]*
1.123	0.6437	1	23.21
2.037	0.8129	2	20.87
3.458	0.7176	3	22.91
4.272	0.7561	4	21.36

Further on, the implementation of Pareto method is illustrated for the optimization of number of sizes on the marker. The mathematical development uses the results of the simulation described in the previous paragraph.

The following steps are followed:

- **Formulation of the problem.** The aim is to optimize the production design for a batch of 20 products of a given model, for which the marker design simulations provided the data in Table 42.2.

- **Defining the criteria and functions associated to them.** The CU and Total Length optimization criteria are defined per batch. CU is a function to be maximized, L total a function to be minimized. The number of sizes on the sheet is considered as an independent variable:

$$\text{Criterion 1}: \ \text{f1(x)} = \text{CU(n)}, \mathbf{max}\text{(CU)}, \tag{42.2}$$

$$\text{Criterion 2}: \ \text{f2(x)} = \text{Lt(n)}, \mathbf{min}\text{(Lt)}. \tag{42.3}$$

- **Analytical formulation of the functions** associated to the criteria and determination of their extremes. For the analytical expression of the CU and Lt functions, the point pairs in Table 42.2 are used. Figure 42.3 shows the graphs of the functions and the approximation polynomials of third degree (for which the correlation coefficient is almost unitary).

It is easy to see from the graphical representation the areas where the functions are extreme. If CU and Lt functions are written analytically as:

$$\text{CU} = 0.0664\text{n}^3 - 0.5306\text{n}^2 + 1.2962\text{n} - 0.1883, \tag{42.4}$$

$$\text{Lt} = -0.1326\text{n}^3 + 1.0142\text{n}^2 - 2.3487\text{n} + 3.788, \tag{42.5}$$

the first derivatives and their roots can be deduced (Table 42.3).

- **Interpretation of results** and establishment of the Pareto front. One can notice that the maximum of the CU function corresponds to the abscissa x2CU = 1.90,

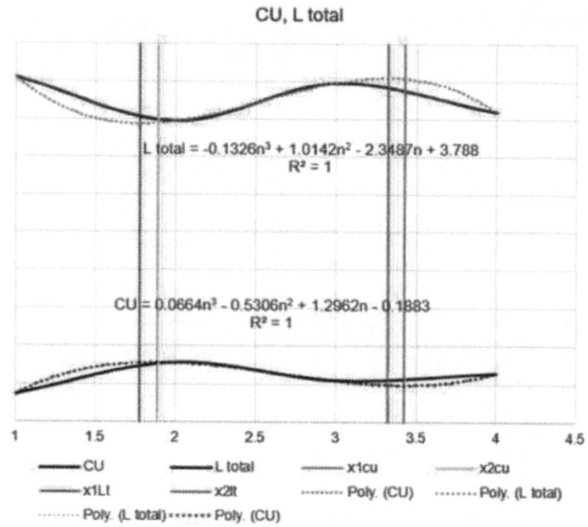

Fig. 42.3 Graphs of the criteria CU and Lt and their extreme points

Table 42.3 Roots of the first derivative of functions CU and Lt

x *1 CU*	x *2 CU*	x *1 Lt*	x *2 Lt*
3.43	1.90	1.78	3.32

and the minimum of the Lt function to the abscissa x1Lt = 1.78. As a result, the optimal solution, for which CU is maximum and Lt minimum, is nearby the abscissa n = 2. Given that n must be an integer, n = 2 is considered the solution to the problem.

For a continuous variation of the independent variable the Pareto front can be drawn, in a representation $f_1(f_2)$ or $f_2(f_1)$. Figure 42.4 shows the relation Lt(CU), which from a theoretical point of view offers a set of optimal solutions.

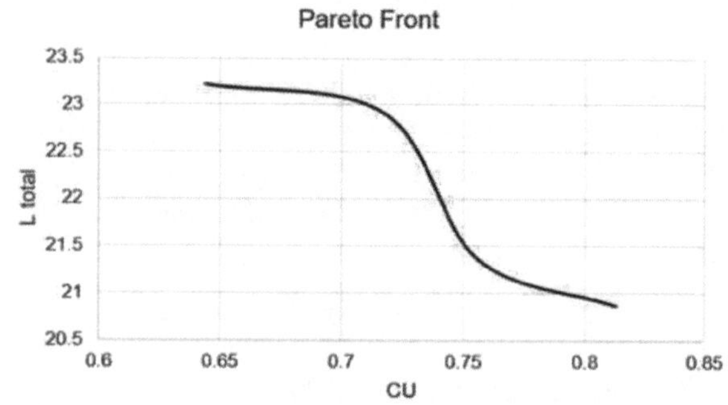

Fig. 42.4 Pareto front

From a practical point of view, in the formulated problem, the lower side of the front is of interest, from which the solution corresponding to the abscissa n = 2 was chosen.

The numerical example above generated a mathematical support for the experimental conclusion that the most favorable design of the marker is obtained for the combination of two sizes.

42.3 Conclusions

Analytical approach in the design of markers used in garment industry is difficult and not yet practiced on large scale. CAD programs provide fast computation of geometrical items, such as lengths and areas, and the coefficient of utility of fabric. This parameter is important in the evaluation of the efficiency of fabric use and the waste of fabric after the manufacturing of a batch of products. The coefficient of utility depends on many parameters, among which the number of sizes on one marker is essential. The content of this paper shows that the mathematical approach of this problem is possible, and the obtained results are validated by skilled designers, who got expertise through a long practical work. The problem is approached in two manners: numerical simulation and multi-objective optimization. The result is useful for the designers in the garment industry directly, and may be used in software enhancement in the future versions.

Acknowledgements The authors acknowledge the support received from the company S.C. Prodconf R.line S.R.L. of Oradea, which provided garment models and specific software.

References

1. Alhammadi, H.Y., Romagnoli, J.A.: Process design and operation: Incorporating environmental, profitability, heat integration and controllability considerations. Comput. Aided Chem. Eng. **17**, 264-305 (2004)
2. Datta, D.B., Seal, P.: Various approaches in pattern making for garment sector. J. Textile Eng. Fashion Technol. **4**(1), 29–34 (2018)
3. Elsahida, K., Fauzi, A.M., Sailah, I., Siregar, I.Z.: Sustainable production of natural textile dyes industry. In: International Conference on Innovation In: Technology And Management For Sustainable Agroindustry (ITAMSA 2019), vol. 472: Art. 012036 (2020)
4. Esparza, A.: How to calculate marker efficiency in 3 simple steps 2020. https://www.smartpatternmaking.com/blogs/articles/how-to-calculate-marker-efficiency-in-3-simple-steps
5. Khattab, T.A., Abdelrahman, M.S., Rehan, M.: Textile dyeing industry: environmental impacts and remediation. Environ. Sci. Pollut. Res. **27**(4), 3803–3818 (2020)
6. Kunz, G.I., Glock, R.E.: Apparel Manufacturing: Sewn Product Analysis, 4th edn. Pearson, US (2005)
7. *** Cum dorește Uniunea Europeană să ajungă la o economie circulară până în 2050? https://www.europarl.europa.eu/news/ro/headlines/priorities/economia-circulara/20210128STO96607/cum-doreste-ue-sa-obtina-o-economie-circulara-pana-in-2050

8. *** The impact of textile production and waste on the environment 2021. https://www.europarl.europa.eu/news/en/headlines/society/20201208STO93327/the-impact-of-textile-production-and-waste-on-the-environment-infographic
9. *** Marker making efficiency|Calculation of Marker 2017. http://diantextile.blogspot.com/2017/02/marker-making-efficiency-calculation-of.html
10. https://geminicad.com/products/nest-expert/